The 6th International Congress of CiiEM—Immediate and Future Challenges to Foster One Health

The 6th International Congress of CiiEM—Immediate and Future Challenges to Foster One Health

Editors

José Brito
Nuno Taveira
Ana I. Fernandes

MDPI

Basel • Beijing • Wuhan • Barcelona • Belgrade • Novi Sad • Cluj • Manchester

Editors
José Brito
Egas Moniz School of Health & Science
Monte de Caparica, Portugal

Nuno Taveira
Egas Moniz School of Health & Science
Monte de Caparica, Portugal

Ana I. Fernandes
Egas Moniz School of Health & Science
Monte de Caparica, Portugal

Editorial Office
MDPI
St. Alban-Anlage 66
4052 Basel, Switzerland

This is a reprint of articles from the Proceedings published online in the open access journal *Medical Sciences Forum* (ISSN 2673-9992) (available at: https://www.mdpi.com/2673-9992/22/1).

For citation purposes, cite each article independently as indicated on the article page online and as indicated below:

Lastname, A.A.; Lastname, B.B. Article Title. *Journal Name* **Year**, *Volume Number*, Page Range.

ISBN 978-3-0365-9174-2 (Hbk)
ISBN 978-3-0365-9175-9 (PDF)
doi.org/10.3390/books978-3-0365-9175-9

Contents

About the Editors

José Brito

José Brito is a Full Professor and Coordinator of the WDXRFLab at the Egas Moniz School of Health & Science, Portugal, where he has been teaching a variety of courses in Biophysics and Biostatistics to students enrolled in Health and Biomedical Sciences programmes. He holds degrees in Mathematics (University of Lisbon, PT) and Ergonomics (Technical University of Lisbon, PT) and a PhD in Biophysics (University of Lisbon, PT). The general perspective of his research activity is the application of models and methods, from exact sciences to the study of current problems in public, occupational, and environmental health. His early research focused on the use of in vivo X-ray fluorescence spectrometry techniques for the development of PBKMs for bone-seeking elements in humans. He is currently involved in studies on the relationships between environmental exposure to chemical and physical pollutants and mortality and morbidity in the Portuguese population. He has been principal investigator or collaborator in several projects funded by Portuguese (FCT) and Canadian (NSERC, OIT and CIHR) agencies, and has published several papers in peer-reviewed international journals.

Nuno Taveira

Nuno Taveira is Full Professor and Coordinator of the PhD Programme in Biomedical Sciences at the Egas Moniz School of Health & Science, Portugal. He holds a degree in Pharmaceutical Sciences and a PhD in Pharmacy (University of Lisbon, PT). He is a Research Associate and Head of Research Unit at the Research Institute for Medicines (iMed.ULisboa), Faculty of Pharmacy, Lisbon, Portugal. He is a collaborator of the Global (GBD) and Local (LBD) Burden of Diseases, Injuries and Risk Factors study, led by the Institute of Health Metrics and Evaluation, Washington, USA. His main research interests are the development and preclinical evaluation of antiviral drugs, vaccines, and microbicides; the epidemiology, drug resistance, and evolution of HIV, HBV, and HCV; and HIV-2 infection (diagnosis, pathogenesis, natural history, neutralising antibody response). He is an Associate Editor, Editorial Board Member, or Review Editor for several international journals. He is also the author of more than 350 research publications, including 2 patents, 4 books, 22 book chapters, and 122 papers.

Ana I. Fernandes

Ana I. Fernandes is an Associate Professor of Pharmaceutics and Head of the PharmSci Lab at the Egas Moniz School of Health & Science, Portugal. She holds a degree in Pharmaceutical Sciences (University of Lisbon, PT) and a Ph.D. in Drug Delivery (University of London, UK). She has been involved in the study of polymers and polymeric systems with biomedical applications, namely in the delivery of therapeutic proteins and conventional drugs. Her current research is related to formulations in paediatrics, drug solubility enhancement by co-amorphization, and the 3D printing of pharmaceuticals, as well as nutraceuticals and the usage of lifestyle drugs. Over the years, she has been the principal investigator or collaborator in several externally financed projects (National Funding Agency – FCT), a scientific consultant, an editor for numerous journals, and has published in several international revered journals.

medical sciences forum

MDPI

Editorial

Preface of the 6th Congress of the Egas Moniz Centre for Interdisciplinary Research—Immediate and Future Challenges to Foster One Health

José Brito *, Nuno Taveira and Ana I. Fernandes *

Egas Moniz Center for Interdisciplinary Research (CiiEM), Egas Moniz School of Health & Science, 2829-511 Almada, Portugal; ntaveira@ff.ulisboa.pt
* Correspondence: jbrito@egasmoniz.edu.pt (J.B.); aifernandes@egasmoniz.edu.pt (A.I.F.)

1. Preface

The 6th Congress of the Egas Moniz Centre for Interdisciplinary Research was held in Caparica from 5 to 7 July under the theme "Immediate and future challenges to foster One Health". Aligned with the key strategic orientations and expected impacts of research and innovation on society, economy, and science, as defined in the Horizon Europe strategic plan, the Congress provided an opportunity for researchers and health professionals to present and discuss the expected impact of their works in different areas. These areas included: (1) access to innovative, sustainable, and high-quality health care; (2) the use of new tools, technologies, and digital solutions for a healthy society; (3) the promotion of sustainable and circular management of natural resources while tackling pollution; and (4) the security of sustainable food and nutrition systems.

There were more than 400 registered participants, including 9 international invited keynote speakers, 35 oral communications and 150 poster presentations, after a thorough peer-review process (89% acceptance rate).

In line with the mission of the Egas Moniz School of Health and Science to improve the translation of research activities into educational practice, the Congress has also emerged as an opportunity to disseminate the research work of undergraduate and postgraduate students enrolled in the different graduation programs of the institution. A significant contribution to the training of young researchers has been made by the Proceedings Scientific Committee (Table 1) during the peer-review process, providing invaluable guidance on how to meet the standards of scientific writing and publication of research papers.

Citation: Brito, J.; Taveira, N.; Fernandes, A.I. Preface of the 6th Congress of the Egas Moniz Centre for Interdisciplinary Research—Immediate and Future Challenges to Foster One Health. *Med. Sci. Forum* **2023**, 22, 22. https://doi.org/10.3390/msf2023022022

Published: 7 August 2023

Table 1. Proceedings Scientific Committee.

Ana Isabel Fernandes (Coordinator)	Madalena Salema Oom
Fernanda Loureiro	Renata Ramalho
Íris Almeida	Sónia Vicente

2. Statement of Peer-Review

Works were submitted online, until 2 June 2023, through the conference submission management system (Wordpress Plugin WP Abstracts Pro v.2.6.1 Abstracts & Manuscripts Submission) as abstracts (max. 500 words) and, optionally, also as short proceeding papers (up to 4 pages) to be published in this dedicated volume of MDPI's *Medical Sciences Forum* journal. A total of 208 abstracts and 63 proceeding papers were submitted.

A thorough double-blind peer-review of the proceedings was conducted by a total of 73 reviewers (Table 2). The nominal list of those involved in the edition and reviewing process is included in the acknowledgments. To ensure the high publication standards of the journal, proceedings were reviewed by two independent scholars. Reviews were conducted by expert referees to the professional and scientific standards expected of a proceedings journal, and considered (1) the aim, relevance, and originality of the work; (2) materials and methods; (3) ethical considerations; (4) scientific soundness; and (5) writing—clarity, text organization and formatting, referencing, and English language and style. Narrative literature reviews, study protocols, and systematic reviews without meta-analyses were not considered.

Table 2. Board of Reviewers.

Aida Duarte	Diego González	Maria João Cebola
Aida Simões	Dora Ladislau	Miguel Correia
Alexandra Rodrigues	Evgenia Bekman	Miguel Garcia
Alexandre Santos	Gonçalo Martins Pereira	Nataniel Rosa
Alzira Cavacas	Graciliana Lopes	Nathalie Ferreira
Amândio Dias	Guilhermina Moutinho	Nuno Coelho
Ana Clara Ribeiro	Hélder Costa	Nuno Taveira
Ana Cristina Cardoso Neves	Helena Barroso	Patricia Cavaco Silva
Ana Cristina Manso	Inês Caldeira Fernandes	Paula Ravasco
Ana Cristina Pestana Neves	Isabel Barahona	Paulo Mascarenhas
Ana Paula Serro	Joana Costa	Paulo Oliveira
Ana Raquel Barata	Joana Garrido	Pedro Brandão
Ana Vieira	João Aguiar	Ricardo Alves
André Almeida	João Albernaz Neves	Ricardo Baúto
António Delgado	João Botelho	Sara Raposo
Carla Ascenso	João Couvaneiro	Sergio Félix
Carla Martinho Neto	João Dias	Sílvia Almeida
Carlos Zagalo	Jorge Caldeira	Sofia Jordão
Catarina Almeida	José Feliz	Susana Monteiro
Catarina Ramos	José Grillo	Teresa Costa
Cátia Caneiras	Júlia Antunes	Tiago Ferro
Cecília Rozan	Lucinda Bessa	Vanessa Machado
Cidália Castro	Mara Guerreiro	Xavier Melo
Cristina Soeiro	Margarida Miranda	
Deolinda Auxtero	Maria do Rosário Dias	

The authors were notified of the reviewers' comments and asked for revision and resubmission until 21 June 2023. The Proceedings Scientific Committee was ultimately responsible for checking the corrections and making the final decision. Authors received a notification about the acceptance or rejection of their work by 26 June 2023.

The 49 accepted proceeding papers (77.8% acceptance rate) were presented at the conference (16 oral and 33 poster communications) and checked for plagiarism with Turnitin (www.turnitin.com) before being sent for publication. The volume editors also double-checked compliance with formatting rules.

Acknowledgments: We acknowledge the input of the experts across the different scientific areas who contributed to more scientifically sound proceeding papers.

Conflicts of Interest: The authors declare no conflict of interest.

medical sciences forum

MDPI

Proceeding Paper

Exploring Genetic Variations and Psychological Factors in Alcohol and Drug Consumption in a Portuguese Female Sample †

Joana Couceiro [1], Sara Oliveira [2], Catarina Godinho [2], Inês Alves [2], Juliana Fonte [2], Ricardo Baúto [1], Alexandra Maia e Silva [1], Iris Almeida [1], Cristina Soeiro [1,*] and Alexandre Quintas [1,*]

1 Egas Moniz Center for Interdisciplinary Research (CiiEM), Egas Moniz School of Health & Science, Caparica, 2829-511 Almada, Portugal; joana.m.couceiro@gmail.com (J.C.); rbauto@egasmoniz.edu.pt (R.B.); amaiaesilva@egasmoniz.edu.pt (A.M.e.S.); ialmeida@egasmoniz.edu.pt (I.A.)

2 Egas Moniz School of Health & Science, Caparica, 2829-511 Almada, Portugal; sara.oliveira99@gmail.com (S.O.); kattarinagodinho@hotmail.com (C.G.); ines.alves-13@hotmail.com (I.A.); julianafonte4@gmail.com (J.F.)

* Correspondence: csoeiro@egasmoniz.edu.pt (C.S.); alexandre.quintas@gmail.com (A.Q.)

† Presented at the 6th International Congress of CiiEM—Immediate and Future Challenges to Foster One Health, Almada, Portugal, 5–7 July 2023.

Abstract: While in some countries, the possession of psychoactive substances leads to prison sentences or execution, other countries, like Portugal, follow an avenue leading to a drug-tolerant culture. However, there is a lack of empirical data on Portuguese populations to measure the drug-use trend. The present study uses multidisciplinary approaches to explore the prevalence of alcohol and drug consumption on a Portuguese student population (N = 81, ages ranged from 17 to 40 years), associating it with psychological and genetic factors. The results show a prevalence of cannabis consumption higher than what is reported by the EMCDDA, and suggests that carriers of the minor allele of 5-HTTLPR have a higher propensity for addiction.

Keywords: genetic variations; psychological factors; consumptions; drugs; alcohol

Citation: Couceiro, J.; Oliveira, S.; Godinho, C.; Alves, I.; Fonte, J.; Baúto, R.; Silva, A.M.e.; Almeida, I.; Soeiro, C.; Quintas, A. Exploring Genetic Variations and Psychological Factors in Alcohol and Drug Consumption in a Portuguese Female Sample. *Med. Sci. Forum* **2023**, *22*, 34. https://doi.org/10.3390/msf2023022034

Academic Editors: José Brito, Nuno Taveira and Ana I. Fernandes

Published: 16 August 2023

1. Introduction

Drug abuse is a ubiquitous historical phenomenon that largely depends on the cultural and political environment. In Portugal, the primary drug law is Decree-Law 15/93, which frames the legal regime applicable to drug trafficking and consumption of psychoactive substances. Decree-Law 130-A/2001 changed the 1993 framework by decriminalizing the consumption, acquisition and possession of drugs for personal consumption. Notwithstanding this, people possessing substances of abuse quantities below the amounts stated on the current drug Decree-Law and not under suspicion of drug trafficking are evaluated by the local Commission for Dissuasion of Drug Addiction. In extreme cases, punitive sanctions can be applied, but the main purpose is to promote addiction treatment. The Portuguese case for drug control frequently receives focus due to the encouraging results regarding drug use prevalence, which remains reasonably low compared to other European countries. However, the statistics compiled by the European Monitoring Centre for Drugs and Drug Addiction (EMCDDA) indicate that between 2001 and 2007, lifetime prevalence rates for cannabis, cocaine, amphetamines, ecstasy, and LSD use rose for the Portuguese general population (ages 15–64) and for the 15–34 age group [1].

The Portuguese government's investment in drug safety, which involves assessment, drug checking, and wastewater analysis, among other analytical tools to monitor the drug consumption trend, is the lowest in Europe, with a value of 0.003% of GDP [2], since most of the drug-related public budget is expended on the health consequences of drug abuse. Consequently, the lack of empirical data is a significant problem that may bias the reports' numbers. The latest report by the EMCDDA indicates a lifetime prevalence of 8% of cannabis use among the Portuguese population aged 15–34. However, the survey used on this report was from 2016. The figures for the abuse of other drugs are much lower, with a lifetime prevalence of 0.3% for cocaine and 0.2% for MDMA. Finally, among individuals seeking treatment, 33.9% use cannabis and 39.4% use opioids [1].

Independently of the success or lack of success of public policies, psychoactive substance use must be studied and debated based on its very foundations without any prejudice. Literature data on psychoactive substance use suggest evolutionary, environmental, neurobiological, personality [3], and genetic causes for drug addiction. However, multi- and transdisciplinary approaches are scarce. Thus, a holistic approach for drug addiction is needed. This assumes a higher relevance on cannabis and alcohol use due to their consumption prevalence and people enrolling in treatment programs.

Factors such as the progression of consumption patterns, the initiation age, the substances used, the routes of administration, concurrent substance consumption, relapse circumstances, previous cessation attempts, and drug-related health problems are significant in comprehending the psychosocial aspects and the accompanying psychopathological impact of drug use [4]. From a biological perspective, some studies suggest that hereditary influences may contribute to a predisposition for substance use; however, these effects can be mitigated by the presence of adequate parental monitoring and supervision, further underscoring the impact of psychosocial development [5]. Several genes have been identified and associated with addiction, including the serotonin transporter gene (*SLC6A4*). Serotonin (5-HT) is a neurotransmitter that, after release, is taken up again from synaptic spaces by SLC6A4, localized in presynaptic neuronal membranes, terminating the synaptic actions of 5-HTT and recycling it into the neurotransmitter pool. The *SLC6A4* is a 31 kb gene on chromosome 17 (17q11.2) (OMIM-182138). *5-HTTLPR*, its most common polymorphism, is a 44-bp indel at the promotor region modifying SLC6A4 activity. It has two major alleles: the long (L) and the short (S) form [6]. The S form is associated with lower basal transcriptional efficiency of the *SLC6A4* gene promotor, resulting in lower serotonin uptake activity comparing with the L form [7]. Several case–control and family-based association studies try to link variants in *SLC6A4* with alcohol or drug dependence. The results are contradictory, as some of these studies reported positive findings, while many others found no evidence of association. A 2013 meta-analysis based on 55 studies supports the associations of *5-HTTLPR* with alcohol, heroin, cocaine, and methamphetamine dependence and abuse [7]. However, the authors suggest that further studies are needed. According to several studies *SLC6A4/5-HTTPLR* is one of the four genes associated with addiction, particularly in the European population [8]. The present research aims to explore the intricate interplay between genetic factors and the psychosocial development of individuals with a history of substance abuse.

2. Materials and Methods

This pilot study examined a sample of 81 female university students with ages ranging from 17 to 40 years, with an average age of 21.04 years (SD = 3.16). DNA was obtained from a buccal swab and the psychological variables were collected through three self-report instruments assessing substance consumption, adverse childhood experiences, aggression and personality. For the analysis of the polymorphism *SLC6A4/5-HTTLPR*, DNA was obtained from the buccal swab. DNA extraction was performed with the QIAamp DNA investigator Kit® (QIAGEN), according to the manufacturer's instructions. *5-HTTLPR* genotyping was performed via PCR, using the following primers: 5′ GGCGTTGCCGCTCT-GAATGC 3′ (forward) and 5′ GAGGGACTGAGCTGGACAACCAC 3′ (reverse). In this process, 25 µL PCR mix reactions were used, containing approximately 50 ng of genomic DNA, 1X Horse-Power™ Taq DNA polymerase MasterMix® (Canvax), 5 pmol of each primer, and 2% DMSO. Reactions were carried out with the following program: 94 °C for 2 min, followed by 35 cycles at 94 °C for 10 s, 61 °C for 30 s, and 68 °C for 30 s. The amplification products were analyzed on a 2.5% (*w/v*) agarose gel; random products were confirmed via Sanger sequencing. Regarding psychological variables and substance consumption, The Adverse Childhood Experiences Questionnaire (ACE; Felitti & Anda, 1998; Portuguese Version: Maia & Silva, 2007), The Reactive-Proactive Aggression Questionnaire (RPQ; Raine et al., 2006; Portuguese Version: Pechorro et al., 2017) and the NEO Five-Factor Inventory (NEO-FFI, McCrae and Costa in 2004) were used to collect data.

3. Results

The results indicate that 11.1% (n = 9) of the subjects reported never consuming alcoholic beverages. The remaining 88.9% was divided between "sometimes" consuming alcohol, accounting for 49.4% (n = 40), and "frequent" consumption, representing 39.5% (n = 32). Regarding cannabinoid consumption, 61.7% (n = 50) reported never consuming cannabinoids, 28.4% (n = 23) reported consuming cannabinoids "sometimes", and the remaining 9.9% (n = 8) reported consuming them frequently. As for heavier drug consumption, including ecstasy, cocaine, or heroin, 91.4% (n = 74) reported never consuming these drugs, while 3.7% (n = 3) reported consuming them "sometimes", and 4.9% (n = 4) reported consuming them "frequently", meaning that nearly 9% of all subjects were engaged in hard drug use. Regarding the Five Factor Model (Personality), we also found a weak positive correlation between neuroticism and alcohol consumption, which we identified as the most relevant (Spearman) correlation ($r = 0.22$, $p < 0.01$). This result provides a linkage that suggests that genetic factors can influence the regulation of neurotransmitters and neural pathways associated with reward and stress response, making individuals with higher neuroticism scores more vulnerable to developing problematic alcohol consumption patterns. Concerning *SLC6A4/5-HTTLPR*, the subjects were separated only between consumers and non-consumers. *5-HTTLPR* allele distribution was consistent with the Hardy–Weinberg equilibrium. The major allele frequency was L = 91(56.2%), which was consistent with the values observed in previous studies for the European population (48–71%). The genotype frequencies were analyzed, and there was no evidence of a significant association between *5-HTTLPR* and consumption for all substances under analysis (alcohol, cannabinoids, and ecstasy, cocaine, and heroin). For alcohol abuse, 18.1% of the 'consumption' group had the SS-genotype, while in the 'no consumption' group, no one with the same genotype was found. The frequency of the LL-genotype was 27.8% and 33.3% for the 'consumption' and 'no consumption' groups, respectively. The heterozygous genotype was present in 54.2% of the consumers and in 66.7% of the non-consumers. For cannabinoid consumption analysis, the genotype distribution in the consumers was 54.8%, 22.6%, and 22.6% for the LS-, LL-, and SS-genotype, respectively. In the 'no consumption' group, the distribution was 56%, 32%, and 12%, respectively. Finally, for ecstasy, cocaine and heroin abuse, this trend is even higher. The frequency of homozygous individuals in the 'consumption' group was 28.6% for the S allele and 14.3% for the L allele. On the contrary, in non-consumers, this distribution was 14.9% for the SS-genotype and 29.7% for the LL-genotype. The heterozygous individuals were evenly distributed between the two groups (57.1% and 55.4%, correspondingly). Although the results only show trends, without statistical significance, they are in line with what is described in the literature. The *5-HTTLPR* S-allele results in a less efficient transporting function, leading serotonin to remain longer in the synapse, reducing its recycling period, thus resulting in a reduction in circulating serotonin [8]. This is the reason why *5-HTTLPR* has been linked to higher levels of anxiety and a negative mood in healthy individuals, and consequent consumption/addiction.

4. Discussion

Based on the presented research, deepening the studies on genetic variability and psychological variables is of utmost importance for developing a better understanding of substance use. Exploring these factors contributes to a more personalized approach to treating and preventing addiction, considering individual differences and specific intervention responses. Furthermore, analyzing the female gender in this context is crucial, as women have biological, psychosocial, and cultural particularities that influence alcohol and drug consumption. The inclusion of studies focused on the female gender fills significant gaps and provides a solid foundation for the implementation of health policies and appropriate interventions. Therefore, continuing research in this area is encouraged, aiming to understand the risk factors, consumption patterns, and specific needs of women to promote effective and comprehensive interventions in addressing alcohol and drug consumption.

Author Contributions: Conceptualization, A.Q. and C.S.; methodology, J.C., A.Q and C.S.; software, J.C., C.S., S.O., C.G., I.A. (Inês Alves) and J.F.; validation, C.S. and A.Q.; formal analysis, C.S., J.C., A.M.e.S. and A.Q.; investigation, J.C., C.S., S.O., C.G., I.A. (Inês Alves) and J.F.; resources, A.Q. and C.S.; data curation, C.S., R.B. and J.C.; writing—original draft preparation, J.C. and R.B.; writing—review and editing, A.Q.; supervision, A.Q., C.S. and J.C.; project administration, I.A. (Iris Almeida); funding acquisition, I.A. (Iris Almeida), R.B. and A.M.e.S. All authors have read and agreed to the published version of the manuscript.

Funding: Research funding by CiiEM Investiga 2018 (CEI/24/2018) by Egas Moniz Center for Interdisciplinary Research (CiiEM).

Institutional Review Board Statement: The study was conducted in accordance with the Declaration of Helsinki and approved by the Institutional Review Board (or Ethics Committee) of Egas Moniz School of Health and Science for studies involving humans.

Informed Consent Statement: We rigorously adhered to the highest ethical standards, including the principles outlined in the Helsinki Declaration. Furthermore, we obtained informed consent from all participants, ensuring their autonomy and well-being throughout the scientific study.

Data Availability Statement: Not applicable.

Acknowledgments: The authors acknowledge Egas Moniz Center for Interdisciplinary Research (CiiEM) from Egas Moniz School of Health and Science for the financial support to CiiEM Investiga 2018 (CEI/24/2018).

Conflicts of Interest: The authors declare no conflict of interest.

References

1. European Monitoring Centre for Drugs and Drug Addiction. *European Drug Report 2021: Trends and Developments*; Publications Office of the European Union: Luxembourg, 2021.
2. Prieto, L. Labelled drug-related public expenditure in relation to gross domestic product (gdp) in Europe: A luxury good? *Subst. Abus. Treat. Prev. Policy* **2010**, *5*, 9. [CrossRef]
3. Terracciano, A.; Löckenhoff, C.E.; Crum, R.M.; Bienvenu, O.J.; Costa, P.T. Five-factor model personality profiles of drug users. *BMC Psychiatry* **2008**, *8*, 1–10. [CrossRef] [PubMed]
4. Waaktaar, T.; Torgersen, S. Genetic and environmental causes of variation in perceived loneliness in young people. *Am. J. Med. Genet.* **2012**, *159B*, 580–588. [CrossRef] [PubMed]
5. Gelernter, J.; Kranzler, H.; Cubells, J.F. Serotonin transporter protein (SLC6A4) allele and haplotype frequencies and linkage disequilibria in African- and European-American and Japanese populations and in alcohol-dependent subjects. *J. Hum. Genet.* **1997**, *101*, 243–246. [CrossRef] [PubMed]
6. Heils, A.; Teufel, A.; Petri, S.; Stöber, G.; Riederer, P.; Bengel, D.; Lesch, K.P. Allelic variation of human serotonin transporter gene expression. *J. Neurochem.* **1996**, *66*, 2621–2624. [CrossRef] [PubMed]
7. Cao, J.; Hudziak, J.J.; Li, D. Multi-cultural association of the serotonin transporter gene (SLC6A4) with substance use disorder. *Neuropsychopharmacology* **2013**, *38*, 1737–1747. [CrossRef] [PubMed]
8. Lopez-Leon, S.; González-Giraldo, Y.; Wegman-Ostrosky, T.; Forero, D.A. Molecular genetics of substance use disorders: An umbrella review. *Neurosci. Biobehav. Rev.* **2021**, *124*, 358–369. [CrossRef] [PubMed]

medical sciences forum

MDPI

Proceeding Paper

Sunscreening and Photosensitizing Therapy in the Elderly †

Maria Deolinda Auxtero *, José Brito and Isabel Margarida Costa

Egas Moniz Center for Interdisciplinary Research (CiiEM), Egas Moniz School of Health & Science, 2829-511 Caparica, Portugal; jbrito@egasmoniz.edu.pt (J.B.); isabelc@egasmoniz.edu.pt (I.M.C.)
* Correspondence: mauxtero@egasmoniz.edu.pt; Tel.: +351-212946753
† Presented at the 6th International Congress of CiiEM—Immediate and Future Challenges to Foster One Health, Almada, Portugal, 5–7 July 2023.

Abstract: People aged 65+ are generally polymedicated. In this age group, it is important to be aware of the potential for drug-induced photosensitivity. Sunscreen use is an effective way to prevent sun damage to the skin. Our study aimed to evaluate sunscreen use habits in 104 non-institutionalized elderly people aged 65 years and older and the taking/use of photosensitizing medications. It was concluded that all but one of the elderly take photosensitizing drugs but do not use sunscreen regularly. Although all therapies were prescribed or advised by health professionals, virtually all the elderly stated that they did not receive advice on this matter.

Keywords: drug-induced photosensitivity; drug-induced photoallergy; drug-induced phototoxicity; elderly; sunscreen protection

Citation: Auxtero, M.D.; Brito, J.; Costa, I.M. Sunscreening and Photosensitizing Therapy in the Elderly. *Med. Sci. Forum* **2023**, 22, 4. https://doi.org/10.3390/msf2023022004

Academic Editors: Nuno Taveira and Ana I. Fernandes

Published: 9 August 2023

1. Introduction

Photodermatoses are a group of skin problems induced by ultraviolet radiation and, in some cases, visible light. Drug-induced photosensitivity includes photoallergy and phototoxicity and is one of the most common photodermatoses [1]. The rate of cases of photosensitization has been steadily increasing. This may be associated with excessive sun exposure to achieve a socially valued suntan, but also with an increase in exogenous triggers present in foods, dietary supplements, cosmetics, and pharmaceuticals. Medicines represent a large percentage of photosensitizing substances, leading to an increased risk of sunburn, rashes, and other types of skin lesions. Hundreds of drugs administered topically or systemically are implicated in photosensitivity reactions by means of complex mechanisms [2]. These include antimicrobial, cardiovascular, antihyperlipidemic, psychotropic, antineoplastic [2], and non-steroidal anti-inflammatory [3] drugs. Photosensitivity is a skin-related adverse drug reaction which can lead to additional complications demanding hospitalization and the restriction of the use of some medications. The elderly population are particularly prone to drug-induced photosensitivity due to multiple pathologies and polypharmacy-related problems; complex therapeutic regimens, mainly in industrialized societies, result in a burden on healthcare systems.

Regular use of sunscreen is one of the main methods of preventing photodermatoses. However, compliance with sun protection guidelines is highly variable and tends to be poor and gender dependent [4,5]. The prevalence of sunscreen non-usage reported in the literature ranges from 9.9% to 68% [6,7]. In 2020, the percentage of seniors over 65 who always used sunscreen was 13.6% in men and 30.9% in women [5]. Nevertheless, little is known about sun protection habits in the medicated elderly, especially those living in a familiar environment and subject to some daily sun exposure. Thus, our study aimed to analyze the sun protection practices of non-institutionalized elderly people medicated with at least two medications and to evaluate the relationship between these practices and the use of photosensitizing drugs.

2. Materials and Methods

A cross-sectional study was conducted in May 2023 with elderly people over 65 years old, taking at least two medicines and living alone or with relatives, in Portugal and abroad. Upon acceptance of the participants, informed consent was signed. Participants were requested to show the packages of all medicines they were taking at the time of the interview, and a series of demographic and anthropometric data were collected, such as age, gender, nationality, and composition of the family household. Additionally, habits of sun protection were assessed through the question "Do you use sunscreen?". Respondents had four options, namely: "never", "rarely", "only in summer", and "regularly".

Medicines taken by the elderly were characterized in terms of ATC code, route of administration, prescriber, and photosensitizing potential. Drug-induced photosensitivity appears to be underreported, mainly due to issues of difficult clinical recognition and scarce documentation in public databases. For the present work, the potential for drug photosensitivity was assessed through an extensive literature search, including reviews, clinical cases, summary of product characteristics (SmPC), package leaflets (PL), and databases such as Medscape [8].

Since variables of interest are measured on a nominal or ordinal scale, the association between them was assessed using the Spearman correlation coefficient or contingency tables and the chi-square test of independence. All statistical tests were applied at the 5% level of significance, using SPSS 28 (Armonk, NY, USA: IBM Corp.). An association between habits of sun protection and nationality was not assessed due to the scarcity of data per cell.

3. Results

3.1. Subjects

Demographic, Anthropometric, and Social Behavior Data

A total of 104 seniors with an average age of 78.4 years old (range 65 to 96 years) were included in the study. Table 1 lists the frequency of the variables of interest. The percentage of individuals who do not use sunscreen is undistinguishable from those who use it only in summer (sig. = 0.305), but it is higher than those who rarely (sig. = 0.002) or regularly use it (sig. < 0.001).

Table 1. Demographic and social behavior data of participants.

Gender	Frequency	Percentage
Male	36	34.6
Female	68	65.4
Citizenship		
Portugal	62	59.6
France	32	30.8
Spain	6	5.8
Other	4	3.8
Household		
Live alone	36	34.6
Live with other elderly	49	47.1
Live with younger adults	19	18.3
Use of sunscreen		
Never	43	41.3
Rarely	19	18.3
Only in summer	34	32.7
Regularly	8	7.4

3.2. Therapeutics

3.2.1. Therapeutic Groups

A total of 489 drugs were taken by the participants (average 4.7 drugs/elderly), corresponding to 168 different molecules from 14 therapeutic groups. The top three groups were cardiovascular system (28.7%); nervous system (19.5%); and alimentary tract and metabolism (14.9%).

3.2.2. Route of Administration

More than 93% of the drugs were administered via the oral route (Figure 1).

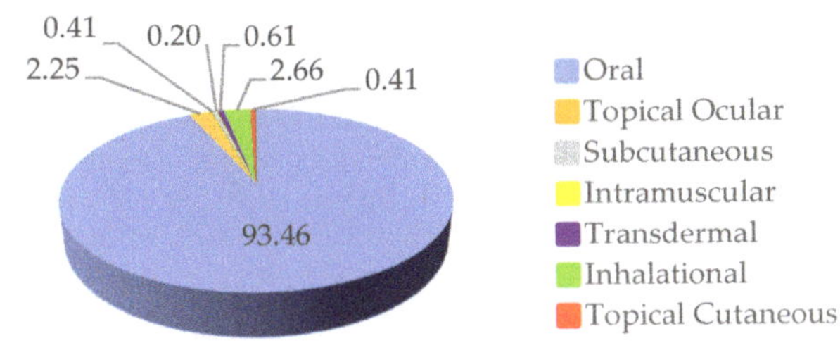

Figure 1. Routes of drug administration used by the elderly in the study.

3.2.3. Photosensitizing Drugs

Ninety-nine percent of the elderly people who participated in the study use between 1 and 13 photosensitizing drugs (Figure 2). Only one elderly person was not taking any photosensitizing medicine.

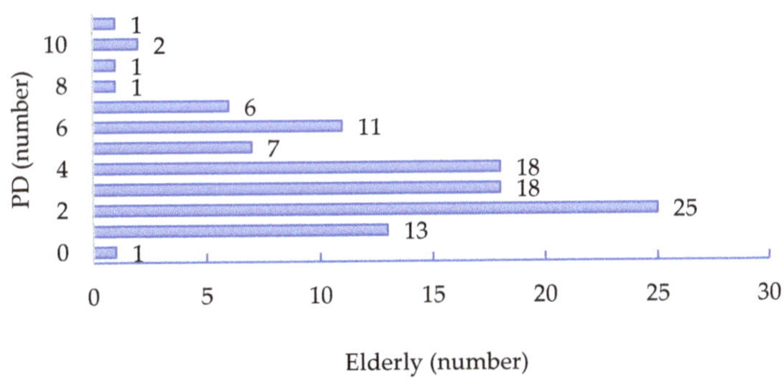

Figure 2. Number of photosensitizing drugs (PD) taken by the elderly.

3.3. Associations between Use of Sunscreen Habits and Variables of Interest

3.3.1. Gender

There is no association between gender and use of sunscreen habits according to the chi-square test for independence (sig. = 0.470).

3.3.2. Use of Photosensitizing Drugs

A Spearman correlation coefficient of −0.104 was estimated from the data of n = 104 subject, with an estimated significance of 0.157. In conclusion, no association exists between skin protection and the use of photosensitizing medicines.

3.3.3. Household

The association with 'living alone' was assessed using contingency tables and the chi-square test of independence, which showed that the variables are not associated (sig. = 0.297). The same lack of association was observed between skin protection behavior and living with younger adults (<65 y.o.) (sig. = 0.428).

4. Discussion and Conclusions

Nearly all the elderly people in this study took several photosensitizing drugs, but they did not have regular habits of sun protection. In fact, only 7.4% of the participants used sunscreen regularly. This value is lower than 13.6% and 30.9% previously reported for men and women in the same age group [5]. To our knowledge, there are no published data on the sun protection habits of the elderly (institutionalized or in a home environment) who take photosensitizing medications. In the present study, no questions were asked about the eventual symptoms of photodermatoses. Moreover, the duration of the treatments is not known. Although all the medicines were prescribed by a doctor (99.6%) or advised to be taken by a pharmacist (0.4%), only in two cases did a participant report being informed of the importance of using sunscreen during the therapy. To minimize the risk of skin lesions caused by photosensitizing drugs, it is of the utmost importance that the elderly be advised to use sunscreen whenever therapy includes such medication.

Author Contributions: Conceptualization, M.D.A. and I.M.C.; formal analysis, J.B. and M.D.A.; investigation, M.D.A. and I.M.C.; writing—original draft preparation, M.D.A.; writing—review and editing, M.D.A., I.M.C. and J.B. All authors have read and agreed to the published version of the manuscript.

Funding: This research received no external funding.

Institutional Review Board Statement: The study was conducted in accordance with the Declaration of Helsinki and approved by the Ethics Committee of Egas Moniz School of Health and Science (protocol code 1230, 25th of May 2023).

Informed Consent Statement: Informed consent was obtained from all subjects involved in the study.

Data Availability Statement: The data presented in this study are available on request from the corresponding author. The data are not publicly available due to privacy restrictions.

Acknowledgments: The authors acknowledge the third-year students of the Integrated Masters in Pharmaceutical Sciences of IUEM, Egas Moniz School of Health and Science for their help with data collection.

Conflicts of Interest: The authors declare no conflict of interest.

References

1. Blakely, K.M.; Drucker, A.M.; Rosen, C.F. Drug-Induced Photosensitivity-An Update: Culprit Drugs, Prevention and Management. *Drug Saf.* **2019**, *42*, 827–847. [CrossRef] [PubMed]
2. Kowalska, J.; Rok, J.; Rzepka, Z.; Wrześniok, D. Drug-Induced Photosensitivity-From Light and Chemistry to Biological Reactions and Clinical Symptoms. *Pharmaceuticals* **2021**, *14*, 723. [CrossRef] [PubMed]
3. Blanca-Lopez, N.; Soriano, V.; Garcia-Martin, E.; Canto, G.; Blanca, M. NSAID-induced reactions: Classification, prevalence, impact, and management strategies. *J. Asthma Allergy* **2019**, *12*, 217–233. [CrossRef] [PubMed]
4. Bocquier, A.; Fressard, L.; Legleye, S.; Verger, P.; Peretti-Watel, P. Social Differentiation of Sun-Protection Behaviors: The Mediating Role of Cognitive Factors. *Am. J. Prev. Med.* **2016**, *50*, e81–e90. [CrossRef] [PubMed]
5. Rep, M.M.M. *QuickStats: Percentage of Adults Aged ≥18 Years Who Always Use Sunscreen When Outside for >1 Hour on a Sunny Day, by Sex and Age Group—National Health Interview Survey, United States, 2020*; Centers for Disease Control and Prevention (CDC): Hyattsville, MD, USA, 2022; Volume 71, p. 747. [CrossRef]
6. Bertolin, M.; Cercatto, M.C.; Requena, C.; Serra-Guillen, C.; Llombart, B.; Sanmartin, O.; Guillen, C.; Nagore, E. Awareness, Attitude, and Adherence to Preventive Measures in Patients at High Risk of Melanoma. A Cross-Sectional Study on 185 Patients. *J. Cancer Educ.* **2015**, *30*, 552–566. [CrossRef] [PubMed]

7. Da Silva, E.S.; Dumith, S.C. Non-use of sunscreen among adults and the elderly in southern Brazil. *An. Bras. Dermatol.* **2019**, *94*, 567–573. [CrossRef] [PubMed]
8. Drugs & Diseases. Available online: https://reference.medscape.com/ (accessed on 16 May 2023).

medical sciences forum

MDPI

Proceeding Paper

Hyposalivation and Xerostomia: Prevalence and Associated Factors in the Elderly †

Inês Caetano Santos [1,*], Ana Rita Dias [1], Joana Maximiano [1], Ana Cristina Manso [1], Mário Polido [1], Luís Proença [1], José João Mendes [1] and Helena Canhão [2]

[1] Egas Moniz Center for Interdisciplinary Research (CiiEM), Egas Moniz School of Health & Science, 2829-511 Almada, Portugal; ritafonseca.1999@gmail.com (A.R.D.); joanamaximiano6@gmail.com (J.M.); mansocristina@gmail.com (A.C.M.); mpolido@egasmoniz.edu.pt (M.P.); luisfapro@gmail.com (L.P.); jmendes@egasmoniz.edu.pt (J.J.M.)

[2] NOVA Medical School, 1169-056 Lisboa, Portugal; helena.canhao@nms.unl.pt

* Correspondence: isantos@egasmoniz.edu.pt

† Presented at the 6th International Congress of CiiEM—Immediate and Future Challenges to Foster One Health, Almada, Portugal, 5–7 July 2023.

Abstract: In the elderly, many chronic diseases and systemic conditions, medications, and salivary dysfunctions, mostly caused by therapeutic radiation and chemotherapy, are associated with hyposalivation and xerostomia, affecting quality of life. The purpose of this study was to identify and evaluate potential factors associated with xerostomia and/or hyposalivation in a local elderly population. Almost half of the participants had hyposalivation and some reported xerostomia. There was a significant association between hyposalivation and xerostomia. Hyposalivation was most common among women, those taking one or more types of medication simultaneously, those taking antidepressants, and participants with rheumatoid arthritis and depression. There was a tendency for individuals with poorer oral-health-related quality of life to exhibit hyposalivation.

Keywords: xerostomia; hyposalivation; geriatrics

Citation: Santos, I.C.; Dias, A.R.; Maximiano, J.; Manso, A.C.; Polido, M.; Proença, L.; Mendes, J.J.; Canhão, H. Hyposalivation and Xerostomia: Prevalence and Associated Factors in the Elderly. *Med. Sci. Forum* **2023**, *22*, 33. https://doi.org/10.3390/msf2023022033

Academic Editors: José Brito, Nuno Taveira and Ana I. Fernandes

Published: 16 August 2023

1. Introduction

In the elderly, several chronic diseases and systemic conditions, such as autoimmune diseases, polymedication, and salivary dysfunction, mostly caused by therapeutic radiation and chemotherapy, are associated with reduced salivary flow [1]. Xerostomia is defined as the perception of dry mouth, whereas hyposalivation is the objective reduction in salivary flow [2]. However, both conditions can cause signs or symptoms that affect quality of life [1,3]. The purpose of this study was to identify and evaluate potential factors associated with xerostomia and/or hyposalivation in a local elderly population.

2. Materials and Methods

A total of 150 elderly participants, aged 65 years or older, were included in this study, which was carried out at the Egas Moniz Dental Clinic between December 2021 and May 2022. Sociodemographic data, general health status data, Summated Xerostomia Inventory (SXI-PL) scores, and Geriatric Oral Health Assessment Index (GOHAI) scores were obtained via a questionnaire. Hyposalivation was assessed by calculating the rate of stimulated (<0.7 mL/min) and unstimulated (<0.1 mL/min) salivary flow using sialometry. The data were statistically analyzed using descriptive and inferential methodologies (chi-square and Mann–Whitney tests), by means of IBM SPSS Statistics v.27.0. A significance level of 5% was used in all inferential analyses.

3. Results

The majority of participants were female (60.0%) and the mean age was 72.9 (±4.0) years. Arterial hypertension was the most common systemic disease (60.7%). More than half were taking medication (86.0%), with antihypertensives being the most common (59.0%). Most participants were taking two or more types of medication at the same time (67.0%). Almost half of the participants had hyposalivation (48.7%) and 31.3% reported a 'dry mouth perception'. A significant association was found between hyposalivation and xerostomia perception ($p < 0.05$). Hyposalivation was more common in women (52.2%); those simultaneously taking 1, 2, 3, or 5+ types of medication (41.9%, 53.3%, 60.6%, and 62.5%, respectively); and those taking antidepressants (57.1%). Participants with rheumatoid arthritis and depression were also more likely to have hyposalivation (60.0% and 66.7%, respectively). Although no significant association was found, there was a tendency for individuals with poorer oral-health-related quality of life to have hyposalivation (57.7%).

4. Discussion and Conclusions

Xerostomia and hyposalivation semiologically allow us to indicate the clinical and subclinical state of a patient. Hyposalivation was a very common condition in this study, which may be explained by the co-existence of some chronic diseases, such as rheumatoid arthritis or depression, the concomitant use of medication, or the atrophy of salivary gland tissue associated with ageing [1,2]. Some types of medication, such as antidepressants, can cause vasoconstriction in the salivary glands, altering their hydro-electrolyte balance, leading to changes in acinar cell function or glandular tubule structure [4]. Hyposalivation in female participants can be explained by the decrease in progesterone and estrogen due to the menopause, resulting in the salivary glands being unable to absorb these hormones [5]. Hyposalivation is a condition that promotes poorer quality of life in this age group and should be diagnosed and treated in order to improve the well-being of the elderly [3].

Author Contributions: Conceptualization, I.C.S., H.C. and A.C.M.; methodology, I.C.S., H.C. and A.C.M.; validation, M.P. and J.J.M.; formal analysis, I.C.S., A.R.D. and J.M.; investigation, I.C.S., A.R.D. and J.M.; resources, I.C.S.; data curation, L.P.; writing—original draft preparation, I.C.S. and A.R.D.; writing—review and editing, I.C.S., A.R.D., A.C.M. and J.J.M.; supervision, A.C.M., J.J.M. and H.C.; project administration, A.C.M., J.J.M., M.P. and H.C. All authors have read and agreed to the published version of the manuscript.

Funding: This study received no external funding.

Institutional Review Board Statement: The study was conducted in accordance with the Declaration of Helsinki and approved by the Ethics Committee of Egas Moniz School of Health and Science (protocol code 1013, approved on 27 January 2022).

Informed Consent Statement: Informed consent was obtained from all subjects involved in the study.

Data Availability Statement: The data presented in this study are available on request from the corresponding author. The data are not publicly available as they are part of an ongoing study.

Conflicts of Interest: The authors declare no conflict of interest.

References

1. Kim, Y.J. Xerostomia and Its Cellular Targets. *Int. J. Mol. Sci.* **2023**, *24*, 5358. [CrossRef]
2. López-Pintor, R.M.; Casañas, E.; González-Serrano, J.; Serrano, J.; Ramírez, L.; de Arriba, L.; Hernández, G. Xerostomia, Hyposalivation, and Salivary Flow in Diabetes Patients. *J. Diabetes Res.* **2016**, *2016*, 4372852. [CrossRef] [PubMed]
3. Barbe, A.G.; Schmidt, P.; Bussmann, M.; Kunter, H.; Noack, M.J.; Röhrig, G. Xerostomia and hyposalivation in orthogeriatric patients with fall history and impact on oral health-related quality of life. *Clin. Interv. Aging* **2018**, *13*, 1971–1979. [CrossRef]

4. Närhi, T.O.; Meurman, J.H.; Ainamo, A. Xerostomia and Hyposalivation: Causes, consequences and treatment in the elderly. *Drugs Aging* **1999**, *15*, 103–116. [CrossRef] [PubMed]
5. Eliasson, L.; Carlén, A.; Laine, M.; Birkhed, D. Minor gland and whole saliva in postmenopausal women using a low potency oestrogen (oestriol). *Arch. Oral Biol.* **2003**, *48*, 511–517. [CrossRef] [PubMed]

medical sciences forum

MDPI

Proceeding Paper

Influence of Color Vision Impairments on Shade Matching among Dental Students †

Ana Mano Azul [1,*], Ana Monteiro [1], António H. S. Delgado [1,2] and Helena Valente [3]

1 Egas Moniz Center for Interdisciplinary Research (CiiEM), Egas Moniz School of Health & Science, 2829-511 Almada, Portugal; anapgmonteiro26@gmail.com (A.M.); asalesdelgado@egasmoniz.edu.pt (A.H.S.D.)
2 Division of Biomaterials and Tissue Engineering, Eastman Dental Institute, University College London (UCL), Royal Free Hospital, London NW3 2PF, UK
3 Egas Moniz Hospital, Rua da Junqueira, 1349-019 Lisbon, Portugal; mhvalente@mail.telepac.pt
* Correspondence: aazul@egasmoniz.edu.pt
† Presented at the 6th International Congress of CiiEM—Immediate and Future Challenges to Foster One Health, Almada, Portugal, 5–7 July 2023.

Abstract: This study examined the impact of color vision deficits on shade selection accuracy among 119 Portuguese dental students. Only males displayed color vision deficiencies (4/119). Participants completed an Ishihara color vision screening test and a shade selection test using the VITAPAN classical scale. Overall, 59.4% of students correctly matched shades. No significant gender difference was found. Matching shades was easier in the order of C > B > A > D, with darker shades posing more difficulty. Color vision deficiencies did not significantly affect accurate tooth shade selection. However, males showed a higher prevalence of color vision deficiencies, and shade matching ease varied based on matice.

Keywords: color; color vision deficits; shade selection

Citation: Azul, A.M.; Monteiro, A.; Delgado, A.H.S.; Valente, H. Influence of Color Vision Impairments on Shade Matching among Dental Students. *Med. Sci. Forum* **2023**, 22, 10. https://doi.org/10.3390/msf2023022010

Academic Editors: José Brito, Nuno Taveira and Ana I. Fernandes

Published: 9 August 2023

1. Introduction

In the past few decades, there has been an increasing demand from the general public for better dental aesthetics [1]. Although patients lack dental-specific knowledge, they are able to distinguish if a restoration has a biomimetic appearance or if it looks artificial and unaesthetically pleasing [2]. For this reason, accurate shade selection in the restorative treatment is crucial for its success. A visual method is commonly used by dentists to perform this selection. However, not all dentists are able to accurately undertake this selection, as the method is subjective and they might suffer from color vision deficits [2]. For this reason, all dentists should have some basic training in the science of color and carry out screening tests to assess their visual acuity.

2. Materials and Methods

The sample for this study comprised 119 dental students from Instituto Universitário Egas Moniz (IUEM, Caparica, Portugal), who were in their final year. The participants were predominantly females (75%), within an age range of 22–42 years old and with a reported mean age of 23. The evaluation process consisted of two stages. Firstly, a color vision test was employed for screening, utilizing the Ishihara test. This test was administered under standardized lighting conditions and involved participants reading numbers displayed inside colored circles. Secondly, a shade selection test was carried out using the VITAPAN classical scale (Vita Zahnfabrik, Bad Säckingen, Germany) for posterior comparison. During this last phase, each student was provided with three teeth from the VITAPAN classical scale and was required to match the shade of each tooth with a complete VITAPAN scale, while being blind to the original shade. Regarding statistical data treatment, descriptive statistics were performed utilizing SPSS v. 26.0 software.

3. Results

Among the participants in the study, color vision deficiency was observed exclusively in males, with 4 out of 119 males incorrectly identifying all discs in the Ishihara test. When it came to matching shades on the VITAPAN classical scale, the overall accuracy rate was 59.4% among all students. Specifically, male students achieved a 54.8% accuracy rate, while female students exhibited a slightly higher accuracy rate of 61%, although this difference was not statistically significant. In terms of shade matching ease, students found it relatively easier to correctly match shades in the following order of matices: C, B, A and D. Darker shades were generally more challenging to match accurately compared to lighter shades on average.

4. Discussion

The topic presented in this study holds significance as accurate shade selection is vital for successful restorative dental procedures. The study aimed to contribute to the existing literature by examining the correlation between color vision deficiencies and shade matching using a commercial tooth shade scale, among a random pool of dental students. The results of this study align with previous research conducted by Khosla et al. (2017) [1] and Ethell et al. (2006) [3], which concluded that color vision deficiencies do not significantly affect color matching. However, contrasting findings have been reported by Davison et al. (1990) and Barna et al. (1981) [4,5], who found that color vision deficiencies can impact color discrimination and may necessitate assistance when using color scales.

Overall, while some studies support the findings of the present study that color vision deficiencies do not significantly impact color matching, other studies have reported differences in color discrimination abilities. These differences could potentially be influenced by variations in lighting conditions and environmental factors during the tests. It is important for dental professionals with color vision deficiencies to be aware of their limitations and seek appropriate support when using color scales in their practice.

5. Conclusions

In conclusion, the findings of this study indicate that color vision deficiencies do not appear to significantly impact the accurate identification and selection of tooth shades using a commercial scale, specifically the VITAPAN classical scale. However, it is noteworthy that a higher prevalence of color vision deficiencies was observed among male participants. Furthermore, the ease of shade matching on the tooth shade scale seems to depend on the hue. These results align with the notion that individuals with color vision deficiencies can still effectively perform shade selection tasks in dental practice using appropriate tools and techniques.

Author Contributions: Conceptualization, A.M. and A.M.A.; methodology, H.V.; investigation, A.M.; resources, H.V. and A.M.A.; data curation, A.H.S.D.; writing—original draft preparation, A.M. and A.H.S.D.; writing—review and editing, A.M.A., A.H.S.D. and A.M.; supervision, A.M.A. and H.V. All authors have read and agreed to the published version of the manuscript.

Funding: This research received no external funding.

Institutional Review Board Statement: The study was conducted in accordance with the Declaration of Helsinki and approved by the Ethics Committee of Egas Moniz School of Health and Science (protocol code 743—08.05.2019).

Informed Consent Statement: Not applicable.

Data Availability Statement: Data are available upon reasonable request from the author.

Conflicts of Interest: The authors declare no conflict of interest.

References

1. Khosla, A.; Maini, A.P.; Wangoo, A.; Singh, S.; Mehar, D.K. Prevalence of Colour Vision Anomalies Amongst Dental Professionals and its Effect on Shade Matching of Teeth. *J. Clin. Diagn. Res.* **2017**, *11*, ZC33–ZC36. [CrossRef] [PubMed]
2. Naik, A.V.; Pai, R.C. Color Blindness in Dental Students and Staff—An obstacle in shade selection for restorations. *Ann. Essences Dent.* **2010**, *2*, 25–28.
3. Ethell, J.; Jarad, F.D.; Youngson, C.C. The effect of colour defective vision on shade matching accuracy. *Eur. J. Prosthodont. Restor. Dent.* **2006**, *14*, 131–136. [PubMed]
4. Davison, S.P.; Myslinski, N.R. Shade selection by color vision-defective dental personnel. *J. Prosthet. Dent.* **1990**, *63*, 97–101. [CrossRef] [PubMed]
5. Barna, G.J.; Taylor, J.W.; King, G.E.; Pelley, G.B., Jr. The influence of selected light intensities on color perception within the color range of natural teeth. *J. Prosthet. Dent.* **1981**, *46*, 450–453. [CrossRef]

medical sciences forum

MDPI

Proceeding Paper

Microbiological Analysis of Borehole Water Quality †

Catarina Silva [1], Telma Costa [1], Nádia Silva [1], Sérgio Valério [2], Helena Barroso [2], Maria João Hilário [2,*] and Sara Alves [2]

[1] Egas Moniz School of Health & and Science, 2829-511 Almada, Portugal; catarinasilva96@hotmail.com (C.S.); telmafiliparc@gmail.com (T.C.); nadia.solene2001@gmail.com (N.S.)
[2] Egas Moniz Center for Interdisciplinary Research (CiiEM), Egas Moniz School of Health & and Science, 2829-511 Almada, Portugal; sfcvalerio@gmail.com (S.V.)
* Correspondence: mjhilario1973@gmail.com
† Presented at the 6th International Congress of CiiEM—Immediate and Future Challenges to Foster One Health, Almada, Portugal, 5–7 July 2023.

Abstract: Groundwater is often used as a primary resource by those who have boreholes inside their properties; however, this has caused concerns among health professionals. This water may contain microorganisms or substances that are harmful. The main objective of this study was to microbiologically analyse the quality of the water coming from boreholes in the village of Santo Ovídio, Setúbal, by searching for bacteria that indicate faecal contamination, —such as total and faecal coliforms, *Escherichia coli*, faecal Enterococci, *Clostridium* and *Pseudomonas aeruginosa*—and quantifying them via the membrane filtration method. The research method was a quantitative, simple, descriptive, level I study with a sample size of 20 participants. It was found that 60% of the samples contained at least one of these microorganisms.

Keywords: pathogenic microorganisms; water quality; faecal contamination; membrane filtration method

Citation: Silva, C.; Costa, T.; Silva, N.; Valério, S.; Barroso, H.; Hilário, M.J.; Alves, S. Microbiological Analysis of Borehole Water Quality. *Med. Sci. Forum* **2023**, *22*, 11. https://doi.org/10.3390/msf2023022011

Academic Editors: José Brito, Nuno Taveira and Ana I. Fernandes

Published: 9 August 2023

1. Introduction

The presence of local sanitation systems, human activity, urbanization, industrialization, and sewage makes groundwater vulnerable to contamination by organisms. Populations who have boreholes or wells often use groundwater as their main source of water. Many people assume water is safe to drink because it looks good and is crystal clear. However, this water may contain microorganisms or substances harmful to human health and be unsuitable for human consumption. The transmission of diseases through water mainly involves the faecal–oral route. To protect and preserve water quality, and therefore public health, it is essential to perform physicochemical, microbiological, and ecotoxicological analyses of water.

Escherichia coli (*E. coli*) and other faecal coliforms are used as the health parameters for monitoring water quality. Coliforms include all aerobic and facultative anaerobic organisms, as well as Gram-negative, non-sporulated, and rod-shaped bacteria, which ferment lactose with gas formation. Within the group of total coliforms, a subgroup of thermotolerant coliforms can be observed. These are the ones that indicate faecal pollution since they are restricted to the gastrointestinal tract. These are characterized by the presence of the enzymes ß-D-galactosidase and ß-glucuronidase and the ability to ferment lactose and mannitol with gas production, in a medium containing bile salts or other surfactants [1]. Within the thermotolerant coliforms, *E. coli* is the microorganism considered to be the best indicator of faecal contamination. The Chromogenic Coliform Agar medium is used for the isolation and identification of all these microorganisms, producing pink colonies for non-faecal coliforms and blue colonies for faecal coliforms [2]. Other bacteria, such as *Pseudomonas aeruginosa* (*P. aeruginosa*) and Enterococci can also be isolated in water. *P. aeruginosa* is Gram-negative aerobic that is non-sporulated and positive for catalase and oxidase. In CN Agar medium, it produces greenish-coloured fluorescent colonies due to the production of the pigments, pyoverdine and pyocyanin [2,3]. Faecal Enterococci are non-spore-forming, facultative anaerobes and are negative for catalase and oxidase [4]. The selective medium used for its isolation is Slanetz and Bartley agar. This is based on the ability of this bacterium to reduce TTC (2,3,5-triphenyl tetrazolium chloride—colourless dye) to formazin, which leads to typical dark pink to red colonies [2,4]. Another Gram-positive bacterium present in the intestinal tract is *Clostridium perfringens* (*C. perfringens*). This bacillus is a strict anaerobe and is the most important sulfite-reducing bacteria within the genus *Clostridium*, forming heat-resistant spores. *C. perfrigens* does not multiply in aquatic environments, but its spores can survive in water for months and, consequently, its presence may indicate long-term faecal pollution [2,5]. The medium used to detect it is Tryptone–sulfite–cycloserine. The anaerobic sulfite-reducing spores of the bacterium can reduce the sulfite in the medium to sulphide, which in turn reacts with the ferric salt of the medium, thus producing black colonies [2,3].

The International Organization for Standardization is responsible for establishing international standards known as ISO standards for the purpose of improving quality. There is an ISO standard for each surveyed microorganism. The method of choice for the research of each microorganism is the membrane filtration method. To confirm the presence of specific microorganisms in each medium, biochemical confirmation tests were performed, such as the use of Hajna Kliger medium, as well as catalase and oxidase tests. The Hajna Kliger medium is used for the confirmation and identification of *Enterobacteriaceae*. This medium allows can differentiate between dextrose- lactose and/or sucrose-fermenting bacteria and detect hydrogen sulfide and gas production. The catalase test is used to evaluate the presence of the enzyme, catalase [6]. Catalase is present in most aerobic and facultative anaerobic bacteria. The presence of this enzyme makes it possible to differentiate microorganisms of the genus *Staphylococcus* spp. (catalase-positive) from other non-catalase-producing Gram-positive cocci. The oxidase test allows the presence of the cytochrome, oxidase enzyme, to be determined. The oxidase test is extremely important for differentiating between *Enterobacteriaceae* and *Pseudomonas* [2,4,7].

2. Materials and Methods

The main objective of the present study was to microbiologically analyse the quality of water coming from boreholes in the locality of Santo Ovídio, Setúbal. This was carried out through a microbiological analysis of 20 water samples from boreholes belonging to study participants, who were previously informed about the study via leaflets. As a procedure, samples were collected in sterile containers and kept refrigerated until they reached the laboratory. Microbiological analysis was performed within 24 h of collection. Then, 100 mL was filtered for each analysis/quantification. Filtration was carried out through a membrane filter using a vacuum pump. After filtration, the filters were transferred to the appropriate culture medium. Subsequently, additional confirmation tests were performed. According to the variables, the sample, and the objectives of the study, no statistical tests were used; only a simple statistical analysis was performed using SPSS. The results are organized in tables and circle graphs.

3. Results

The obtained results are shown in Table 1. Figure 1 shows a pie chart with the total number of samples that tested positive for at least one of the investigated bacteria. These samples (60%of the total) were considered contaminated.

Table 2 shows the percentages of all microorganisms found in the tested samples. As there were poly-contaminated samples, the total number of positives in the study represented in this table is higher than 20 (total number of samples).

Table 1. The number of colonies present in each sample.

Samples /Bacteria	Total Coliforms	Faecal Coliforms (*E. coli*)	*Intestinal Enterococci*	*Pseudomonas aeruginosa*	*Clostridium* Sulfite-Reducing
SO1	0 UFC/100 mL	0 UFC/100 mL	0 UFC/100 mL	0 UFC/200mL	0 UFC/100 mL
SO2	0 UFC/100 mL	0 UFC/100 mL	0 UFC/100 mL	0 UFC/200mL	0 UFC/100 mL
SO3	1 UFC/100 mL	0 UFC/100 mL	0 UFC/100 mL	0 UFC/200mL	0 UFC/100 mL
SO4	0 UFC/100 mL	0 UFC/100 mL	0 UFC/100 mL	0 UFC/200mL	19 UFC/100 mL
SO5	0 UFC/100 mL	0 UFC/100 mL	0 UFC/100 mL	0 UFC/200mL	1 UFC/100 mL
SO6	12 UFC/100 mL	2 UFC/100 mL	0 UFC/100 mL	0 UFC/200mL	0 UFC/100 mL
SO7	>100 UFC/100 mL	6 UFC/100 mL	18 UFC/100 mL	5 UFC/200mL	1 UFC/100 mL
SO8	0 UFC/100 mL	0 UFC/100 mL	0 UFC/100 mL	0 UFC/200mL	0 UFC/100 mL
SO9	>100 UFC/100 mL	0 UFC/100 mL	0 UFC/100 mL	0 UFC/200mL	100 UFC/100 mL
SO10	0 UFC/100 mL	0 UFC/100 mL	0 UFC/100 mL	0 UFC/200mL	0 UFC/100 mL
SO11	>100 UFC/100 mL	0 UFC/100 mL	0 UFC/100 mL	0 UFC/200mL	40 UFC/100 mL
SO12	>100 UFC/100 mL	68 UFC/100 mL	2 UFC/100 mL	0 UFC/200mL	>100 UFC/mL
SO13	101 UFC/100 mL	1 UFC/100 mL	1 UFC/100 mL	0 UFC/200mL	40 UFC/100 mL
SO14	>100 UFC/100 mL	15 UFC/100 mL	13 UFC/100 mL	104 UFC/200mL	20 UFC/100 mL
SO15	30 UFC/100 mL	1 UFC/100 mL	1 UFC/100 mL	0 UFC/200mL	15 UFC/100 mL
SO16	0 UFC/100 mL	0 UFC/100 mL	0 UFC/100 mL	0 UFC/200mL	>15 UFC/mL
SO17	0 UFC/100 mL	0 UFC/100 mL	0 UFC/100 mL	0 UFC/200mL	0 UFC/100 mL
SO18	0 UFC/100 mL	0 UFC/100 mL	0 UFC/100 mL	0 UFC/200mL	0 UFC/100 mL
SO19	0 UFC/100 mL	0 UFC/100 mL	0 UFC/100 mL	0 UFC/200mL	0 UFC/100 mL
SO20	0 UFC/100 mL	0 UFC/100 mL	0 UFC/100 mL	0 UFC/200mL	0 UFC/100 mL

Table 2. Percentages of all microorganisms found in the tested samples.

		Positive	
		N	Percentage
Bacterium	Total Coliforms	9	28.1%
	Faecal Coliforms/*E. coli*	6	18.8%
	Enterococci	5	15.6%
	Pseudomonas	2	6.3%
	Sulphite Reducing *Clostridium*	10	31.3%
	Total positive samples in the study	32	100.0%

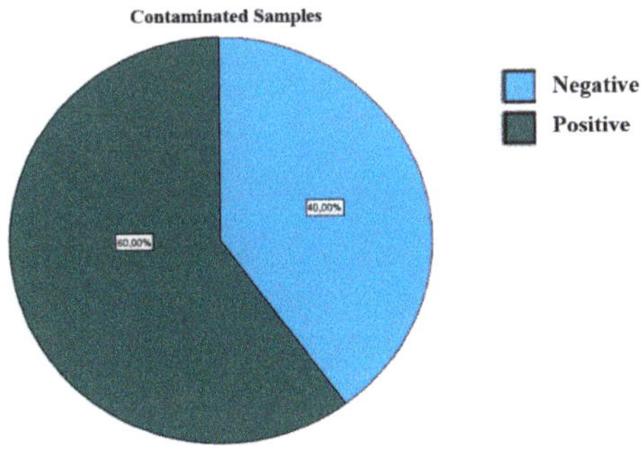

Figure 1. Percentages of contaminated samples vs. non-contaminated samples.

4. Discussion

The objective of this study was to access if the water derived from boreholes was of good microbial quality. Contamination by bacteria was observed in 60% of the samples. The analysis of the positive samples revealed the presence of different bacteria: total coliforms (28.1%), *E. coli* (18.8%), *Enterococci* (15.6%), *Pseudomonas* (6.3%), and *Clostridium* (31.3%) were the most common. All of these results were confirmed by biochemical tests and validated by controls. The study confirms that some waters are highly contaminated, presenting all the bacteria investigated as observed in SO07 and SO14. Only SO19 did not present any microorganisms; however, we cannot confirm good microbiological water quality, since not all microorganisms were investigated. All samples that were positive for all investigated bacteria ended up exceeding the bacterial limits permitted by law, which states that these microorganisms should not be present in groundwater.

5. Conclusions

Considering that residents are senior citizens (a more vulnerable population) and the safety of groundwater was not confirmed, consuming this contaminated water could increase health problems. An alert was given to the population in danger of consuming water from untreated boreholes, and they were informed of all the obtained results. With this study, we hope to provide a starting point for future studies to address this topic in more locations and with a larger sample, contributing to the safer use of groundwater.

Author Contributions: Conceptualization, C.S.; methodology, C.S. and T.C.; software, S.V.; validation, C.S., T.C. and H.B.; formal analysis, C.S. and T.C.; investigation, C.S. and T.C.; resources, C.S., T.C. and H.B.; data curation, C.S. and T.C.; writing—original draft preparation, T.C.; writing—review and editing, C.S., T.C., N.S., H.B. and S.V.; visualization, C.S., T.C., N.S. and H.B.; supervision, S.V., S.A., M.J.H. and H.B.; project administration, S.V. and S.A. All authors have read and agreed to the published version of the manuscript.

Funding: This research received no external funding, but all the required materials were provided by Egas Moniz School of Health & Science.

Institutional Review Board Statement: The study was conducted in accordance and approved by the Institutional Review Board (or Ethics Committee) of Egas Moniz School of Health & Science (protocol code MP.EM.PE.18_02 and date of 7 June 2022).

Informed Consent Statement: Informed consent was obtained from all subjects involved in the study.

Data Availability Statement: Data is available in Egas Moniz School of Health and Science, but cannot be accessed due to privacy and ethical restrictions.

Conflicts of Interest: The authors declare no conflict of interest.

References

1. Souto, J.P.; Sarmento Lira, A.G.; Figueira, J.d.; Nascimento da Silva, A.; Silva, E.S. Poluição Fecal da Água: Microrganismos Indicadores. In Proceedings of the VI Congresso Brasileiro de Gestão Ambiental, Porto Alegre, RS, Brazil, 23–26 November 2015.
2. Alves Cunha, M.C. Execução de Ensaios Microbiológicos nas Áreas Alimentar, Ambiental e Técnica em Contexto Empresarial. Master's Dissertation, Faculdade de Ciências da Universidade do Porto, Porto, Portugal, 2017.
3. Mendes, B.; Santos Oliveira, J. *Qualidade da Água para Consumo Humano*, 1st ed.; Lidel: Lisboa, Portugal, 2004.
4. Valente Bernardo, M.M. Comparação dos Métodos Aplicados na Deteção de Bactérias Coliformes, *Escherichia coli* e *Enterococcus* sp. em Águas para Fins Recriativos. Master's Dissertation, Universidade do Algarve, Faro, Portugal, 2007.
5. Quintela, S.D. Qualidade da Água para Consumo Humano: Riscos Associados à Presença de Biofilmes em Torneiras de Espaços Escolares. Master's Dissertation, Escola Superior de Tecnologia da Saúde do Porto, Porto, Portugal, 2016.
6. Barroso, H.; Taveira, N.; Meliço-Silvestre, A. *Microbiologia Médica*; Lidel: Lisboa, Portugal, 2014.
7. Rosa Candeias, G.M. Qualidade Microbiológica do Gelo Usado em Estabelecimentos de Restauração e de Bebidas. Master's Dissertation, Instituto Superior de Ciências da Saúde Egas Moniz, Almada, Portugal, 2014.

medical sciences forum

MDPI

Proceeding Paper

Personality and Aggressive Behavior in a Domestic Violence Suspects Sample †

Iris Almeida [1,*], Ana Carolina Pires [2], Carolina Nobre [3], Daniela Vicente [2] and Joana Marques [3]

1 Laboratório de Ciências Forenses e Psicológicas Egas Moniz (LCFPEM), Egas Moniz Center for Interdisciplinary Research (CiiEM), Egas Moniz School of Health & Science, 2829-511 Almada, Portugal
2 Egas Moniz School of Health & Science, 2829-511 Almada, Portugal; acpires8@hotmail.com (A.C.P.); danielavicente00@gmail.com (D.V.)
3 Victims Information and Assistance Office (GIAV), Egas Moniz School of Health & Science, 2829-511 Almada, Portugal; carolinanobrepsic@gmail.com (C.N.); joanaromba@hotmail.com (J.M.)
* Correspondence: ialmeida@egasmoniz.edu.pt
† Presented at the 6th International Congress of CiiEM—Immediate and Future Challenges to Foster One Health, Almada, Portugal, 5–7 July 2023.

Abstract: The goal of this study is to show the relationship between the 5-factor model of personality and aggression in 54 suspects of domestic violence, between 23 and 68 years old, assessed in the Victims Information and Assistance Office (GIAV) within the Public Prosecutor's Office. The results show us a positive correlation between neuroticism and physical aggression, anger, hostility, and total aggression score; a negative correlation between extraversion and physical aggression; a negative correlation between agreeableness and physical aggression, anger, and total aggression score; and a negative correlation between conscientiousness and physical aggression, anger, and total aggression score. Our results show the importance of studying the relationship between personality and aggressive behavior and allow us to understand and find assessment strategies (e.g., personality and aggressive behavior assessment) and prevention strategies for domestic violence. Therefore, we reinforce the relevance of continuing the study of this topic, which could strengthen a closer relationship between forensic psychology and law.

Keywords: personality; aggressive behavior; five-factor model; domestic violence

Citation: Almeida, I.; Pires, A.C.; Nobre, C.; Vicente, D.; Marques, J. Personality and Aggressive Behavior in a Domestic Violence Suspects Sample. *Med. Sci. Forum* **2023**, *22*, 14. https://doi.org/10.3390/msf2023022014

Academic Editors: José Brito, Nuno Taveira and Ana I. Fernandes

Published: 9 August 2023

1. Introduction

The five-factor model (FFM) is the most modern and widely acknowledged method of describing and evaluating personality. According to the FFM, the five basic domains that incorporate personality are neuroticism, extraversion, openness to experience, agreeableness, and conscientiousness [1]. According to Costa and McCrae (1992), neuroticism characterizes people who frequently experience negative emotions (e.g., anxiety, resentment, and grief), have irrational beliefs and excessive desire and cravings, and are more likely to experience emotional instability and psychological distress [2]. People who have a high score on this dimension are examples of anxious and insecure people, as opposed to people who score poorly, who typically are at ease and pleased with themselves [3]. The level of activity, the demand for outside stimulation, the quantity and quality of social engagement, and the quality of delight are all accounted for by the extraversion domain. People that perform well on this dimension can be characterized as talkative and social. People who score poorly on the dimension are contrasted with them and portrayed as restrained and serious [3]. The ability to be open to experience is measured by one's proactive pursuit of appreciation and tolerance for experience for its own sake, as well as one's exploration of the unfamiliar. People that perform well on this factor are considered to be unconventional and curious. Contrarily, persons with low scores are conventional and realist [3]. Agreeableness measures the degree of interpersonal orientation toward others along a continuum ranging from sympathy and compassion to antagonistic ideas, feelings, and behaviors. People who perform well on this dimension are thought to be kind and sincere, whereas those who perform poorly are thought to be agitated and manipulative [3]. Conscientiousness refers to a person's ability to regulate their impulsivity. Those who exhibit high levels of conscientiousness are viewed as reliable and scrupulous [2]. Low scorers, in contrast, are unreliable and reckless [3].

According to personality and aggression theorists, personality traits play a significant role in predicting aggressive behavior [4]. The understanding of this relationship was improved by the use of the FFM [2,5,6] and was used by a few researchers [7–9] to investigate the relationship between aggression and several personality traits. While developing the Aggressive Questionnaire, Buss and Perry (1992) assumed that this behavior could be categorized into four factors: verbal aggression (e.g., yelling, cursing, threatening, insulting); physical aggression (e.g., hitting, striking, breaking things); anger (e.g., physiological arousal); and hostility (e.g., feelings of ill will and injustice) [10]. Agreeableness, which is negatively correlated with both self-reported and peer-reported aggressive behavior and violence [7], is the strongest personality predictor of this type of behavior [11]. Anger and agreeableness were found to be negatively correlated [12]; therefore, it may be claimed that those who have low levels of agreeableness are more likely to experience rage when they are provoked. Apart from the measure of agreeableness, all the individual variables have a statistically significant impact on aggressive behavior [3], whereas neuroticism is positively correlated with hostility and anger. Neuroticism was found to have the highest individual impact on aggression, implying that the more stable your emotions, the less aggressive you are. This phenomenon can be explained by the fact that emotionally unstable people (those with high levels of neuroticism) are more prone to experience negative emotions, and as a result, they have lower levels of patience [3]. Some authors [13], by studying the relationship between personality traits and the Aggressive Questionnaire, concluded that, first, there is a correlation between neuroticism and all four measures of the Aggression Questionnaire. Second, hostility and verbal aggression were mostly unrelated to (poor) conscientiousness, although physical aggression and anger were highly related to it. On the other hand, the correlations between the component extroversion and violent conduct are contradictory: the link between self-reported physical violence and extroversion, for instance, was shown to be negative [11], although another study [13] discovered a positive correlation between both. Regarding the last factor of the FFM, openness to experience, it appears to be unrelated to any aggressive behavior [7]. Low agreeableness, low conscientiousness, and high neuroticism have previously been associated with aggression [14,15]. Higher BPAQ scores were positively associated with neuroticism and negatively associated with agreeableness and conscientiousness [16]. In a study with 38 suspects of domestic violence [16], the authors found a positive correlation between neuroticism and hostility, a negative correlation between openness to experience and overall aggression, a negative correlation between agreeableness and physical aggression and anger, and a negative correlation between conscientiousness and anger.

2. Materials and Methods

The sample is composed of 54 suspects of domestic violence ($n_{\text{men}} = 46$ (85.2%); $n_{\text{women}} = 8$ (14.8%)) between 23 and 68 years old ($M = 45.39$, $sd = 10.36$) assessed in the Victims Information and Assistance Office (GIAV) within the Public Prosecutor's Office. Most of the sample were only suspects ($n = 35$), and others were victims and suspects simultaneously ($n = 19$). A large proportion of the sample had middle school education (27.8% ($n = 15$)), followed by high school education (25.9% ($n = 14$)) and B.Sc. degrees (24.1% ($n = 13$)). Our sample was mostly experts in the intellectual and scientific professions (24.1% ($n = 13$)) and industrial, agricultural, and fishing workers (16.7% ($n = 9$)). The relationship between victims and our sample were: 19 married; 14 ex-boyfriends/girlfriends; 11 divorced; 3 boyfriends/girlfriends; 3 partners; 3 ex-partners; 1 lover. Regarding criminal history, 38.9% ($n = 21$) had previous contacts with the Criminal Justice System, including convictions in 13.0% ($n = 7$) of cases. Data were collected from lawsuits, semi-structured interviews, collateral information, and clinical and forensic assessment tools such as NEO-PI-R [2] and BPAQ [10].

3. Results and Discussion

The goal of this study is to show the relationship between the five-factor model of personality (NEO-PI-R) and aggression (BPAQ).

Table 1 shows a positive correlation between neuroticism and physical aggression, anger, hostility, and total aggression score; a negative correlation between extroversion and physical aggression; a negative correlation between agreeableness and physical aggression, anger, and total aggression score; and a negative correlation between conscientiousness and physical aggression, anger, and total aggression score.

Table 1. Relationship between personality and aggression.

	Neuroticism	Extraversion	Agreeableness	Conscientiousness
Physical aggression	0.431 **	−0.296 *	−0.289 *	−0.342 *
Anger	0.620 **	−0.135	−0.502 **	−0.402 **
Hostility	0.441 **	−0.168	−0.188	−0.165
Total aggression	0.589 **	−0.203	−0.413 **	−0.338 *

** $p < 0.01$; * $p < 0.05$.

These results are consistent with the findings of the empirical studies mentioned above [7,11,13–17], namely, the relation between FFM and aggression. People with higher levels of neuroticism are more likely to experience negative emotions and engage in aggressive behaviors, and people with higher levels of agreeableness and conscientiousness are more likely to experience positive emotions and are less likely to engage in aggressive behaviors.

Our results show the importance of studying the relationship between personality and aggressive behavior and allow us to understand and find assessment strategies (e.g., personality and aggressive behavior assessment) and prevention strategies for domestic violence. Therefore, we reinforce the relevance of continuing the study of this topic, which could strengthen a closer relationship between forensic psychology and law and a better understanding of domestic violence offenders' characteristics.

Author Contributions: Conceptualization, I.A., A.C.P., C.N., D.V. and J.M.; methodology, I.A., A.C.P., C.N., D.V. and J.M.; software, I.A., A.C.P., C.N., D.V. and J.M.; validation, I.A., A.C.P., C.N., D.V. and J.M.; formal analysis, I.A., A.C.P., C.N., D.V. and J.M.; investigation, I.A., A.C.P., C.N., D.V. and J.M.; resources, I.A., A.C.P., C.N., D.V. and J.M.; data curation, I.A., A.C.P., C.N., D.V. and J.M.; writing—original draft preparation, I.A., A.C.P., C.N., D.V. and J.M.; writing—review and editing, I.A., A.C.P., C.N., D.V. and J.M.; visualization, I.A., A.C.P., C.N., D.V. and J.M.; supervision, I.A.; project administration, I.A., A.C.P., C.N., D.V. and J.M. All authors have read and agreed to the published version of the manuscript.

Funding: This research received no external funding.

Institutional Review Board Statement: This study is part of a protocol established by the Portuguese Public Prosecutor's Office and Egas Moniz School of Health & Science to assess and analyze the characteristics of victims and offenders in the field of violence. The strictness of ethical and deontological principles is safeguarded once criminal records have been restricted access by law (including judicial secrecy). Therefore, all assessed subjects gave their informed consent, and their data were processed anonymously. This study was conducted by the Declaration of Helsinki, and all ethical standards of scientific research were respected, as well as the Code of Ethics of the Order of Portuguese Psychologists and the General Data Protection Regulation.

Informed Consent Statement: All ethical issues were considered due to the sensitive nature of the detailed data, the respective informed consent, confidentiality limits, and information about the ethics and technician's impartiality. Informed consent was obtained from all subjects involved in the study. The participants signed an informed consent term, which contained the goal of the evaluation, the limits of confidentiality, and information about the ethics and impartiality of the technicians. Written informed consent has been obtained.

Data Availability Statement: Data sharing is not applicable because some of the information derives from criminal records.

Conflicts of Interest: The authors declare no conflict of interest.

References

1. McCrae, R.R.; Costa, P.T., Jr. A contemplated revision of the NEO Five-Factor Inventory. *Personal. Individ. Differ.* **2004**, *36*, 587–596. [CrossRef]
2. Costa, P.T.; McCrae, R.R. *Revised NEO Personality Inventory (NEO-PI-R) and NEO Five-Factor Inventory (NEO-FFI)*; Psychological Assessment Resources: Odessa, FL, USA, 1992.
3. Pervin, A.L.; John, P.O. *Personality: Theory and Research*, 7th ed.; John Wiley & Sons: Hoboken, NJ, USA, 1997.
4. Anderson, C.A.; Huesmann, R.L. Human aggression: A social-cognitive view. In *The Sage Handbook of Social Psychology*; Hogg, M.A., Cooper, J., Eds.; Sage Publications: London, UK, 2003; pp. 259–287.
5. Jensen-Campbell, L.A.; Graziano, W.G. Agreeableness as a moderator of interpersonal conflict. *J. Personal.* **2001**, *69*, 323–362. [CrossRef] [PubMed]
6. Miller, J.D.; Lynam, D.; Leukefeld, C. Examining antisocial behavior through the five-factor model of personality. *Aggress. Behav.* **2003**, *29*, 497–514. [CrossRef]
7. Gleason, K.A.; Jensen-Campbell, L.A.; Richardson, D.S. Agreeableness as a predictor of aggression in adolescence. *Aggress. Behav.* **2004**, *30*, 43–61. [CrossRef]
8. Graziano, W.G.; Jensen-Campbell, L.A.; Hair, E.C. Perceiving interpersonal conflict and reacting to it: The case for Agreeableness. *J. Personal. Soc. Psychol.* **1996**, *70*, 820–835. [CrossRef] [PubMed]
9. Suls, J.; Martin, R.; David, J.P. Person-environment fit and its limits: Agreeableness, neuroticism, and emotional reactivity to interpersonal conflict. *Personal. Soc. Psychol. Bull.* **1998**, *24*, 88–98. [CrossRef]
10. Buss, A.H.; Perry, M.P. The Aggression Questionnaire. *J. Personal. Soc. Psychol.* **1992**, *63*, 452–459. [CrossRef] [PubMed]
11. Sharpe, J.P.; Desai, S. The revised neo personality inventory and the MMPI-2 psychopathology five in the prediction of aggression. *Personal. Individ. Differ.* **2001**, *31*, 505–518. [CrossRef]
12. Martin, R.; Wan, C.K.; David, J.P.; Wegner, E.L.; Olson, B.D.; Watson, D. Style of anger expression: Relation to expressivity, personality, and health. *Personal. Soc. Psychol. Bull.* **1999**, *25*, 1196–1207. [CrossRef]
13. Gallo, L.C.; Smith, T.W. Construction validation of health-related personality traits: Interpersonal circumplex and Five-Factor model analysis of the Aggression Questionnaire. *Int. J. Behav. Med.* **1998**, *5*, 129–147. [CrossRef] [PubMed]
14. Jones, S.E.; Miller, J.D.; Lynam, D.R. Personality, antisocial behavior, and aggression: A meta-analytic review. *J. Crim. Justice* **2011**, *39*, 329–337. [CrossRef]
15. Jiang, X.; Li, X.; Dong, X.; Wang, L. How the Big Five personality traits related to aggression from perspectives of the benign and malicious envy. *BMC Psychol.* **2022**, *10*, 203. [CrossRef] [PubMed]
16. Dam, V.H.; Hjordt, L.V.; da Cunha-Bang, S.; Sestoft, D.; Knudsen, G.M.; Stenbaek, D.S. Trait aggression is associated with five-factor personality traits in males. *Brain Behav.* **2021**, *11*, e02175. [CrossRef] [PubMed]
17. Baúto, R.V.; Carreiro, A.F.; Pereira, M.; Guarda, R.; Almeida, I. Personality and Aggressive Behavior: The Relation between the Five-Factor and Aggression Models in a Domestic Violence Suspects Sample. *Med. Sci. Forum* **2021**, *5*, 18.

medical sciences forum

MDPI

Proceeding Paper

Aggression, Genetics, and Adverse Childhood Experiences in a University Sample †

Sara Oliveira [1,*], Cristina Soeiro [1,2], Alexandre Quintas [1,3] and Joana Couceiro [1,3]

1 Egas Moniz School of Health & Science, 2829-511 Almada, Portugal; c.soeiro@netcabo.pt (C.S.); alexandre.quintas@gmail.com (A.Q.); joana.m.couceiro@gmail.com (J.C.)
2 Laboratório de Psicologia (LabPSI), Centro de Investigação Interdisciplinar Egas Moniz (CiiEM), 2829-511 Almada, Portugal
3 Laboratório de Patologia Molecular e Bioquímica Forense, Centro de Investigação Interdisciplinar Egas Moniz (CiiEM), 2829-511 Almada, Portugal
* Correspondence: sara.oliveira99@gmail.com

† Presented at the 6th International Congress of CiiEM—Immediate and Future Challenges to Foster One Health, Almada, Portugal, 5–7 July 2023.

Abstract: The literature shows that aggression in adulthood is associated with adverse childhood experiences and genetics. This research aims to study the phenomenon of aggression in adulthood and analyse its relationship with genetics (dopamine and serotonin polymorphisms) and adverse childhood experiences. The sample was collected as part of the research project "Aggression and Genetics in a University Context" and consists of 93 individuals, 81 women, and 12 men. Participants completed a protocol consisting of a sociodemographic questionnaire, the Reactive–Proactive Aggression Questionnaire (RPQ), the Short Form of the Buss–Perry Aggression Questionnaire (BPAQ-SF), and the Childhood History Questionnaire (ACE). The most important results indicate that adverse childhood experiences are correlated with aggression in adulthood.

Keywords: aggression; adverse childhood experiences; serotonin; dopamine

Citation: Oliveira, S.; Soeiro, C.; Quintas, A.; Couceiro, J. Aggression, Genetics, and Adverse Childhood Experiences in a University Sample. *Med. Sci. Forum* **2023**, *22*, 40. https://doi.org/10.3390/msf2023022040

Academic Editors: José Brito, Nuno Taveira and Ana I. Fernandes

Published: 18 August 2023

1. Introduction

Human aggression is a controversial concept, with different perspectives on its origins. Aggression is a maladaptive expression of human behaviour directed at others intending to harm, injure, or harass [1,2]. According to the General Model of Aggression (GAM), human aggression is influenced by knowledge structures with implications in various sociocognitive phenomena, including perception, interpretation, decision making, and behaviour [2]. Knowledge structures are formed from experience and influence perception at different levels, from simple object perception to complex interpersonal perception. GAM thus integrates a social, cognitive, and biological aspect of the development of aggression, divided into two main processes: distal and proximal processes. Distal processes underlie each episode of proximal processes and explain, through modifiers, how genetic and environmental factors can influence personality through changes in knowledge structures. The proximal processes describe how personal and situational factors influence aggressive thoughts, feelings of anger, and arousal levels, which in turn affect appraisal and decision-making processes, resulting in aggressive or non-aggressive behaviour [2].

The study of human aggression is very complex. It is crucial to understand how aggression is influenced at the genetic level [3]. Our behaviour can be influenced by two neurotransmitters, dopamine (DA) and serotonin (5-hydroxytryptamine or 5-HT) [4]. The proper functioning of neurotransmitters depends on the balance of [3], which, when changed at the genetic level, significantly dysregulate the DA and 5-HT system [5]. These changes at the genetic level are called genetic polymorphisms, which differ from genetic mutations in that, to be considered polymorphic, they must have a frequency of more than

1% in the chromosomes of the general population [6]. The changes at the genetic level may appear during childhood and may occur in individuals with a history of adverse experiences at this stage of life [7].

Adverse childhood experiences (ACEs) result from a series of adverse situations to which individuals are exposed that, over time, become a limitation to normative human development [8]. Changes at the genetic level can occur during, and especially in, childhood, and can occur in individuals who exhibit ACE at this stage of life [2], which in turn is considered a predictor of human aggression [8]. This study is relevant to psychology and genetics as it attempts to study a complex phenomenon like aggression by combining two areas to understand the human aggression. The present research aims to study aggression in adulthood, its relation with genetics, through the identification of serotonin and dopamine polymorphisms (5-HTTLPR and DAT1) and with adverse childhood experiences.

2. Materials and Methods

The sample consists of university students (M = 20.95; DP = 2.98). The data in this study were collected as part of the research project "Aggression and Genetics in a university context". It is, therefore, a sample of volunteer university students who participated in the overall project.

The sample consists of 93 individuals, of whom 12 are male (12.9%) and 81 are female (87.1%). Data were collected online through a battery of psychological tests. The instruments used were as follows: (1) Buss–Perry Aggression Questionnaire—Short Form (BPAQ-SF), adapted for the Portuguese population [9]. It assesses four aggression subscales: physical aggression, verbal aggression, anger, and hostility. (2) The Reactive–Proactive Aggression Questionnaire (RPQ), adapted for the Portuguese population [10], is a self-report measure that distinguishes between reactive and proactive aggression. (3) Adverse Childhood Experiences (ACEs), Portuguese version [11], is a self-report questionnaire for adults that aims to assess the occurrence of adverse childhood experiences. It can be grouped into 2 major categories: (1) experiences against the individual and (2) the family environment. To calculate the total, it is necessary to create a variable: adversity total. First, it is necessary to recalculate all the categories, i.e., each category is transformed into a dichotomous value category, where the value 'zero' is given if the subject does not report this form of adversity, or the value 'one' if this adversity is reported [11].

The analysis of the genetic material revealed two polymorphisms, one of serotonin (5-HTTLPR) and the other of dopamine (DAT1). The 5-HTTLPR polymorphism has been divided into three different groups: heterozygous (ID); homozygous for insertion (II); and homozygous for deletion (DD). When we refer to homozygous for insertion or homozygous for deletion, it means that there has been an insertion or deletion of one or more nucleotides in a genetic sequence.

3. Results

Chi-squared tests were run to analyse the association between the polymorphisms (5-HTTLPR and DAT1) with the ACE subscales. A significant association was obtained between the 5-HTTLPR (serotonin polymorphism) and the emotional abuse scale (χ^2 = 6.756; p = 0.034). Thus, heterozygotes (ID) (63.6%) are the ones with more indicators of the presence of a history of the emotional abuse scale, followed by deletion homozygotes (DD) (27.3%) and insertion homozygotes (II) (9.1%).

To verify the relationship between the aggression scales' scores (BPAQ-SF and RPQ) and ACE scores, Pearson correlation analysis was used, as shown in Table 1.

Table 1. Correlations between the indicators of aggression (BPAQ-SF and RPQ) and ACE ($n = 93$).

	Anger	Hostility	Total BPAQ	Reactive Aggression	Proactive Aggression	Total RPQ
Experiences Against Individual	0.22 *	0.45 **	0.37 **	0.30 **	0.25 *	0.30 **
Dysfunctional Family Environment	-	0.23 *	-	-	-	-
Total ACE	-	0.39 **	0.30 **	0.22 *	-	0.22 *

* $p < 0.05$; ** $p < 0.01$.

4. Discussion

It was possible to verify the existence of a significant relationship between the 5-HTTLPR polymorphism and emotional abuse, with the percentage of individuals who are heterozygous (ID) having a higher number of indicators of the presence of a history of emotional abuse compared to those who are homozygous. This polymorphism is mainly involved in inhibiting impulsive aggression [12,13]. This finding is supported in the literature, with several studies showing that adverse childhood experiences and the 5-HTTLPR are related [12].

No association was found between dopamine and the other variables. It was not possible to confirm, as reported in the literature, that the presence of the DAT1 polymorphism in individuals with ACEs presents higher levels of aggression [11]. Studies conducted in university students on the differences between men and women showed that the effects of polymorphisms on aggression were significant only for men [14].

It was also found that the greater the number of experiences against the individual (e.g., types of abuse and neglect), the greater the level of aggression in adulthood. This finding is consistent with what has been explained in the literature [15].

It was found that the dysfunctional environment was only correlated with hostility, i.e., the more significant the exposure to the dysfunctional environment (e.g., experiencing domestic violence), the higher the level of hostility in the future. A meta-analysis study shows that both anger and hostility are associated with exposure to an event such as domestic violence [16].

Finally, the fourth outcome found that adverse childhood experiences were not associated with anger or proactive aggression, so this finding does not support the literature [17].

When examining the results of the present research, it is possible to identify an association with the distal processes of the GAM. In this case, environmental modifiers assessed as adverse childhood experiences correlate with the forms and functions of aggression as explained by the model. Biological modifiers, assessed in this research as serotonin and dopamine, were only correlated with environmental modifiers and, according to the GAM, these act together to influence personality, thus modifying personal (and situational) factors and consequently influencing aggressive thoughts and feelings of anger, which in turn affect appraisal and decision-making processes, resulting in aggressive or non-aggressive behaviour [18].

The research carried out has several limitations. The first is related to the sample, as it is small. The second limitation relates to the lack of heterogeneity (there are more women than men), which is reflected in the low level of aggression, as men show higher levels of aggression.

5. Conclusions

In summary, the results show that the significant relationship between serotonin polymorphism and adverse childhood experiences are correlated with aggression in adulthood. Through this result, we can verify the importance of how a childhood free of adverse experiences can be associated with a low risk of being aggressive in the future.

Author Contributions: All authors contributed to the realisation of this article at the same level of importance. All authors have read and agreed to the published version of the manuscript.

Funding: This research received no external funding.

Institutional Review Board Statement: The study was conducted in accordance with the Declaration of Helsinki and approved by the Institutional Review Board (or Ethics Committee) of Egas Moniz School of Health and Science for studies involving humans.

Informed Consent Statement: Informed consent was obtained from all subjects involved in the study.

Data Availability Statement: The data presented in this study are available on request from the corresponding author. The data are not publicly available due to privacy and ethical restrictions.

Acknowledgments: We would like to express a special thanks to the Laboratory of Psychology Egas Moniz—Labpsi and to the partnership with the Laboratório de Patologia Molecar e Bioquímica Forense for helping in this project with the genetics.

Conflicts of Interest: The authors declare no conflict of interest.

References

1. Allen, J.J.; Anderson, C.A. Aggression and Violence: Definitions and Distinctions. In *The Wiley Handbook of Violence and Aggression*; John Wiley & Sons, Ltd.: Chichester, UK, 2017; pp. 1–14. [CrossRef]
2. Allen, J.J.; Anderson, C.A.; Bushman, B.J. The General Aggression Model. *Curr. Opin. Psychol.* **2018**, *19*, 75–80. [CrossRef]
3. Sapolsky, R. *Comportamento: A Biologia No Nosso Melhor E Pior*; Círculo Leitores: Lisboa, Portugal, 2018.
4. Daw, N.D.; Kakade, S.; Dayan, P. Opponent Interactions between Serotonin and Dopamine. *Neural Netw.* **2002**, *15*, 603–616. [CrossRef] [PubMed]
5. De Simoni, M.G.; Dal Toso, G.; Fodritto, F.; Sokola, A.; Algeri, S. Modulation of Striatal Dopamine Metabolism by the Activity of Dorsal Raphe Serotonergic Afferences. *Brain Res.* **1987**, *411*, 81–88. [CrossRef] [PubMed]
6. Willard, H. Diversidade genética humana: Mutação e polimorfismo. In *Genética Médica*; Nussubaum, R., Mcinnes, R., Huntington, F., Eds.; Guanabara Koogan: Rio de Janeiro, Brazil, 2002; pp. 274–293.
7. Twardosz, S.; Lutzker, J.R. Child Maltreatment and the Developing Brain: A Review of Neuroscience Perspectives. *Aggress* **2010**, *15*, 59–68. [CrossRef]
8. Felitti, V.J.; Anda, R.F.; Nordenberg, D.; Williamson, D.F.; Spitz, A.M.; Edwards, V.; Koss, M.P.; Marks, J.S. Relationship of Childhood Abuse and Household Dysfunction to Many of the Leading Causes of Death in Adults. *Am. J. Prev. Med.* **1998**, *14*, 245–258. [CrossRef] [PubMed]
9. Pechorro, P.; Barroso, R.; Poiares, C.; Oliveira, J.P.; Torrealday, O. Validation of the Buss–Perry Aggression Questionnaire-Short Form among Portuguese Juvenile Delinquents. *Int. J. Law Psychiatry* **2016**, *44*, 75–80. [CrossRef] [PubMed]
10. Pechorro, P.; Ray, J.V.; Raine, A.; Maroco, J.; Gonçalves, R.A. The Reactive–Proactive Aggression Questionnaire: Validation Among a Portuguese Sample of Incarcerated Juvenile Delinquents. *J. Interpers. Violence* **2017**, *32*, 1995–2017. [CrossRef] [PubMed]
11. Silva, S.; Maia, Â. *Versão Portuguesa do Family ACE Questionnaire (Questionário da História de Adversidade na Infância)*; Psiquilibrios Edicões: Braga, Portugal, 2008.
12. Reif, A.; Rösler, M.; Freitag, C.M.; Schneider, M.; Eujen, A.; Kissling, C.; Wenzler, D.; Jacob, C.P.; Retz-Junginger, P.; Thome, J.; et al. Nature and Nurture Predispose to Violent Behavior: Serotonergic Genes and Adverse Childhood Environment. *Neuropsychopharmacol* **2007**, *32*, 2375–2383. [CrossRef] [PubMed]
13. Volavka, J. The neurobiology of violence: An update. *J. Neuropsychiatry Clin. Neurosci.* **1999**, *11*, 307–314. [CrossRef] [PubMed]
14. Caspi, A.; Sugden, K.; Moffitt, T.E.; Taylor, A.; Craig, I.W.; Harrington, H.; McClay, J.; Mill, J.; Martin, J.; Braithwaite, A.; et al. Influence of Life Stress on Depression: Moderation by a Polymorphism in the 5-HTT Gene. *Science* **2003**, *301*, 386–389. [CrossRef] [PubMed]
15. Verona, E.; Joiner, T.E.; Johnson, F.; Bender, T.W. Gender Specific Gene–Environment Interactions on Laboratory-Assessed Aggression. *Biol. Psychol.* **2006**, *71*, 33–41. [CrossRef] [PubMed]
16. Anderson, C.A. Human Aggression. *Annu. Rev. Psychol.* **2002**, *53*, 27–51. [CrossRef] [PubMed]
17. Mumford, E.A.; Taylor, B.G.; Berg, M.; Liu, W.; Miesfeld, N. The social anatomy of adverse childhood experiences and aggression in a representative sample of young adults in the U.S. *Child Abus. Negl.* **2019**, *88*, 15–27. [CrossRef] [PubMed]
18. Manuck, S.B.; Flory, J.D.; Ferrell, R.E.; Dent, K.M.; Mann, J.J.; Muldoon, M.F. Aggression and Anger-Related Traits Asso-ciated with a Polymorphism of the Tryptophan Hydroxylase Gene. *Biol. Psychiatry* **1999**, *45*, 603–614. [CrossRef] [PubMed]

medical sciences forum

MDPI

Proceeding Paper

Cyberaggression, Personality and Genetics †

Catarina Godinho [1,*], Cristina Soeiro [1,2], Alexandre Quintas [1,3] and Joana Couceiro [1,3]

1 Egas Moniz School of Health & Science, 2829-511 Almada, Portugal; c.soeiro@netcabo.pt (C.S.); alexandre.quintas@gmail.com (A.Q.); joana.m.couceiro@gmail.com (J.C.)
2 Laboratório de Psicologia (LabPSI), Centro de Investigação Interdisciplinar Egas Moniz (CiiEM), 2829-511 Almada, Portugal
3 Laboratório de Patologia Molecular e Bioquímica Forense, Centro de Investigação Interdisciplinar Egas Moniz (CiiEM), 2829-511 Almada, Portugal
* Correspondence: kattarinagodinho@hotmail.com
† Presented at the 6th International Congress of CiiEM—Immediate and Future Challenges to Foster One Health, Almada, Portugal, 5–7 July 2023.

Abstract: Cyber aggression can be defined as a form of aggression where the perpetrator uses digital media to harm a person or group of people. The literature has linked aggression to personality and genetics. This study aimed to show the relationship between cyber aggression and personality and polymorphisms associated with aggression-related serotonergic and dopaminergic processes. The sample consisted of 93 individuals, 81 women (87.1%) and 12 males (12.9%) with a mean age of 20.95 (SD = 2.98), and it was collected as part of the research project "Aggressiveness and Genetics in a University Context". The most important results indicate that cyber aggression and aggression are related to personality traits, especially agreeableness. An influence of genetic variables was not found.

Keywords: aggression; cyber aggression; personality; genetics; serotonin; dopamine

1. Introduction

Human aggression can be defined as any behaviour performed by an individual with intention to cause harm to another person or group of people. The perpetrator must believe that the behaviour will cause harm to the victim and the victim must be motivated to avoid that harm [1].

With the expansion of communication technologies in recent years and their integration into everyday life, new benefits and problems have emerged, including a new form of aggression, cyber aggression, amplified by internet communication [2,3]. This includes defamation, social exclusion, swearing and comments that indirectly apply to an individual [2,4]. Some forms of cyber aggression are similar to verbal aggression, involving hostile words and insults to cause psychological harm to the victim [3,5].

Several variables are used to study aggression, and therefore cyber aggression, and one of these variables is personality. Personality traits such as agreeableness, neuroticism [6] and conscientiousness [7] are usually studied to understand their relationship with aggression, and studies have shown a relationship between personality traits and aggression. Looking at these personality dimensions and their relationship with aggression, the agreeableness trait is the one that is most negatively associated with aggression, i.e., people with higher scores on this personality dimension show less aggressive behaviour [6], as is conscientiousness, which also tends to be negatively associated with aggression [7]. On the other hand, neuroticism is positively associated with aggressive behaviour. Openness to experience is not associated with aggression [6], and extroversion has had mixed results, with studies finding negative relationships between extroversion and physical aggression [7] and other studies finding a positive relationship between the two [8].

Personality and aggression are also linked to some genetic markers. Some studies on this issue have found a relationship between the serotoninergic and dopaminergic systems

Citation: Godinho, C.; Soeiro, C.; Quintas, A.; Couceiro, J. Cyberaggression, Personality and Genetics. *Med. Sci. Forum* **2023**, 22, 41. https://doi.org/10.3390/msf2023022041

Academic Editors: José Brito, Nuno Taveira and Ana I. Fernandes

Published: 18 August 2023

and aggression. These genetic markers impact the manifestation of hostile behaviour, anger and other antecedents of aggressive behaviour [9–11] and personality, thus influencing human behaviour [12]. According to some studies on the General Model of Aggression, one of the models that tries to explain aggression, it has been possible to verify a link between aggression and serotonin [13].

This paper aimed to show the relationship between cyber aggression and personality traits and some polymorphisms related to serotoninergic and dopaminergic processes.

2. Materials and Methods

The sample consists of 93 individuals, including 81 females (87.1%) and 12 males (12.9%), aged between 17 and 40 years, with a mean age of 20.95 (SD = 2.98). First, genetic material was collected to find two polymorphisms related to serotonin and dopamine. The collection was performed by collecting DNA from mouth cells obtained from swabs. The volunteers extracted their own mouth cells, which were properly stored for later use in the Laboratory of Molecular Pathology and Forensic Biochemistry. Participants completed an online battery of psychological instruments: (1) the short version of the Buss-Perry Aggression Questionnaire (BPAQ-SF) [14], which assesses four forms of aggression (Physical and Verbal Aggression, Anger and Hostility); (2) the Reactive-Proactive Aggression Questionnaire (RPQ) [15], which assesses reactive and proactive aggression; (3) the Cyber-Aggression Typology Questionnaire (CATQ) [16], which assesses the different dimensions of cyber, impulsive-appetitive, impulsive-aversive, controlled-appetitive and controlled-aversive aggression; (4) and the NEO Five-Factor Inventory-3 (NEO-FFI-3) [17], which assesses the five dimensions of personality (Openness, Conscientiousness, Extroversion, Agreeableness and Neuroticism).

The present research is integrated in the scope of the project "Aggressiveness and Genetics in a University Context" which was previously approved in the IUEM's ethics committee. All of the participants signed an informed consent form, and their data were coded to ensure the privacy and confidentiality of each individual participant.

3. Results and Discussion

Concerning the cyber aggression behaviours, weak to moderate, positive, statistically significant correlations were found between the cyber aggression subscales and the other forms and functions of aggression. The results presented in Table 1 indicate that the agreeableness dimension showed significant and negative correlations with all the subscales that define cyber aggression, noting that participants with higher scores in agreeableness showed lower scores in all the subscales that assess cyber aggression. Another study [18] supports this finding by showing that agreeableness is negatively correlated with the perpetration of cyber aggression. No significant correlations were found between the other personality dimensions and cyber aggression.

Table 1. Relationship between cyber aggression and personality.

	Neuroticism	Conscientiousness	Extraversion	Agreeableness	Openness
Impulsive-Aversive	0.09	−0.09	0.06	−0.35 **	−0.09
Controlled-Aversive	−0.00	0.00	0.08	−0.32 **	−0.09
Controlled-Appetitive	0.03	−0.05	0.03	−0.27 **	−0.02
Impulsive-Appetitive	−0.02	−0.05	0.05	−0.37 **	0.02
Total	0.06	−0.07	0.06	−0.37 **	−0.07

** $p < 0.05$.

Regarding the dopaminergic and serotoninergic processes related to cyber aggression, the results showed no significant differences between the two genetic polymorphisms in relation to cyber aggression. Concerning the dopaminergic and serotoninergic processes involved in personality, the results showed no significant differences between the categories defining the two genetic polymorphisms and personality. According to studies on the General Model of Aggression, it has been possible to verify relationships between personality and aggression, especially with the serotonin polymorphism, which was not verified in this study [13].

4. Conclusions

In summary, this study presented significant results concerning the relationship between cyber aggression and personality. The study between cyber aggression and genetics did not show any significant results, even though the literature found relationships between the two variables, which may be related to this study's limitations. These are related to the sample, which is small and presents a much lower number of men than women, which may have significantly influenced the results obtained. This is important because studies have reported higher aggression scores in men than women [19,20]. It is therefore essential to have a large and more representative sample of the university population. With this study, we tried to find and confirm information about cyber aggression, personality and genetics that can contribute to the prevention of and intervention in cases of cyber aggression.

Author Contributions: All authors contributed to the realization of this article as the same level of importance. All authors have read and agreed to the published version of the manuscript.

Funding: This research received no external funding.

Institutional Review Board Statement: The study was conducted in accordance with the Declaration of Helsinki and approved by the Institutional Review Board (or Ethics Committee) of Egas Moniz School of Health and Science for studies involving humans.

Informed Consent Statement: Informed consent was obtained from all subjects involved in the study.

Data Availability Statement: The data presented in this study are available on request from the corresponding author. The data are not publicly available due to privacy and ethical restrictions.

Acknowledgments: We would like to give a special thanks to the Laboratory of Psychology Egas Moniz-Labpsi and to the partnership with the Laboratório de Patologia Molecar e Bioquímica Forense for helping in this project with the genetics.

Conflicts of Interest: The authors declare no conflict of interest.

References

1. Anderson, C.A. Human Aggression. *Annu. Rev. Psychol.* **2002**, *53*, 27–51. [CrossRef] [PubMed]
2. Grigg, D.W. Cyber-Aggression: Definition and Concept of Cyberbullying. *Aust. J. Guid. Couns.* **2010**, *20*, 143–156. [CrossRef]
3. Langos, C. Cyberbullying: The Challenge to Define. *Cyberpsychol. Behav. Soc. Netw.* **2012**, *15*, 285–289. [CrossRef] [PubMed]
4. Buss, A.; Perry, M. The aggression questionnaire. *J. Pers. Soc. Psychol.* **1992**, *63*, 452–459. [CrossRef] [PubMed]
5. DeMarsico, D.; Bounoua, N.; Miglin, R.; Sadeh, N. Aggression in the Digital Era: Assessing the Validity of the Cyber Motivations for Aggression and Deviance Scale. *Assessment* **2022**, *29*, 764–781. [CrossRef] [PubMed]
6. Gleason, K.A.; Jensen-Campbell, L.A.; South Richardson, D. Agreeableness as a Predictor of Aggression in Adolescence. *Aggr. Behav.* **2004**, *30*, 43–61. [CrossRef]
7. Sharpe, J.P.; Desai, S. The Revised Neo Personality Inventory and the MMPI-2 Psychopathology Five in the Prediction of Aggression. *Pers. Individ. Differ.* **2001**, *31*, 505–518. [CrossRef]
8. Gallo, L.C.; Smith, T.W. Construct Validation of Health-Relevant Personality Traits: Interpersonal Circumplex and Five-Factor Model Analyses of the Aggression Questionnaire. *Int. J. Behav. Med.* **1998**, *5*, 129–147. [CrossRef] [PubMed]
9. Fernàndez-Castillo, N.; Cormand, B. Aggressive Behavior in Humans: Genes and Pathways Identified through Association Studies. *Am. J. Med. Genet.* **2016**, *171*, 676–696. [CrossRef]
10. Sapolsky, R. *Comportamento: A Biologia No Nosso Melhor e Pior*; Círculo Leitores: Lisboa, Portugal, 2018.
11. Wang, M.; Chen, P.; Li, H.; Kemp, A.H.; Zhang, W. Catechol-O-Methyltransferase Gene Val158Met Polymorphism Moderates the Effect of Social Exclusion and Inclusion on Aggression in Men: Findings From a Mixed Experimental Design. *Front. Psychol.* **2021**, *11*, 622914. [CrossRef] [PubMed]
12. Takano, A.; Arakawa, R.; Hayashi, M.; Takahashi, H.; Ito, H.; Suhara, T. Relationship between Neuroticism Personality Trait and Serotonin Transporter Binding. *Biol. Psychiatry* **2007**, *62*, 588–592. [CrossRef]
13. Allen, J.J.; Anderson, C.A. Aggression and Violence: Definitions and Distinctions. In *The Wiley Handbook of Violence and Aggression*; John Wiley & Sons, Ltd.: Chichester, UK, 2017; pp. 1–14. [CrossRef]

14. Pechorro, P.; Barroso, R.; Poiares, C.; Oliveira, J.P.; Torrealday, O. Validation of the Buss–Perry Aggression Questionnaire-Short Form among Portuguese Juvenile Delinquents. *Int. J. Law Psychiatry* **2016**, *44*, 75–80. [CrossRef] [PubMed]
15. Pechorro, P.; Ray, J.V.; Raine, A.; Maroco, J.; Gonçalves, R.A. The Reactive–Proactive Aggression Questionnaire: Validation Among a Portuguese Sample of Incarcerated Juvenile Delinquents. *J. Interpers. Violence* **2017**, *32*, 1995–2017. [CrossRef] [PubMed]
16. Runions, K.C.; Bak, M.; Shaw, T. Disentangling Functions of Online Aggression: The Cyber-Aggression Typology Questionnaire (CATQ): Cyber-Aggression Typology Questionnaire. *Aggr. Behav.* **2017**, *43*, 74–84. [CrossRef] [PubMed]
17. Pedroso De Lima, M.; Gonçalves, E.; Salgueira, A.; Gonzalez, A.-J.; Costa, J.J.; Costa, M.J.; Costa, P. A versão portuguesa do NEO-FFI: Caracterização em função da idade, género e escolaridade. *Psicologia* **2014**, *28*, 1–10. [CrossRef]
18. Zhou, Y.; Zheng, W.; Gao, X. The Relationship between the Big Five and Cyberbullying among College Students: The Mediating Effect of Moral Disengagement. *Curr. Psychol.* **2019**, *38*, 1162–1173. [CrossRef]
19. Verona, E.; Joiner, T.E.; Johnson, F.; Bender, T.W. Gender Specific Gene–Environment Interactions on Laboratory-Assessed Aggression. *Biol. Psychol.* **2006**, *71*, 33–41. [CrossRef] [PubMed]
20. Manuck, S.B.; Flory, J.D.; Ferrell, R.E.; Dent, K.M.; Mann, J.J.; Muldoon, M.F. Aggression and Anger-Related Traits Associated with a Polymorphism of the Tryptophan Hydroxylase Gene. *Biol. Psychiatry* **1999**, *45*, 603–614. [CrossRef] [PubMed]

medical sciences forum

MDPI

Proceeding Paper

Alexithymia, Personality, Internet Addiction, and Interpersonal Relationships †

Andreia Matias [1,*], Cristina Soeiro [1], Iris Almeida [1], Renato Pessoa Santos [2], Ricardo Ventura Baúto [1] and Sandra Oliveira [2]

[1] Egas Moniz Center for Interdisciplinary Research (CiiEM), Egas Moniz School of Health & Science, 2829-511 Almada, Portugal; csoeiro@egasmoniz.edu.pt (C.S.); ialmeida@egasmoniz.edu.pt (I.A.); rbauto@egasmoniz.edu.pt (R.V.B.)

[2] Academia Militar, 1169-203 Lisbon, Portugal; santoskami@gmail.com (R.P.S.); sle.oliveira@hotmail.com (S.O.)

* Correspondence: andreiamatias@gmail.com

† Presented at the 6th International Congress of CiiEM—Immediate and Future Challenges to Foster One Health, Almada, Portugal, 5–7 July 2023.

Abstract: This article delves into an explanatory model of interpersonal behavior based on alexithymia and internet dependence, demonstrating how these factors affect interpersonal relationships in a sample of university students. Alexithymia refers to a persistent difficulty in identifying, describing, and expressing emotions, which can hinder emotional connection with others. Internet addiction can lead to a neglect of interactions and social isolation, thereby reducing interpersonal relationships. The results of this study suggest that both alexithymia and internet addiction can have a negative impact on interpersonal relationships by interfering with the expression of emotions, communication and social engagement.

Keywords: alexithymia; internet dependency; personality; interpersonal relationships

1. Introduction

Alexithymia is a clinical concept used to describe people who have significant difficulty, or even inability, to express their emotions and intentions. [1]. Studies suggest that alexithymia is a multifactorial personality trait associated with deficits in cognitive–emotional processing; it is often observed in various disorders involving difficulties in identifying, verbalizing, and experiencing emotional situations [2]. Personality is described by a hierarchical model called the Big Five (neuroticism, extraversion, openness to experience, agreeableness, and conscientiousness). This plays an important role in the lives of subjects who have different personalities, and thus experience their emotions differently. A positive association was found between alexithymia and neuroticism, the latter being considered the greatest predictor of alexithymia. A negative association was found between alexithymia and extraversion, openness to experience and conscientiousness [3]. These domains, along with agreeableness, are associated with lower levels of alexithymia. Openness to experience is, after neuroticism, the greatest predictor of alexithymia [4]. Some authors suggest that alexithymia may play a critical role in the pathogenesis of internet dependency disorders, among others [5]. Existing literature also shows that individuals with alexithymia have difficulty in identifying, expressing, and communicating emotions, and they may use the internet as a tool for social interaction to better regulate their emotions and fulfill their social needs. It is common for individuals with alexithymia to engage in addictive behaviors as a form of escape from their real lives and the negative emotions underlying their condition. Thus, individuals with alexithymia are more likely to be addicted to the internet than individuals without alexithymia, as they use it as a tool for emotional expressing. Online addiction behaviors may impact the way emotions are regulated by enhancing emotion control, gaining validation online, and compensating for disadvantages in real life [6]. It is

Citation: Matias, A.; Soeiro, C.; Almeida, I.; Santos, R.P.; Baúto, R.V.; Oliveira, S. Alexithymia, Personality, Internet Addiction, and Interpersonal Relationships. *Med. Sci. Forum* **2023**, 22, 42. https://doi.org/10.3390/msf2023022042

Academic Editors: José Brito, Nuno Taveira and Ana I. Fernandes

Published: 21 August 2023

becoming essential to study the relationship between the two variables, and how internet use is perceived by individuals with alexithymia. The causal relationship is not yet clear due to the interaction of numerous other variables that may influence this relationship (e.g., environment and family). Alexithymia is related to behavioral dependence and lack of assertiveness, which suggests that individuals with difficulty expressing their emotions assertively become unable to adequately cope with negative emotions and stressful situations. Other research [3,4] has also shown that subjects with high levels of alexithymia tend to have difficulty expressing emotions and understanding the emotions of others, which can negatively affect their interpersonal behaviors. This can include difficulty in showing affection and facing problems communicating with others. Therefore, in interpersonal relationship behaviors, alexithymia can have a significant effect. Thus, difficulties in recognizing and expressing emotions can lead to inadequate communication and less ability to emotionally connect with others. Internet addiction refers to the excessive or uncontrolled use of the internet, which has negative psychological and social effects [7]. This dependency can negatively affect interpersonal relationships as individuals may prioritize virtual interactions over face-to-face contact, resulting in social isolation, decreased social interactions, and difficulties in face-to-face communication. We also know that excessive internet use seriously affects academic development, physical and mental health, and interpersonal relationships. There are studies indicating that factors affecting internet addiction include personality traits, emotions, and knowledge. When analyzing the relationship between internet dependence and personality, specifically related to alexithymia, the main findings indicate that more neurotic, extrovert, less agreeable and less conscientious individuals have a higher degree of internet addiction [8,9]. However, the literature has not yet reached a consensus on the association between neuroticism and excessive internet use. This article aims to study an explanatory model for interpersonal behaviors based on alexithymia, personality, and internet dependency. This research contributes to the development of prevention programs and strategies for training social skills in young adults, considering individual differences and the challenges of the internet use.

2. Methods and Materials

2.1. Participants

The sample consists of 491 young university students, of whom 119 are male (24.2%), and 372 are female (85.8%).

2.2. Instruments

The survey consists of five scales: (a) a sociodemographic questionnaire, aimed at complementing and contextualizing the collected information; (b) the Toronto Alexithymia Scale (TAS-20) by Taylor, Ryan, and Bagby (1985), translated and validated in Portuguese by Prazeres, Parker, and Taylor, 2000); (c) the Internet Addiction Test (IAT, Young, 1998); (d) the Neo-Five Factor Inventory (NEO-FFI, Costa, and McCrae, 1989), (e) the Interpersonal Behavior Scale—Reduced Version (ECIr-translation and adaptation by Vagos, Pereira, and Arrindell, 2014).

3. Results

The results of the TAS scale indicate that the sample presents low scores in all three test subscales, indicating a generally low level of alexithymia. The results obtained via the scale of internet addiction show that 65.2% of students present a low level of severity of internet use and 34.2% a moderate level of severity of internet use. The analysis of the relationship between alexithymia and internet addiction showed low/moderate, positive, and significant correlations between alexithymia subscales and the internet addiction general score (Table 1).

Table 1. Correlations between the indicators of *alexithymia* (TAS) and *internet addiction* (IAT) ($N = 491$).

	TAS-Difficulty in Describing Feelings and Emotions	TAS-Difficulty in Identifying Feelings and Emotions	TAS-Outward-Oriented Thinking Style
(IAT) Internet addiction	0.483 **	0.344 **	0.162 **

** $p < 0.01$.

Table 2 indicates that there is a significant, low/moderate and positive correlation between all alexithymia subscales and neuroticism. On the other hand, moderate negative correlations were observed between alexithymia (difficulty in describing and identifying feelings and emotions) and personality dimensions, agreeableness, and conscientiousness.

Table 2. Correlations between the indicators of *alexithymia* (TAS) and *personality* (NEOFFI) ($N = 491$).

	Neuroticism	Openness to Experience	Extraversion	Agreeableness	Conscientiousness
TAS-Difficulty describing feelings and emotions	0.435 **	−0.351 **		−0.232 **	−0.263 **
TAS-Difficulty identifying feelings and emotions	0.566 **	−0.241 **		−0.181 **	−0.296 **
TAS-Outward-Oriented Thinking Style	0.138 **		0.140 **		

** $p < 0.01$.

Table 3 shows low/moderate correlations between all alexithymia subscales and those of the Interpersonal Behaviors scale, which assesses the intensity of discomfort felt in social events that trigger an assertive response.

Table 3. Correlations between the indicators of alexithymia (TAS) and the *Interpersonal Behavior Scale* (IBS) ($N = 491$).

	IBS-Negative Assertiveness-Discomfort	IBS-Management Expression-Discomfort	IBS-Assertiveness Initiative-Discomfort	IBS-Positive Assertiveness-Discomfort
TAS-Difficulty describing feelings and emotions	0.255 **	0.238 **	0.307 **	0.297 **
TAS-Difficulty identifying feelings and emotions	0.294 **	0.302 **	0.327 **	0.262 **
TAS-Outward-Oriented Thinking Style	0.157 **	0.134 **	0.165 **	0.168 **

** $p < 0.01$.

4. Discussion

Based on the presented research, alexithymia and internet dependency can be negatively associated with interpersonal relationships since they disrupt emotional expression, communication, and social engagement. Also, the association between alexithymia, internet addiction with the personality indicates that more neurotic, less agreeable and less conscientious individuals reveal a higher degree of internet addiction and Alexithymia. The obtained results confirm the data already presented in the literature. Gathering more data and an analysis of gender differences are the next stages of research. Promoting the development of prevention programs and social skills strategies for young adults considering individual differences and the challenges of internet use is the main contribution of this project.

Author Contributions: Conceptualization, A.M., C.S., I.A., R.P.S. and S.O.; methodology, A.M., C.S., I.A., R.P.S., R.V.B. and S.O.; software, A.M., C.S., I.A., R.P.S., R.V.B. and S.O.; formal analysis, A.M., C.S., I.A., R.P.S., R.V.B. and S.O.; investigation, A.M., C.S., I.A., R.P.S., R.V.B. and S.O.; resources, A.M., C.S., I.A., R.P.S., R.V.B. and S.O.; data curation, A.M., C.S., I.A., R.P.S., R.V.B. and S.O.; writing—original draft preparation, A.M., C.S., I.A., R.P.S., R.V.B. and S.O.; writing—review and editing, A.M., C.S., I.A., R.P.S., R.V.B. and S.O.; visualization, A.M., C.S., I.A., R.P.S., R.V.B. and S.O.; supervision, C.S. All authors have read and agreed to the published version of the manuscript.

Funding: This research received no external funding.

Institutional Review Board Statement: The study was conducted in accordance with the Declaration of Helsinki and approved by the Institutional Review Board (or Ethics Committee) of Egas Moniz School of Health and Science for studies involving humans.

Informed Consent Statement: We rigorously adhered to the highest ethical standards, including the principles outlined in the Helsinki Declaration. Furthermore, we obtained informed consent from all participants, ensuring their autonomy and well-being throughout the scientific study.

Data Availability Statement: The data presented in this study are available on request from the corresponding author. The data are not publicly available due to privacy and ethical restrictions.

Acknowledgments: The authors would like to thank the students of the third year of the Psychology course at the Egas Moniz School of Health and Sciences, who contributed to the literature review and sample collection.

Conflicts of Interest: The authors declare no conflict of interest.

References

1. Freire, L. Alexitimia: Dificuldade de expressão ou ausência de sentimento? Uma análise teórica. *Psicol. Teor. Pesqui.* **2010**, *26*, 15–24. [CrossRef]
2. Taylor, G.J.; Bagby, M.; Parker, J.D.A. *Disordes of Affect Regulation: Alexithymia in Medical and Psychiatric Illness*; Cambridge University Press: Cambridge, UK, 1997.
3. Donges, U.S.; Suslow, T. Alexithymia and automatic processing of emotional stimuli: A systematic review. *Rev. Neurosci.* **2017**, *28*, 247–264. [CrossRef] [PubMed]
4. Taylor, G.J.; Bagby, R.M.; Kushner, S.C.; Benoit, D.; Atkinson, L. Alexithymia and adult attachment representations: Associations with the five-factor model of personality and perceived relationship adjustment. *Compr. Psychiatry* **2014**, *55*, 1258–1268. [CrossRef] [PubMed]
5. Heshmati, R.; Pellerone, M. The big five personality traits and dispositional mindfulness as predictors of alexithymia in college students. *Clin. Neuropsychiatry* **2019**, *16*, 98–106. [PubMed]
6. Mahapatra, A.; Sharma, P. Association of internet addiction and alexithymia—A scoping review. *Addict. Behav.* **2018**, *81*, 175–182. [CrossRef] [PubMed]
7. Ahmed, G.K.; Abdalla, A.A.; Mohamed, A.M.; Mohamed, L.A.; Shamaa, H.A. Relationship between time spent playing internet gaming apps and behavioral problems, sleep problems, alexithymia, and emotion dysregulations in children: A multicentre study. *Child Adolesc. Psychiatry Ment. Health* **2022**, *16*, 67. [CrossRef] [PubMed]
8. Grynberg, D.; Luminet, O.; Corneille, O.; Grèzes, J.; Berthoz, S. Alexithymia in the interpersonal domain: A general deficit of empathy? *Personal. Individ. Differ.* **2010**, *49*, 845–850. [CrossRef]
9. Monteiro, A.P.; Sousa, M.; Correia, E. Adição à internet e caraterísticas de personalidade: Um estudo com estudantes universitários portugueses. *Rev. Port. Educ.* **2020**, *33*, 159–176. [CrossRef]

medical sciences forum

MDPI

Proceeding Paper

Symptomatology in Intimate Partner Violence Offenders and Victims †

Iris Almeida [1,*], Alexandre Mateus [2], Guilherme Sena [3], Joana Fialho [2], Maria Beatriz Ribeiro [2], Rafaela Morgado [3] and Ricardo Ventura Baúto [1]

1 Egas Moniz Center for Interdisciplinary Research (CiiEM), Egas Moniz Forensic and Psychological Sciences Laboratory (LCFPEM), Egas Moniz School of Health & Science, 2829-511 Almada, Portugal; rbauto@egasmoniz.edu.pt

2 Egas Moniz School of Health & Science, 2829-511 Almada, Portugal; alex283843@gmail.com (A.M.); joanalcostafialho@gmail.com (J.F.); tiz2001@gmail.com (M.B.R.)

3 Egas Moniz Forensic and Psychological Sciences Laboratory (LCFPEM), Egas Moniz School of Health & Science, 2829-511 Almada, Portugal; guisenacosta@gmail.com (G.S.); rafaela.fmorgado@gmail.com (R.M.)

* Correspondence: ialmeida@egasmoniz.edu.pt

† Presented at the 6th International Congress of CiiEM—Immediate and Future Challenges to Foster One Health, Almada, Portugal, 5–7 July 2023.

Abstract: Intimate partner violence (IPV) consists of any action that holds the intention to inflict physical, sexual and/or psychological damages to the offending person's partner. This study aims to identify the psychopathological symptoms in IPV offenders and victims. Our sample consisted of 59 offenders and 63 IPV victims for whom the assessment was requested by a court order. The information on the evaluation was obtained through the database of the Forensic Psychology Office—Egas Moniz. The results show significative differences in five of the nine symptom dimensions, namely somatization, obsessive-compulsive, anxiety, phobic anxiety, and psychoticism, and victims have a complex presentation of symptoms. The impact of victimization is revealed in the negative repercussions at the level of global functioning in IPV victims rather than in IPV offenders.

Keywords: intimate partner violence; psychopathology; offenders; victims

Citation: Almeida, I.; Mateus, A.; Sena, G.; Fialho, J.; Ribeiro, M.B.; Morgado, R.; Baúto, R.V. Symptomatology in Intimate Partner Violence Offenders and Victims. *Med. Sci. Forum* **2023**, *22*, 43. https://doi.org/msf2023022043

Academic Editors: José Brito, Nuno Taveira and Ana I. Fernandes

Published: 22 August 2023

1. Introduction

Domestic violence is a major social problem with severe consequences for everyone. It consists of every type of violence such as physical, sexual, emotional, psychological, economic, and negligence, perpetrated in the family [1]. According to the World Health Organization (WHO), 30% of women around the world have been victims of domestic violence in their life, and 27% of women between 15 and 49 years old have been battered by their intimate partner. Intimate partner violence (IPV) is a form of domestic violence that consists of any action that holds the intention to inflict physical, sexual and/or psychological damage to the offender's partner [2]. In 2022, Portugal reported a total of 26,073 domestic violence cases (15.8% more than in 2021), where 80% were male offenders and 72.4% were female victims [3]. So IPV continues to represent most crimes committed in Portugal and constitutes a large problem that can have detrimental effects on families as well as on the larger community.

Although there is no direct relationship between psychopathology and IPV, it is common to identify personality and psychiatric disorders in several individuals judged by the criminal justice system [4–6]. For example, in a study carried out by Siria and colleagues [7], differences were identified between groups of offenders with and without psychopathology, with the former having higher levels of violence, anxiety (10.7%) and drug addiction (9.1%). Also, in this study, offenders experienced significantly higher rates in the prevalence of anxiety, bipolar disorder, dysthymia, alcohol dependence, drug dependence, thought disorder and severe depression syndromes. On another level, previous studies have shown that male offenders tended to have a higher probability of having any given psychiatric disorder than non-offenders [8,9].

As mentioned above, IPV is a public health problem with a significant impact on victims' physical and mental health, particularly concerning depression [10,11] and post-traumatic stress disorder [12,13], with this last one being recognized as one of the most prevalent mental health outcomes of IPV [14]. The frequent use of maladaptive coping was also related to the presence of more depressive symptomatology [15]. The literature points out that severe and prolonged interpersonal trauma often results in a complex exhibition of symptoms that also includes marked changes in personality functioning and in the areas of emotional regulation, consciousness, self-perception, identity, perception, interpersonal relations and system of meaning [16,17]. Many studies have also found neuropsychological consequences in IPV female victims in domains such as memory and learning [18], attention, response inhibition [19], and processing speed on tasks requiring executive functioning [20], resulting from physical violence.

Based on studies of IPV offenders and victims' symptomatology, this study aims to identify psychopathological symptoms in IPV offenders and victims.

2. Materials and Methods

Our sample consisted of 59 offenders and 63 IPV victims for whom the assessment was requested by a court order. The information on the evaluation was obtained through the database of the Forensic Psychology Office [Egas Moniz Forensic and Psychological Sciences Laboratory (LCFPEM)].

The IPV victims sample comprised: 58 women and five men, aged between 18 and 86 years old (M = 39.25; SD = 17.50). The IPV offenders sample comprised: 51 men and 9 women, aged between 20 and 82 years old (M = 42.21; SD = 13.84).

We used the Brief Symptom Inventory BSI [21,22], a self-report scale, composed of 53 items, that evaluate psychopathological and psychological symptoms. Respondents rated their symptoms on a five-point Likert scale, ranging from zero (not at all) to four (extremely), based on their experiences during the past week.

In the current study, an array of symptoms is employed in order to self-report psychological symptoms of clinical relevance, encompassing nine distinct dimensions: somatization (e.g., experiencing faintness), obsessive-compulsive (e.g., facing challenges in memory retention), interpersonal sensitivity (e.g., susceptibility to emotional hurt), depression (e.g., harboring thoughts of self-harm), anxiety (e.g., inner nervousness or tremors), hostility (e.g., feeling easily irritated or annoyed), phobic anxiety (e.g., experiencing fear in open spaces or public places), paranoid ideation (e.g., attributing most troubles to others) and psychoticism (e.g., believing others can control one's thoughts). This tool incorporates three comprehensive indices of overall distress: the Global Severity Index, Positive Symptom Distress Index, and Positive Symptom Total. To ensure the reliability of the identified symptom dimensions, Cronbach's alpha was calculated, resulting in an excellent coefficient of 0.98.

3. Results

Based on studies of IPV offenders and victims' symptomatology, this study aims to identify psychopathological symptoms in IPV offenders and victims. Table 1 shows significative differences in five of nine symptom dimensions, namely somatization, obsessive-compulsive, anxiety, phobic anxiety and psychoticism, with a significant impact on victims' mental healt.

Table 1. IPV offenders and victims' symptomatology.

	Victims		Offenders			
	M	SD	M	SD	t	p
Somatization	1.23	0.59	1	0.39	1.609	0.002
Obsessive-compulsive	1.37	0.68	1.07	0.47	1.726	0.001
Anxiety	1.56	0.97	1.47	0.51	3.106	0.002
Phobic anxiety	1.11	0.46	0.96	0.34	3.202	0.001
Psychoticism	1.26	0.56	1.15	0.46	1.708	0.007

4. Discussion

This study demonstrates that victims and offenders exhibit active symptomatology, with significant differences across five psychopathological dimensions, and with a significant impact on victims' mental health. Among the dimensions examined, victims consistently score higher than offenders. Although no similar effect can be inferred, the relevant literature has consistently identified an objective relationship between the vulnerability resulting from exposure to violent behaviors in victims and offenders and the emergence of psychopathological conditions. Furthermore, the symptoms observed align with those commonly described in the literature, where major depression, anxiety and dependence/addiction frameworks are deeply linked with the context and dynamics of abusive situations. In addition to aiding in assessment, these findings significantly contribute to more effective interventions. They assist professionals in adapting existing programs to achieve better outcomes in preventing recidivism. Consequently, interventions can specifically target the factors that increase the risk of recommitting this type of violence. Despite these valuable findings, this scientific investigation has limitations. The generalizability of the results and data analysis might be restricted due to the small sample size. Moreover, relying solely on self-reported measures to assess psychopathological conditions may not cover all aspects of information processing strategies. Therefore, it is crucial to consider these factors when interpreting the implications of the study.

Author Contributions: Conceptualization, I.A., R.M. and R.V.B.; methodology, I.A. and R.V.B.; software, I.A., R.M. and R.V.B.; validation, I.A., R.M. and R.V.B.; formal analysis, I.A. and R.V.B.; investigation, I.A. and R.V.B.; resources, I.A., R.M. and R.V.B.; data curation, A.M., G.S., J.F. and M.B.R.; writing—original draft preparation, I.A., R.M. and R.V.B.; writing—review and editing, I.A., R.M. and R.V.B.; visualization, I.A., R.M. and R.V.B.; supervision, I.A.; project administration, I.A. and R.V.B. All authors have read and agreed to the published version of the manuscript.

Funding: This research received no external funding.

Institutional Review Board Statement: This study was conducted in accordance with the Declaration of Helsinki, all ethical standards of scientific research were respected, as well as the Code of Ethics of the Order of Portuguese Psychologists and the General Data Protection Regulation.

Informed Consent Statement: Oral and written informed consent has been obtained from all participants. The informed consent explains the goal, the confidentiality limits and ethics and the technician's impartiality.

Data Availability Statement: Data sharing is not applicable because part of the information is derived from criminal records.

Conflicts of Interest: The authors declare no conflict of interest.

References

1. World Health Organization; Pan American Health Organization. *Understanding and Addressing Violence against Women: Intimate Partner Violence*; World Health Organization: Geneva, Switzerland, 2012; Available online: https://apps.who.int/iris/handle/10665/77432 (accessed on 1 June 2023).
2. Matos, M.; Machado, A. *Violência Doméstica: Intervenção em Grupos de Mulheres Vítimas—Manual Para Profissionais*; Comissão para a Cidadania e Igualdade de Género: Porto, Portugal, 2011.
3. de Segurança Interna, S. *Relatório Anual de Segurança Interna*; SSI: Lisboa, Portugal, 2023.

4. Craig, R.J. Use of the Millon Clinical Multiaxial Inventory in the psychological assessment of domestic violence: A review. *Aggress. Violent Behav.* **2003**, *8*, 235–243. [CrossRef]
5. Novo, M.; Farina, F.; Seijo, M.D.; Arce, R. Assessment of a community rehabilitation programme in convicted male intimate-partner violence offenders. *Int. J. Clin. Health Psychol.* **2012**, *12*, 219–234.
6. Weber, T.; Bouman, Y.H.A. Intimate partner violence: Variations in perpetrators and treatment allocation. *J. Interpers. Violence* **2020**, *35*, 1012–1034. [CrossRef] [PubMed]
7. Siria, S.; Fernández-Montalvo, J.; Echauri, J.A.; Arteaga, A.; Azkárate, J.M.; Martínez, M. Differential MCMI-III psychopathological profiles between intimate partner violence perpetrators with and without childhood family violence. *Clin. Psychol. Assoc.* **2021**, *28*, 1020–1029. [CrossRef] [PubMed]
8. Okuda, M.; Olfson, M.; Wang, S.; Rubio, J.M.; Xu, Y.; Blanco, C. Correlates of intimate partner violence perpetration: Results from a national epidemiologic survey. *J. Trauma. Stress* **2015**, *28*, 49–56. [CrossRef] [PubMed]
9. Shorey, R.C.; Febres, J.; Brasfield, H.; Stuart, G.L. The prevalence of mental health problems in men arrested for domestic violence. *J. Fam. Violence* **2012**, *27*, 741–748. [CrossRef] [PubMed]
10. Devries, K.M.; Mak, J.Y.; Bacchus, L.J.; Child, J.C.; Falder, G.; Petzold, M.; Astbury, J.; Watts, C.H. Intimate partner violence and incident depressive symptoms and suicide attempts: A systematic review of longitudinal studies. *PLoS Med.* **2013**, *10*, 3–5. [CrossRef] [PubMed]
11. Dillon, G.; Hussain, R.; Loxton, D.; Rahman, S. Mental and Physical Health and Intimate Partner Violence against Women: A Review of the Literature. *Int. J. Fam. Med.* **2013**, *2013*, 313909. [CrossRef] [PubMed]
12. Campbell, J.C. Health consequences of intimate partner violence. *Lancet* **2002**, *359*, 1331–1336. [CrossRef] [PubMed]
13. Torres, A.; Garcia-Esteve, L.; Navarro, P.; Tarragona, M.J.; Imaz, M.L.; Ascaso, C.; Gelabert, E.; Plaza, A.; Subirà, S.; Valdés, M.; et al. Relationship between intimate partner violence, depressive symptomatology, and personality traits. *J. Fam. Violence* **2013**, *28*, 369–379. [CrossRef]
14. Coid, J.; Petruckevitch, A.; Chung, W.S.; Richardson, J.; Moorey, S.; Feder, G. Abusive experiences and psychiatric morbidity in women primary care attenders. *Br. J. Psychiatry* **2013**, *183*, 332–339. [CrossRef] [PubMed]
15. Tonsing, K.N.; Tonsing, J.C.; Orbuch, T. Domestic violence, social support, coping and depressive symptomatology among South Asian women in Hong Kong. *J. Loss Trauma* **2021**, *26*, 134–152. [CrossRef]
16. Herman, J.L. Complex PTSD: A syndrome in survivors of prolonged and repeated trauma. *J. Trauma. Stress* **1992**, *5*, 377–391. [CrossRef]
17. Zucker, M.; Spinazzola, J.; Blaustein, M.; Zucker, M. Dissociative symptomatology in posttraumatic stress disorder and disorders of extreme stress dissociative symptomatology in posttraumatic stress disorder and disorders of extreme stress. *J. Trauma Dissociation* **2008**, *7*, 19–31. [CrossRef] [PubMed]
18. Valera, E.M.; Berenbaum, H. Brain injury in battered women. *J. Consult. Clin. Psychol.* **2003**, *71*, 797–804. [CrossRef] [PubMed]
19. Little, D.M.; Kraus, M.F.; Joseph, J.; Geary, E.K.; Susmaras, T.; Zhou, X.J.; Pliskin, N.; Gorelick, P.B. Thalamic integrity underlies executive dysfunction in traumatic brain injury. *Neurology* **2010**, *74*, 558–564. [CrossRef] [PubMed]
20. Marín Torices, M.I.; Hidalgo-Ruzzante, N.; Daugherty, J.C.; Jiménez-González, P.; Pérez García, M. Validation of neuropsychological consequences in victims of intimate partner violence in a Spanish population using specific effort tests. *J. Forensic Psychiatry Psychol.* **2018**, *29*, 86–98. [CrossRef]
21. Canavarro, M.C. Inventário de Sintomas Psicopatológicos: BSI. In *Testes e Provas Psicológicas em Portugal*; Simões, M.R., Machado, C., Gonçalves, M.M., Almeida, L.S., Eds.; SHO/APPORT: Braga, Portugal, 1999; Volume 2, pp. 87–109.
22. Canavarro, M.C. Inventário de Sintomas Psicopatológicos: Uma revisão crítica dos estudos realizados em Portugal. In *Avaliação psicológica: Instrumentos Validados para a População Portuguesa*; Simões, M.R., Machado, C., Gonçalves, M.M., Almeida, L.S., Eds.; Quarteto Editora: Coimbra, Portugal, 2007; Volume 3, pp. 305–331.

medical sciences forum

MDPI

Proceeding Paper

Risk Factor Prevalence in Handball Athletes with and without Overuse Injury History †

Inês Ventura [1], Sofia Vieira [1], Joana Rosa [1], Margarida Feliciano [1], José Brito [1], Luciano Maia Alves Ferreira [1], Duarte Tavares [2] and Manuel Barbosa Almeida [1,*]

[1] Egas Moniz Center for Interdisciplinary Research (CiiEM), Egas Moniz School of Health & Science, Campus Universitário, Quinta da Granja, Caparica, 2829-511 Almada, Portugal; inespereiraventura21@gmail.com (I.V.)
[2] Regional Health Administration of Lisbon and Tagus Valley, P.I., Av. Estados Unidos da América nº 77, 1749-096 Lisboa, Portugal
* Correspondence: mbalmeida@egasmoniz.edu.pt
† Presented at the 6th International Congress of CiiEM—Immediate and Future Challenges to Foster One Health, Almada, Portugal, 5–7 July 2023.

Abstract: Handball is a demanding sport with risk of overuse injuries. We aimed to compare scapular motion, the external/internal rotator strength ratio, and shoulder rotator range of motion of handball athletes with and without a history of overuse injury. Cross-sectional data were collected from male handball athletes. The results showed no differences between groups, except for dyskinesia in final assessment position and external rotation range of motion. Scapular dyskinesis prevalence was high in both groups. These findings emphasize the importance of identifying and addressing these risk factors in injury prevention and rehabilitation programs for handball athletes.

Keywords: handball; overhead; shoulder; overuse injury; risk factors; prevention

Citation: Ventura, I.; Vieira, S.; Rosa, J.; Feliciano, M.; Brito, J.; Ferreira, L.M.A.; Tavares, D.; Almeida, M.B. Risk Factor Prevalence in Handball Athletes with and without Overuse Injury History. *Med. Sci. Forum* **2023**, 22, 7. https://doi.org/10.3390/msf2023022007

Academic Editors: Nuno Taveira and Ana I. Fernandes

Published: 8 August 2023

1. Introduction

Handball is a team sport characterized by a high playing time, rapid changes in direction, abrupt landings from jumps, repetitive throwing, and frequent contact and collisions between players, making it a demanding sport with a high risk of injuries [1]. Indirect or non-contact injuries are among the most common types of injuries in handball [2]. Overuse injuries arise from multiple cumulative energy transfers resulting in physical complaints, leading to a reduction in training volume, pain, and decreased performance in training and competitions [3]. In this sport, there are various stress movements and positions that the shoulder joint is subjected to, especially during the throwing motion. Therefore, a weakness or imbalance that alters any of the components of the kinetic chain may lead to a dysfunctional shoulder or may put the individual at a high risk of upper limb injuries [4,5]. Since most injuries in this sport are caused by repetitive overuse of structures in a specific motion, deficits in a joint's range of motion (ROM), the ratio of external/internal (ER/IR) rotator strength of the shoulder, and scapular dyskinesis were identified as modifiable intrinsic risk factors, as their effect can be altered through targeted injury prevention programs [6].

The aim of this study is to compare the prevalence of risk factors such as scapular dyskinesis, the shoulder ER/IR strength ratio, rotation range of motion, and associated factors in handball athletes with and without a history of overuse injury.

2. Materials and Methods

2.1. Study Design

This was a cross-sectional study. We used Kforce-Link® pull dynamometer (Kinvent, Montpellier, France) to assess the ER/IR rotator force; a digital inclinometer, through the Clinometer® smartphone application (Plaincode, Stephanskirchen, Germany), to evaluate shoulder ROM; and a measuring tape to assess scapular displacement through the lateral scapular slide test (LSST).

2.2. Participants

Handball clubs from Lisbon and Setúbal districts in the second division were contacted to obtain their permission to develop this study. Volunteer players from the integrating clubs participated in the study if they were registered male handball athletes in senior and under 20 teams, and athletes with a history of a dominant shoulder overuse injury in the last 12 months were allocated to the injury group. The exclusion criteria were underage, any cognitive impairments, orthopedic shoulder surgery in the last 12 months, and traumatic shoulder injury in the last 3 months.

2.3. Procedures

Assessments were made for both sides. For ROM measurement, a digital inclinometer trough clinometer app was used. It was positioned parallel to the midline of the anterior region of the forearm. Passive IR and ER measurements were taken in the supine position with the dominant shoulder at 90° abduction and the elbow flexed at 90°. For strength measurements of the shoulder, three maximum voluntary isometric contractions of 5 s with 30 s pauses between repetitions were performed. For measurement of scapular displacement, a tape measure was used, measuring between the inferior angle of the scapula and D7 in neutral position, 45 and 90 degrees of shoulder abduction, according to the lateral scapular slide test.

2.4. Statistics

If normal distribution was verified, a parametric independent *t*-test was performed to compare the groups; otherwise, a Mann–Whitney U test was used. To investigate the relative contribution of the characterization variables to the neuromuscular profile of the athletes, a binary logistic regression model for each dependent variable was used.

3. Results

A total of 59 participants participated in the study. The healthy group had 39 athletes with an age of 22.8 ± 5.3 years and a body mass index (BMI) of 26.9 ± 4.5 kg/m^2. The unhealthy group had 20 athletes with an age of 24.0 ± 7.0 years and a BMI of 25.1 ± 3.7 kg/m^2. There were no differences between groups in a variable measurement mean comparison except for dyskinesia at 90° shoulder abduction and 45° horizontal abduction, with maximal IR (p = 0.045) and in external rotation ROM (p = 0.035). The prevalence of risk factors for each group is detailed in Table 1.

Table 1. Prevalence of risk factors by group.

	Healthy Group (n = 39)		Injury Group (n = 20)		p
	Risk n (%)	No Risk n (%)	Risk n (%)	No Risk n (%)	
Internal Rotation	11 (28.2%)	28 (71.8%)	7 (35%)	13 (65%)	0.402
External Rotation	17 (43.6%)	22 (56.4%)	3 (15%)	17 (85%)	0.026 *
TROM	11 (28.2%)	28 (71.8%)	4 (20%)	16 (80%)	0.233
Ratio ER/IR	10 (25.6%)	29 (74.4%)	5 (25%)	15 (75%)	0.609
Scapular Dyskinesis	19 (48.7%)	20 (51.3%)	10 (50%)	10 (50%)	0.572

Abbreviations: ER = external rotators; IR = internal rotators; TROM = total range of motion. p-value with chi-square for comparison of occurrences between groups; * p-value < 0.05.

4. Discussion

Handball is a throwing sport that places a high demand on the shoulder joint, making it prone to injuries. Recent studies have confirmed that between 17% and 44% of all overuse injuries in handball are in the shoulder [2,7]. Identification of changes in ROM, strength of rotational movements, and scapular position can be used to identify athletes at risk of injury. Therefore, we compared the neuromuscular profile characteristics of athletes with and without a history of injury to identify the prevalence of identified risk factors in each group and investigated their association with sociodemographic characteristics.

4.1. Range of Motion

There were no differences between the groups regarding internal rotation (IR) range of motion. However, we found that an additional hour of training reduces the probability of having an IR deficit of more than 10° by 37.2%. Functional adaptations in handball result in a shift in the total available ROM, with a reduction in IR and an increase in ER that could be up to 16°, with increased anterior capsular laxity and excessive glenohumeral translation during ER [8,9]. Our results are consistent with this, as we observed significative differences between groups, with an average increase of 14.65 ± 10.5° in athletes with injury and 7.72 ± 14.9° in healthy athletes. These changes may contribute to the selective atrophy of the infraspinatus muscle, caused by a functional denervation of the suprascapular nerve. A gain in ER ROM of less than 5° has also been identified as a risk factor, as it could indicate a tendency for retraction of IR [10], and our results showed a significantly higher rate of this within the healthy group. These functional alterations may underlie shoulder overuse injuries and are a possible causative factor in the short to medium term, making their prevention extremely important.

A reduced total ROM could increase the probability of experiencing substantial shoulder problems throughout the season [10]. Nonetheless, we observed low rates of athletes presenting a reduced total ROM when compared to the non-dominant side in both groups. Reductions in IR and increases in ER present in the dominant shoulders of uninjured overhead athletes are regarded as normal soft tissue adaptations to repeated throwing [10], and the changes we found are in accordance with that.

4.2. Strength Ratio ER/IR

Although no significant differences in shoulder rotator strength were observed between the groups, the injury group had an ER/IR strength ratio significantly lower than 1. This may be due to the rehabilitation process performed to return to sports, which tends to correct the tendency to a ratio greater than 1 [11]. These findings contradict our hypothesis that athletes with a greater IR strength will be found among the healthy group. Scapular stability is crucial for normal shoulder function, especially in a handball athlete [11]. Our results showed that each additional year of training decreases the risk of a stronger ER than IR by 14.5%, and each added unit within the healthy body mass index (BMI) range decreases the risk of having a strength ratio imbalance in favor of ER by 23.8%.

4.3. Scapular Dyskinesis

The presence of significant scapular movement differences in athletes with injuries may be one of the causes of their injury history, as a combination of reduced scapular adduction/retraction capacity with excessive ER can contribute to an increased risk of anterior glenohumeral instability, which may be the underlying cause of the previously manifested overuse injuries [9]. Despite the fact that we found similar scapular dyskinesis occurrence rates between groups, the injury group had greater scapular displacement on the dominant side compared to healthy athletes in the final evaluation position. Current evidence indicates that most scapular problems in throwing athletes can be attributed to a loss of control of the normal resting scapular position and dynamic scapular movement, resulting in scapular protraction that can result in increased loads, altered movements, and reduced muscle activations that may be associated with decreased performance and increased injury risk [12]; our results are in line with this knowledge.

4.4. Limitations

We assessed the ROM with participants in supine with the shoulder at 90° abduction and the elbow at 90° flexion to replicate a shooting position, but this could promote measurement errors through compensations. The cross-sectional design of our study prevented the determination of our findings as causes or consequences of overuse injuries. A longitudinal setting and improved strategies to control measurement errors are recommended for future studies.

5. Conclusions

Analyzing the rotation ROM, scapular displacement, and ER/IR strength ratio in handball athletes with and without a history of injury should be standard practice for the development of injury prevention programs and targeted rehabilitation for athletes with shoulder overuse injuries, since both groups present high rates of identified risk factors.

Author Contributions: Conceptualization, I.V., S.V., J.R., M.F., D.T., J.B. and L.M.A.F. and M.B.A. methodology, I.V., S.V., J.R. and M.F.; formal analysis, I.V., S.V., J.R. and M.F.; investigation, I.V., S.V., J.R. and M.F; data curation, L.M.A.F., M.B.A., J.B. and D.T.; writing—original draft preparation, I.V., S.V., J.R. and M.F.; writing—review and editing, L.M.A.F., D.T., J.B. and M.B.A.; supervision, M.B.A.; project administration, I.V., S.V., J.R. and M.F. All authors have read and agreed to the published version of the manuscript.

Funding: This research received no external funding.

Institutional Review Board Statement: Egas Moniz Ethics Commission approval CE-1088.

Informed Consent Statement: Informed consent was obtained from all participants in the study.

Data Availability Statement: Mendeley Data, V1, doi: 10.17632/sm66hkjvf4.1, available online: https://data.mendeley.com/datasets/sm66hkjvf4/1 (accessed on 8 August 2023).

Acknowledgments: We would like to thank the Laboratório de Avaliação Física e Funcional em Fisioterapia (LAFFFi) by Sausport for the availability of material, including the KINVENT and all the support in the realization of this project and to FCT/MCTES for the financial support to CiiEM (UIDB/04585/2020) through national funds.

Conflicts of Interest: The authors declare no conflict of interest.

References

1. Raya-González, J.; García-Esteban, S.; Castillo, D.; de Ste Croix, M. Injury profile in professional handball players during 4 consecutive seasons according to playing positions: A longitudinal study. *Sports Health* **2022**, *14*, 273–282. [CrossRef] [PubMed]
2. Achenbach, L.; Luig, P. Epidemiologie und Verletzungsprävention im Handball. *Sport. Sportschaden* **2020**, *34*, 129–135. [CrossRef] [PubMed]
3. Karlsson, J.; Heijne, A.; von Rosen, P. Handball and movement screening–can non-contact injuries be predicted in adolescent elite handball players? A 1-year prospective cohort study. *Physiother. Theory Pract.* **2021**, *37*, 1132–1138. [CrossRef] [PubMed]
4. Kibler, W.B.; Kuhn, J.E.; Wilk, K.; Sciascia, A.; Moore, S.; Laudner, K.; Ellenbecker, T.; Thigpen, C.; Uhl, T. The disabled throwing shoulder: Spectrum of pathology—10-year update. *Arthroscopy* **2013**, *29*, 141–161.e126. [CrossRef] [PubMed]
5. Landreau, P.; Zumstein, M.A.; Lubiatowski, P.; Laver, L. Shoulder Injuries in Handball. In *Handball Sports Medicine: Basic Science, Injury Management and Return to Sport*; Laver, L., Landreau, P., Seil, R., Popovic, N., Eds.; Springer: Berlin/Heidelberg, Germany, 2018; pp. 177–195.
6. Pozzi, F.; Plummer, H.A.; Shanley, E.; Thigpen, C.A.; Bauer, C.; Wilson, M.L.; Michener, L.A. Preseason shoulder range of motion screening and in-season risk of shoulder and elbow injuries in overhead athletes: Systematic review and meta-analysis. *Br. J. Sports Med.* **2020**, *54*, 1019–1027. [CrossRef] [PubMed]
7. Aasheim, C.; Stavenes, H.; Andersson, S.H.; Engbretsen, L.; Clarsen, B. Prevalence and burden of overuse injuries in elite junior handball. *BMJ Open Sport Exerc. Med.* **2018**, *4*, e000391. [CrossRef] [PubMed]
8. Tooth, C.; Gofflot, A.; Schwartz, C.; Croisier, J.-L.; Beaudart, C.; Bruyère, O.; Forthomme, B. Risk factors of overuse shoulder injuries in overhead athletes: A systematic review. *Sports Health* **2020**, *12*, 478–487. [CrossRef] [PubMed]
9. Ouellette, H.; Kassarjian, A.; Tretreault, P.; Palmer, W. Imaging of the overhead throwing athlete. *Semin. Musculoskelet. Radiol.* **2005**, *9*, 316–333. [CrossRef] [PubMed]
10. Clarsen, B.; Bahr, R.; Andersson, S.H.; Munk, R.; Myklebust, G. Reduced glenohumeral rotation, external rotation weakness and scapular dyskinesis are risk factors for shoulder injuries among elite male handball players: A prospective cohort study. *Br. J. Sports Med.* **2014**, *48*, 1327–1333. [CrossRef] [PubMed]

11. Wilk, K.E.; Arrigo, C.A.; Hooks, T.R.; Andrews, J.R. Rehabilitation of the overhead throwing athlete: There is more to it than just external rotation/internal rotation strengthening. *PM&R* **2016**, *8*, S78–S90.
12. Hickey, D.; Solvig, V.; Cavalheri, V.; Harrold, M.; Mckenna, L. Scapular dyskinesis increases the risk of future shoulder pain by 43% in asymptomatic athletes: A systematic review and meta-analysis. *Br. J. Sports Med.* **2018**, *52*, 102–110. [CrossRef] [PubMed]

medical sciences forum

MDPI

Proceeding Paper

Narratives of Hope—The Temporal Dimension in the Ontological Manipulation of the Human Embryo †

Luís Gouveia * and Catarina Delaunay

Interdisciplinary Centre of Social Sciences, Faculty of Social Sciences and Humanities, University Nova de Lisboa, 1070-312 Lisbon, Portugal; catarinadelaunay@fcsh.unl.pt
* Correspondence: lcgouveia86@gmail.com

† Presented at the 6th International Congress of CiiEM—Immediate and Future Challenges to Foster One Health, Almada, Portugal, 5–7 July 2023.

Abstract: The in vitro human embryo is a liminal being towards which the beneficiaries of assisted reproductive technology (ART) can ascribe different moral statuses. Moreover, this ontological manipulation can be combined with a temporal dimension. Namely, a set of narratives of hope can be identified, in terms of how patients project the future considering the accomplishment of the parental project. In turn, these narratives of hope will have repercussions at the ontological level, i.e., on the meaning-making processes around the human embryos created during ART treatments. A Portuguese research project provides empirical data (from semi-directive interviews) to map these plural conceptualizations of human embryos.

Keywords: human embryo; moral status; meaning-making processes; ontological manipulation; narratives of hope

Citation: Gouveia, L.; Delaunay, C. Narratives of Hope—The Temporal Dimension in the Ontological Manipulation of the Human Embryo. *Med. Sci. Forum* **2023**, *22*, 49. https://doi.org/10.3390/msf2023022049

Academic Editors: José Brito, Nuno Taveira and Ana I. Fernandes

Published: 6 September 2023

1. Introduction

The in vitro human embryo constitutes a liminal entity that gives rise to plural conceptualizations [1]. Namely, it is characterized by an ambiguity of its moral status, oscillating between different states—with the boundaries among these states being also fluctuant and difficult to define [2].

Within this ontological manipulation [3], the discourses of ART beneficiaries show that the way embryos are perceived oscillates throughout the therapeutic path [4]. In fact, these conceptualizations range from a *functional* understanding of this being, as a *tool/means* in service of a plan of action (for achieving pregnancy) to an appropriation according to general equivalence classes/categories [3]. That is, the case of classifications such as *valuable biological material*, *genetic heritage*, *offspring*, *potential child*, a *gift* for other couples or to science, etc. [5].

Moreover, this apprehension in generality gives place to a gradual *rise in singularity* [3]: a progressive emotional attachment to the embryo as a singular entity—giving it an irreplaceable character. This singularization of the embryo, as a *child*, manifests in particular emotional ambiences—such as love towards it and grief in case of loss, but also through symbolic gestures like giving a name [3]. And in case of embryo loss, that same name is not given to another embryo—thus expressing that irreplaceable character, different from an understanding in generality.

However, the ontological manipulation of the human embryo presents further complexity. Namely, it is also necessary to consider that the moral statuses do not only vary between different beneficiaries [5]. The meaning-making processes can also change along the therapeutic trajectory of each beneficiary. Modifications in the way ontological manipulation develops—such as the stages of treatment at which the emotional attachment begins—may occur during the therapeutic trajectory of an individual patient.

Concerning these ontological fluctuations, an element of patients' experience emerges, attached to the uncertainty [6] of a clinical treatment's outcome: *hope* regarding the achievement of the therapeutic goal. Indeed, hope is a vector of ontological manipulation regarding different aspects of the clinical trajectory [7]. In the concrete case of ART treatments, it can shape the meanings built around the embryo, triggering fluctuations in the ontological markers that determine the transitions between statuses assigned to that entity [3].

Hope can be characterized as a lens that guides the actor's perspective towards a particular phenomenon situated in the future. Namely, it involves a relationship between *knowledge*—which provides a certain degree of predictability about the outcome of a plan of action (such as a clinical protocol)—and *uncertainty*—about the accomplishment of the plan's goal [8,9].

Moreover, hope is multidimensional, depending on the greater or lesser degree of tangibility of the objective projected into the future. Namely, what emerges from patients' discourses is that *hope* can oscillate within a *continuum* between *hope as expectation* and *hope as desire* [7,9]. *Hope as expectation* means hope as anticipating what is likely and concrete in the future (in this case, achieving pregnancy)—supported, for example, by biomedical information or past experiences. *Hope as desire* constitutes a focus on a more remote/generic possibility—therefore, a *looser* relationship with the objective that guides the therapeutic plan.

In turn, these plural ways of operationalizing hope will have repercussions on an ontological level, particularly in terms of oscillations in the moral status attributed to the human embryo. It is this dynamic between the two dimensions—how narratives of hope performed by ART beneficiaries are a driving force of the ontological manipulation of the embryo—that we seek to develop in a preliminary exploration.

2. Methods

The data presented in this text was collected from a research project conducted in Portugal from 2017 to 2022. The project is named *ETHICHO—Ethico-ontological Choreographies: Forms of objectification and evaluation of the human embryo* in vitro *in the context of Assisted Reproductive Technologies and Scientific Research*. The project relies on a mixed methods approach, encompassing an online questionnaire and semi-directive interviews. However, considering its objectives, the explorative analysis developed in this text rests only on data from the latter technique.

For a brief methodological description, 69 semi-directive interviews were conducted with ART beneficiaries/patients between September 2019 and January 2021. Potential interviewees were contacted through non-probabilistic convenience and snowball sampling, relying predominantly on social networks, along with the support of Portuguese associations for people with infertility. One fundamental methodological criterion for selecting potential respondents was outlined: to have undergone at least one cycle of *second-line treatment*, i.e., a treatment cycle involving the creation of in vitro embryos: In vitro Fertilization (IVF) or Intracytoplasmic Injection technique (ICSI).

All interviews were recorded and fully transcribed, and the qualitative analysis was carried out with the technical support of the software MaxQDA (2018 version). All study participants provided written or verbal informed consent prior to data collection. Ethical approval was obtained from the host institution of the research project before data collection began. In addition, respondents were given pseudonyms to ensure anonymity.

3. Results

Concerning the temporal dimension of analysis, a set of narratives of hope can be identified in beneficiaries' discourses, with patients describing changes in the format of hope from the beginning of treatment and throughout the therapeutic process. These variations in the management of hope can take different forms:

(a) *Oscillations between concrete and vague plans.* In the face of uncertainties and failures throughout the clinical trajectory, a coping mechanism in the present emerges: a (partial) subtraction of the future, in terms of a detachment from *hope as expectancy* to *hope as a desire*;

(b) *Tensions/ambivalences between projecting and dimming the future.* Namely, patients' discourses can show situations of tension/ambivalence between the suspension (with different degrees) of hope as expectation (about the accomplishment of the parental project) in favour of hope as desire, but in which moments of focusing on a concrete future emerge (e.g., fantasizing with elements of the parental project);

(c) *Combinations of expectation and desire.* Patients can also manifest *narratives of hope* in the form of building expectations on short-term stages (such as, for example, obtaining viable embryos), while, at the same time, maintaining a focus on the fulfilment of the parental project in more general forms (thus, as a desire).

In turn, these narratives of hope will have repercussions at an ontological level—i.e., on the meaning-making processes around the human embryos generated during the ART treatments. To illustrate this temporal and ontological dynamic, an excerpt from an interviewed couple is presented.

At the time of the interview, two treatment cycles characterize their clinical path. The couple's discourse illustrates how the failure to achieve pregnancy in this first cycle led to a reconfiguration in the format of engagement in the treatment (and with the embryos) in the second treatment cycle. Namely, a deferment in the process of emotional bonding takes place:

Denise: "[...] we were already warned, we tried not to emotionally bond with them, because if it was already painful the first time, this time there was an emotional barrier. [...] I tried not to create too much affection in the first few months because I was always thinking "Oh, oh, oh, I'm going to lose the baby", "Oh, oh, oh, this is going to be really bad". So, I tried to abstract myself a little bit, because I already knew the pain. I tried to think "This is not happening". I mean, I did all the mandatory medical treatment, but my emotional part was always very contained there, until that ultrasound".

Leonard: "Indeed, the image, the image is very strong. It precedes the contact itself, when you begin to feel moving and the first connection.

Firstly, concerning the temporal dimension, the couple's discourse highlights how the loss of pregnancy in the first cycle is a determinant factor in the change in how engagement in the new treatment cycle takes place. As the female interviewee describes, the involvement in the therapeutic trajectory remains ("I did all the mandatory medical care"). However, there is a change in terms of how hope is operationalized. The narrative of hope associated with the engagement in a therapeutic plan suffers a partial suspension, being managed in the direction of a desire—that is, as a more general form of hope.

Thus, if expectation encompasses a vulnerability for the one who produces/supports this format of hope—considering the emotional impact in case the objective fails—the (partial) suppression of the future (in the form of expectation) becomes a coping mechanism for the couple ("I already knew the pain", "I tried to abstract myself a little bit"). Thus, there is a change in how the female interviewee projects herself into the future concerning the goal of the treatment (i.e., obtaining a full-term pregnancy).

On the other hand, this operationalization of hope also leads to reconfigurations at the ontological level. Namely, the beginning of the process of singularization of the embryo, associated with the gradual construction of an emotional bond, is relegated to a later stage in the trajectory. Specifically, it is the echogram image that stands out as a moment of oscillation in the meaning assigned—instead of the *blastocyst* stage (i.e., the last stage of the development of an embryo in vitro) in the previous treatment cycle. As described, there was an attempt to "not connect emotionally" with this being, in contrast to the first treatment cycle. Thus, at the ontological level, there is a transfer of the oscillations in the meanings attributed to the embryo—in particular, the beginning of the process of singularization—to a subsequent moment of the therapeutic trajectory.

4. Discussion and Conclusions

The different stages of the therapeutic trajectory of ART beneficiaries, which may affect their *hope* regarding the outcome of the treatment, are likely to consubstantiate moments of reconfiguration of the meanings attributed to the embryos and the attached emotional ambiences [1]. These modes of (re)configuration in the meaning-making processes favour; thus, a double pluralism in the ontological constructions around this entity, according to the vicissitudes and circumstances associated with the therapeutic trajectory of patients resorting to ART. In particular, the ontological markers [3] that punctuate the clinical path of the interviewees can, thus, metamorphose according to different moments of success, failure, setbacks and restarts that characterize the therapeutic trajectory [6].

Author Contributions: Writing—original draft preparation, L.G.; writing—review and editing, L.G., C.D.; project administration, C.D.; funding acquisition, C.D; conceptualization, C.D. All authors have read and agreed to the published version of the manuscript.

Funding: This research was funded by FCT—Foundation for Science and Technology, I.P., grant number PTDC/SOC-SOC/29764/2017.

Institutional Review Board Statement: The study was conducted according to the guidelines of the Declaration of Helsinki, and approved by the Ethics Committee of the Universidade NOVA de Lisboa (25 September 2019).

Informed Consent Statement: Informed consent was obtained from all participants.

Data Availability Statement: The data presented in this study are available on request from the project administrator.

Conflicts of Interest: The authors declare no conflict of interest. The funders had no role in the design of the study, in the collection, analyses, or interpretation of data, in the writing of the manuscript, or in the decision to publish the results.

References

1. Giraud, A.-S. Les Statuts de L'être Anténatal: Un Processus D'humanisation 'Relationnel': Assistance Médicale à la Procréation et Mort Périnatale. Ph.D. Thesis, École des Hautes Études en Sciences Sociales, Paris, France, 2015.
2. Provoost, V.; Pennings, G.; De Sutter, P.; Dhont, M. "Something of the two of us". The emotionally loaded embryo disposition decision making of patients who view their embryo as a symbol of their relationship. *J. Psychosom. Obstet. Gynecol.* **2012**, *33*, 45–52. [CrossRef] [PubMed]
3. Boltanski, L. *The Foetal Condition. A Sociology of Engendering and Abortion*; Polity Press: Cambridge, UK, 2013.
4. Delaunay, C.; Santos, M.J.; Gouveia, L. In vitro metaphors: ART beneficiaries' meaning-making on human embryos in the context of IFV in Portugal. Reprod. *Biomed. Soc. Online* **2021**, *13*, 62–74. [CrossRef] [PubMed]
5. Haimes, E.; Porz, R.; Scully, J.L.; Rehmann-Sutter, C. "So, what is an embryo?" A comparative study of the views of those asked to donate embryos for hESC research in the UK and Switzerland. *New Genet. Soc.* **2008**, *27*, 113–126. [CrossRef] [PubMed]
6. Delaunay, C. L'insupportable et l'incertain. Proximité et détachement dans l'assistance médicale à la procréation avec tier donneur. *Anthropol. St.* **2017**, *15*. [CrossRef]
7. Brown, P.; Graaf, S.; Hillen, M. The inherent tensions and ambiguities of hope: Towards a post-formal analysis of experiences of advanced-cancer patients. *Health* **2015**, *19*, 207–225. [CrossRef] [PubMed]
8. Thévenot, L. *L'action au Pluriel. Sociologie des Régimes d'engagement*; Éditions La Découverte: Paris, France, 2006.
9. Peterson, A.; Tanner, C. Between hope and evidence: How community advisors demarcate the boundary between legitimate and illegitimate stem cell treatments. *Health* **2015**, *19*, 188–206. [CrossRef] [PubMed]

medical sciences forum

MDPI

Proceeding Paper

New Compound Combining an Integrase-Targeting Aptamer and a Small Interfering RNA Targeting the Trans-Activation Response/Poly A Region of HIV-1 Potently Suppresses HIV-1 Replication †

Inês Moranguinho [1], Pedro Borrego [1,2], João Lavrado [1], Rui Moreira [1] and Nuno Taveira [1,3,*]

1 Research Institute for Medicines (iMed.ULisboa), Faculty of Pharmacy, Universidade de Lisboa, 1600-277 Lisbon, Portugal; moranguinho.ines@gmail.com (I.M.); pborrego65@gmail.com (P.B.); jlavrado@gmail.com (J.L.); rmoreira@ff.ulisboa.pt (R.M.)
2 Centro de Administração e Políticas Públicas (CAPP), Instituto Superior de Ciências Sociais e Políticas, Universidade de Lisboa, 1300-666 Lisbon, Portugal
3 Centro de Investigação Interdisciplinar Egas Moniz (CiiEM), Instituto Universitário Egas Moniz, 2829-511 Caparica, Portugal
* Correspondence: ntaveira@ff.ulisboa.pt

† Presented at the 6th International Congress of CiiEM—Immediate and Future Challenges to Foster One Health, Almada, Portugal, 5–7 July 2023.

Abstract: We have developed a novel aptamer-based siRNA delivery system for HIV therapy. Apsi510 was obtained by chemical conjugation of an anti-HIV integrase aptamer and an siRNA sequence targeting the HIV-1 TAR/poly A regions to a dendron [2-((4-(2,5-dioxo-2,5-dihydro-1H-pyrrol-1-yl)phenyl)amino)acetaldehyde]. Apsi510 activity against HIV-1NL4.3 was evaluated in two experimental systems using HeLa CD4+ and TZM-bl cells. Apsi510 activity was dose-dependent and inhibited >95% of viral replication at 50 nM. Apsi510 inhibited HIV-1 replication to a similar extent as siRNA alone, indicating efficient intracellular release of the siRNA molecule. Apsi510 is a promising drug candidate for the treatment and prevention of HIV.

Keywords: HIV integrase aptamer; TAR siRNA; Apsi510 compound; antiviral activity

Citation: Moranguinho, I.; Borrego, P.; Lavrado, J.; Moreira, R.; Taveira, N. New Compound Combining an Integrase-Targeting Aptamer and a Small Interfering RNA Targeting the Trans-Activation Response/Poly A Region of HIV-1 Potently Suppresses HIV-1 Replication. *Med. Sci. Forum* **2023**, 22, 23. https://doi.org/10.3390/msf2023022023

Academic Editors: José Brito and Ana I. Fernandes

Published: 11 August 2023

1. Introduction

The human immunodeficiency virus (HIV) is the causative agent of acquired immunodeficiency syndrome (AIDS), a disease that remains incurable and requires lifelong treatment. The best way to end the HIV/AIDS pandemic by 2030 is to cure the infection. Stem cell transplantation has already cured four HIV patients. Simpler, more accessible therapeutic strategies to cure HIV infection have been explored, but none have yet reached the clinic [1].

RNA interference (RNAi) is a natural antiviral defense mechanism found in plants, fungi and invertebrates [2]. It is mediated by small interfering RNA (siRNA) and microRNA (miRNA), which degrade or repress messenger RNAs (mRNAs), thereby silencing the expression of specific genes. Several studies have reported using siRNA agents for silencing key HIV-1 genes and inhibiting HIV-1 replication [3,4]. HIV has a high rate of mutation; therefore, effective siRNAs should target conserved sequences [4]. Naito et al. described a highly effective targeting of the trans-activation response (TAR) element and the poly A region in the HIV-1 LTR [5]. Nevertheless, the lack of efficient delivery systems and low biostability challenge the potential of siRNA-based therapies. Several strategies may improve siRNA therapeutic efficacy, including the use of chemically modified siRNAs and novel siRNA delivery vehicles such as polymers, lipids, peptides, antibodies, nanoparticles, dendrimers and aptamers [6]. Aptamers (Ap) are short (15–80 nucleotides, DNA or RNA) single-stranded oligonucleotides capable of binding specific molecules (small molecules, proteins, nucleic acids) with high specificity and affinity [7]. They have been used in various biomedical applications including aptamer–drug conjugates as a simple and effective way to deliver therapeutic agents to specific targets. However, concerns have been raised about the clinical effectiveness of the first generation of aptamer–drug conjugates (ApDC) due to their low circulating half-life and low stability [6]. Circulating half-life can be increased by chemical modifications, such as locked nucleic acids and the conjugation of high-molecular-weight polymers [6]. Reversible or irreversible covalent bonds (e.g., amides, imines or disulfides) have been used for conjugation, increasing the stability of ApDCs and allowing drug release in target cells [8].

The HIV integrase enzyme is highly conserved and is crucial for proviral DNA integration [9]. Several successful aptamers with anti-integrase activity have been identified, including T30695—an aptamer consisting of a repeating motif of a d(GGGT)4 forming a parallel-strand G-quadruplex [10]. In humanized mice, dendrimer-siRNA nanoparticles targeting both cellular and HIV-1 (tat/rev) transcripts led to a significant viral suppression [11].

In this study, we developed and evaluated the anti-HIV-1 activity of a novel T30695 aptamer-siRNA510 conjugate, named Apsi510. siRNA510 targets the TAR/poly (A) regions in the HIV-1 LTR promoter, which is the binding site for the Tat protein and many cellular factors involved in transcription initiation [12]. Disruption of the TAR/poly A regions disturbs the recruitment of these viral and cellular factors, affecting the production of new virus particles.

2. Materials and Methods

2.1. Apsi510 Synthesis

The inner core of Apsi510 is the linker 2-((4-(2,5-dioxo-2,5-dihydro-1H-pyrrol-1-yl)phenyl)amino)acetaldehyde, which is conjugated with an Ap targeting the HIV-1 integrase protein (T30695) [13] and one siRNA sequence. The synthesis of this linker (**1**) was attained following the methodology depicted in Figure 1. From commercially available 1-fluro-4-nitrobene (**2**, mono ApDnC) and aminoacetaldehyde dimethyl acetal, intermediate **3** was obtained in good yields by nucleophilic aromatic substitution. After the introduction of the acetal side chain, the nitro moiety was subsequently reduced to amine and reacted with maleic anhydride (**4**). After ring closure to form the maleimide intermediate (**5**), acid hydrolysis of the acetal gave the final aldehyde side chain in **1**.

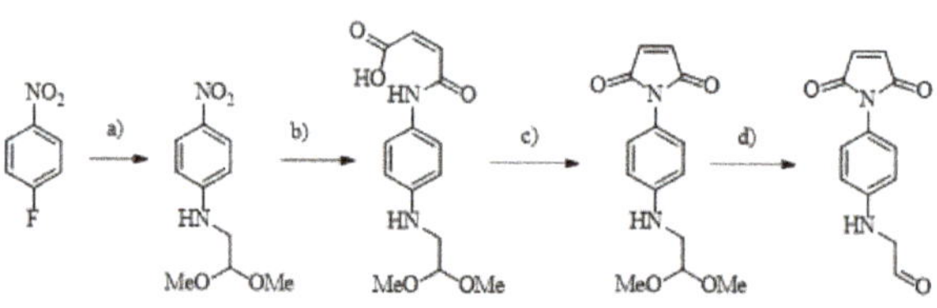

2 3 4 5 1

Figure 1. Synthesis of the linker (**1**) for the aptamer and siRNA510. (**a**) Aminoacetaldehyde dimethyl acetal, DMSO, overnight, r.t; (**b**) (**1**) Pd/C 10%, Et3SiH, MeOH, 1 h, r.t.; (**2**) maleic anhydride, DCM, 3 h, reflux; (**c**) acetic anhydride, sodium acetate, 1 h, r.t.; (**d**) H_2O:TFA (90:10), 10 min, r.t.

The maleimide group in **1** allowed the irreversible covalent conjugation of the thio-modified-HIV-1-integrase-targeting T30695 aptamer. Finally, the acid-labile imine-linker was synthesized by reaction of the commercial amino-modified siRNA510 with the aldehyde side chains to form the acid-labile linkage. The C6-amino-modified siRNA is placed in the 5′-terminus of the sense strand. The imine-acid-labile linkers are expected to release the siRNA content in the acidic endosomal compartments, after cell internalization. Purification of Apsi510 was performed by reversed-phase HPLC.

2.2. Production of Virus and Antiviral Assays

HeLa cells were co-transfected with pNL4.3 or pNL4-3.Luc.R-E- DNA (0.5 μg) and treated with different concentrations of siRNA510 (5–30 nM), Ap(T30695) (25–50 nM) or Apsi510 (25–50 nM). For experiments with pNL4.3, the media were collected after 48 h and virus production in the supernatant was assessed by measuring replication in TZM-bl cells. TZM-bl cells express the luciferase gene under the control of the HIV LTR promoter, which is activated upon HIV infection. For assays with NL4-3.Luc.R-E-virus, firefly luciferase activity was directly measured in HeLa cells 48 h post transfection.

3. Results

3.1. Production of Apsi510

We synthetized the aptamer conjugate (Apsi510) combining a branched aromatic inner core (dendron) and the commercial HIV-1 integrase Ap (T30695) attached by a thioether linkage. This ensemble allowed the attachment of one acid-triggered (acid-labile imine)-releasing linker containing siRNA510.

3.2. Activity of siRNA510, Ap(T30695) and Apsi510 against HIV-1

The activity of different concentrations of siRNA510, Ap (T30695) and Apsi510 was evaluated against the subtype B reference isolate HIV-1NL4.3 in HeLa-CD4+ and TZM-bl cells (Figure 2). siRNA510 and Ap(T30695) (50 nM) alone inhibited >80% of HIV-1 replication in TZM-bl cells (Figure 2A). On the other hand, Apsi510 exhibited a dose-dependent antiviral activity with 50 nM, inhibiting >95% of HIV-1 replication in both experimental systems (Figure 2A,B).

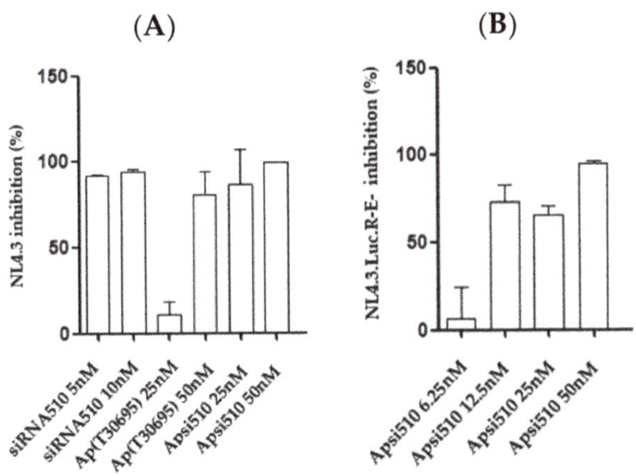

Figure 2. Apsi510 potently suppresses HIV-1 production. (**A**) HeLa cells transfected with pNL4.3 plasmid were treated with different concentrations of siRNA510, Ap(T30695) or Apsi510. Supernatants collected after 48 h were used to infect TZM-bl cells and 48 h after infection, luciferase activity was quantified. (**B**) HeLa cells were transfected with pNL4-3.Luc.R-E-plasmid and treated with Apsi510 (6.25–50 nM). Luciferase activity was quantified 48 h after transfection.

4. Discussion

We successfully synthesized Apsi510—an HIV-1-specific integrase aptamer conjugated to a dendron bound to HIV-1 TAR/poly A-specific siRNA510. Conjugation of siRNA510 with the dendron molecule was expected to protect the siRNA from ribonuclease degradation and to promote cell uptake via endocytosis, thereby increasing the antiviral activity of Apsi510. Apsi510 was a very potent inhibitor of HIV-1 replication in two different experimental systems. The antiviral activity of Apsi510 was higher than those of siRNA510 and Ap(T30695) alone, suggesting an additive effect. A similar inhibitory activity has been obtained with other aptamer-siRNA-based compounds. For example, an anti-gp120 aptamer-anti-tat/rev siRNA chimera had potent anti-HIV activity in vitro and in vivo [14]. Likewise, siRNA aptamers against the HIV protease and the CCR5 co-receptor inhibited HIV-1 replication and cell entry in vitro and in vivo [15]. As siRNA510 has the potential to block RNA expression, Apsi510 may promote HIV latency by the block and lock strategy [1]. On the other hand, as Ap(T30695) inhibits HIV-1 integration, Apsi510 may help to decrease the size of the HIV reservoir and be useful in a shock-and-kill strategy for an HIV cure [1]. These potential activities are being investigated in cell lines that model the HIV reservoir and HIV latency, such as ACH-2 and J-lat E-6. In conclusion, Apsi510 has shown potent activity against HIV-1 and may be useful for the treatment, prevention and cure of HIV infection.

Author Contributions: Conceptualization, J.L., P.B., R.M. and N.T.; investigation, I.M., P.B. and J.L.; formal analysis, I.M., J.L., P.B. and N.T.; methodology, I.M., J.L. and P.B.; writing—original draft, I.M., J.L. and P.B.; writing—review and editing, N.T. All authors have read and agreed to the published version of the manuscript.

Funding: This research was funded by the Foundation for Science and Technology (FCT), I.P., Portugal, through projects UIDB/04138/2020 (iMEd.ULisboa) and UIDB/04585/2020 (CiiEM). Inês Moranguinho was supported by a PhD fellowship from FCT, Portugal (reference SFRH/BD/131062/2017).

Institutional Review Board Statement: Not applicable.

Informed Consent Statement: Not applicable.

Data Availability Statement: Data collected for this study will be shared with publication upon reasonable request, with a signed data access agreement.

Conflicts of Interest: The authors declare no conflict of interest.

References

1. Moranguinho, I.; Valente, S.T. Block-And-Lock: New Horizons for a Cure for HIV-1. *Viruses* **2020**, *12*, 1443. [CrossRef] [PubMed]
2. Schuster, S.; Miesen, P.; van Rij, R.P. Antiviral RNAi in Insects and Mammals: Parallels and Differences. *Viruses* **2019**, *11*, 448. [CrossRef]
3. Adams, D.; Gonzalez-Duarte, A.; O'Riordan, W.D.; Yang, C.C.; Ueda, M.; Kristen, A.V.; Tournev, I.; Schmidt, H.H.; Coelho, T.; Berk, J.L.; et al. Patisiran, an RNAi Therapeutic, for Hereditary Transthyretin Amyloidosis. *N. Engl. J. Med.* **2018**, *379*, 11–21. [CrossRef]
4. Chen, M.J.; Gatignol, A.; Scarborough, R.J. The discovery and development of RNA-based therapies for treatment of HIV-1 infection. *Expert Opin. Drug Discov.* **2023**, *18*, 163–179. [CrossRef]
5. Naito, Y.; Nohtomi, K.; Onogi, T.; Uenishi, R.; Ui-Tei, K.; Saigo, K.; Takebe, Y. Optimal design and validation of antiviral siRNA for targeting HIV-1. *Retrovirology* **2007**, *4*, 80. [CrossRef] [PubMed]
6. Paunovska, K.; Loughrey, D.; Dahlman, J.E. Drug delivery systems for RNA therapeutics. *Nat. Rev. Genet.* **2022**, *23*, 265–280. [CrossRef] [PubMed]
7. Chen, K.; Liu, B.; Yu, B.; Zhong, W.; Lu, Y.; Zhang, J.; Liao, J.; Liu, J.; Pu, Y.; Qiu, L.; et al. Advances in the development of aptamer drug conjugates for targeted drug delivery. *Wiley Interdiscip. Rev. Nanomed. Nanobiotech.* **2017**, *9*, 1438. [CrossRef] [PubMed]
8. Kim, D.H.; Seo, J.M.; Shin, K.J.; Yang, S.G. Design and clinical developments of aptamer-drug conjugates for targeted cancer therapy. *Biomater. Res.* **2021**, *25*, 42. [CrossRef] [PubMed]
9. Rozina, A.; Anisenko, A.; Kikhai, T.; Silkina, M.; Gottikh, M. Complex Relationships between HIV-1 Integrase and Its Cellular Partners. *Int. J. Mol. Sci.* **2022**, *23*, 12341. [CrossRef] [PubMed]
10. Takahashi, M.; Burnett, J.C.; Rossi, J.J. Aptamer-siRNA chimeras for HIV. *Adv. Exp. Med. Biol.* **2015**, *848*, 211–234. [PubMed]
11. Zhou, J.; Neff, C.P.; Liu, X.; Zhang, J.; Li, H.; Smith, D.D.; Swiderski, P.; Aboellail, T.; Huang, Y.; Du, Q.; et al. Systemic Administration of Combinatorial dsiRNAs via Nanoparticles Efficiently Suppresses HIV-1 Infection in Humanized Mice. *Mol. Ther.* **2011**, *19*, 2228–2238. [CrossRef] [PubMed]
12. Das, A.T.; Harwig, A.; Berkhout, B. The HIV-1 Tat protein has a versatile role in activating viral transcription. *J. Virol.* **2011**, *85*, 9506–9516. [CrossRef] [PubMed]
13. Jing, N.; Hogan, M.E. Structure-Activity of Tetrad-forming Oligonucleotides as a Potent Anti-HIV Therapeutic Drug. *J. Biol. Chem.* **1998**, *273*, 34992–34999. [CrossRef]

14. Zhou, J.; Li, H.; Li, S.; Zaia, J.; Rossi, J.J. Novel Dual Inhibitory Function Aptamer–siRNA Delivery System for HIV-1 Therapy. *Mol. Ther.* **2008**, *16*, 1481–1489. [CrossRef]
15. Duclair, S.; Gautam, A.; Ellington, A.; Prasad, V.R. High-affinity RNA Aptamers Against the HIV-1 Protease Inhibit Both In Vitro Protease Activity and Late Events of Viral Replication. *Mol. Ther.-Nucleic Acids* **2015**, *4*, e228. [CrossRef]

medical sciences forum

MDPI

Proceeding Paper

Intra-Oral Halitosis in Periodontitis: The Role of Tongue Coating—A Cross-Sectional Study †

Catarina Izidoro [1,*], João Botelho [1], Vanessa Machado [1], Ana Mafalda Reis [2,3], Luís Proença [1], Ricardo Alves [1] and José João Mendes [1]

[1] Egas Moniz Center for Interdisciplinary Research (CiiEM), Egas Moniz School of Health & Science, Caparica, 2829-511 Almada, Portugal; jbotelho@egasmoniz.edu.pt (J.B.); vmachado@egasmoniz.edu.pt (V.M.); lproenca@egasmoniz.edu.pt (L.P.); ralves@egasmoniz.edu.pt (R.A.); jmendes@egasmoniz.edu.pt (J.J.M.)
[2] Instituto de Ciências Biomédicas Abel Salazar, School of Health and Life Sciences, University of Porto, 4050-313 Porto, Portugal; docmaf@sapo.pt
[3] Neuroradiology Department, Hospital Pedro Hispano, 4464-513 Matosinhos, Portugal
* Correspondence: cizidoro@egasmoniz.edu.pt; Tel.: +351-212-946-737
† Presented at the 6th International Congress of CiiEM—Immediate and Future Challenges to Foster One Health, Almada, Portugal, 5–7 July 2023.

Abstract: Halitosis is an unpleasant breath odor that interferes with self-confidence and with people's professional and social lives. The aim of this cross-sectional study was to evaluate the impact of tongue coating on intra-oral halitosis in patients with periodontitis. The Winkel Tongue Coating Index (WTCI) score was found to be positively and significantly correlated with VSC values (rho = 0.473, $p < 0.001$). WTCI may be associated with levels of volatile sulfur compounds (VSCs) when other causes of extra-oral halitosis are excluded.

Keywords: halitosis; tongue coating; periodontitis

Citation: Izidoro, C.; Botelho, J.; Machado, V.; Reis, A.M.; Proença, L.; Alves, R.; Mendes, J.J. Intra-Oral Halitosis in Periodontitis: The Role of Tongue Coating—A Cross-Sectional Study. *Med. Sci. Forum* **2023**, *22*, 1. https://doi.org/10.3390/msf2023022001

Academic Editors: José Brito, Nuno Taveira and Ana I. Fernandes

Published: 8 August 2023

1. Introduction

Halitosis or oral malodor is a unpleasant breath odor that impacts the quality of life [1,2]. As one of the primary symptoms of periodontitis, halitosis frequently prompts patients to seek treatment [3]. Intra-oral halitosis is caused by pathological conditions (periodontitis and gingivitis) and physiological traits, particularly tongue coating (TC), a grayish-white deposit on the tongue. Several studies have addressed the tongue coating with halitosis, but the number of studies on the tongue coating impact on intra-oral halitosis in periodontitis patients is still scarce. The aim of this cross-sectional study was to evaluate the impact of tongue coating on intra-oral halitosis in patients with periodontitis.

2. Materials and Methods

This research received approval from the Egas Moniz Ethics Committee, following the guidelines of the Helsinki Declaration from 1975, with the 2013 revisions. The study participants were consecutively selected from the Periodontology Department at Egas Moniz Dental Clinic for the purpose of periodontal assessment, spanning from October 2019 to March 2021. To be included in the study, participants needed to have periodontitis and be aged between 18 and 65. They were required to adhere to the recommended methods for halitosis evaluation and provide informed consent. Exclusion criteria encompassed a history of previous periodontal treatment, recent antibiotic usage within the last 4 weeks, a medical history of radiotherapy or chemotherapy, external sources for halitosis, and pregnancy.

A single trained examiner conducted a comprehensive examination of the entire mouth using a manual periodontal CP-12 probe (Hu-Friedy®, Chicago, IL, USA). The definition of periodontitis followed the AAP/EFP 2018 consensus [4]. To diagnose halitosis, a two-step approach was employed: Step (1) A self-reported questionnaire was used to rule out potential causes of extra-oral halitosis; Step (2) Volatile sulfur compounds (VSC) were measured using a device (Halimeter®, Interscan Corp, Chatsworth, CA, USA). A VSC concentration below 80 ppb indicated no detectable odor, while a concentration above 80 ppb indicated the presence of halitosis [5].

The evaluation of tongue coating was performed using the Winkel Tongue Coating Index (WTCI) [6]. In summary, the dorsum (upper surface) of the tongue was divided into six areas: three posterior and three anterior. Tongue coating in each of these sextants was assessed and categorized as follows: 0 = no coating, 1 = light coating, 2 = severe coating. The overall WTCI score was obtained by summing up the scores from all six areas, resulting in a possible range of 0 to 12.

Data were analyzed by descriptive and inferential methodologies. Bivariate correlation was assessed by the Spearman correlation coefficient (rho). A significance level of 5% was established for the inferential analysis.

3. Results

The present study explored the correlation between the intra-oral halitosis (VCS counting) and clinical parameters of periodontitis and tongue coating.

From a total of 71 participants, 51 were evaluated, regarding halitosis status, by VSC counting. From those, 37.3% were diagnosed as exhibiting halitosis (VSC > 80 ppb). The WTCI score ranged from 0 to 9, with a median value of 3 ($\pm$2). WTCI score was found to be positively and significantly correlated with VSC values (rho = 0.473, $p < 0.001$). No significant correlation was identified between WTCI and periodontal clinical parameters Plaque Index (rho = 0.112, $p = 0.437$) and Gingival Index (rho = 0.083, $p = 0.568$).

4. Discussion

Although this study has its limitations, our current discoveries offer supplementary evidence to clarify the impact of tongue coating on intra-oral halitosis in patients with periodontitis. However, it's important to note that these results are derived from a cross-sectional study, and as such, prospective longitudinal studies are required to assess the temporal relationship of these events.

5. Conclusions

Considering the constraints of this observational study, there appears to be a correlation between the Winkel Tongue Coating Index and volatile sulfur compound (VSC) levels in patients with periodontitis, after excluding other causes of extra-oral halitosis. To establish a potential causal relationship, additional intervention studies are necessary. Therefore, it is crucial to incorporate tongue scraping into oral hygiene instructions.

Author Contributions: Conceptualization, C.I., J.B., V.M., A.M.R., L.P., R.A. and J.J.M.; methodology, C.I., J.B., V.M., A.M.R., L.P., R.A. and J.J.M.; validation, C.I., J.B., V.M., A.M.R., L.P., R.A. and J.J.M.; formal analysis C.I., J.B., V.M., A.M.R., L.P., R.A. and J.J.M.; investigation, C.I.; resources, C.I., J.B., V.M., A.M.R., L.P., R.A. and J.J.M.; data curation, C.I., V.M. and J.B.; writing—original draft preparation, C.I., J.B., V.M., A.M.R., L.P., R.A. and J.J.M.; writing—review and editing, C.I., J.B., V.M., A.M.R., L.P., R.A. and J.J.M.; visualization, C.I.; project administration, C.I., J.B., V.M., A.M.R., L.P., R.A. and J.J.M. All authors have read and agreed to the published version of the manuscript.

Funding: This work was financed by national funds through the FCT-Foundation for Science and Technology, I.P., under the project UIDB/04585/2020.

Institutional Review Board Statement: The study was conducted according to the guidelines of the Declaration of Helsinki and approved by Ethics Committee of EGAS MONIZ (No. 781, 26 June 2019).

Informed Consent Statement: Informed consent was obtained from all subjects involved in the study.

Data Availability Statement: Data may be available upon reasonable request.

Acknowledgments: Acknowledgments addressed to the EMDC Periodontology Department.

Conflicts of Interest: The authors declare no conflict of interest.

References

1. Kukkamalla, D.; Cornelio, D.; Mahalinga Bhat, D.; Avadhani, D.; Goyal, D. HALITOSIS—A Social Malady. *IOSR-JDMS* **2014**, *13*, 55–61. [CrossRef]
2. Azodo, C.; Umoh, A. Self-perceived oral malodour among periodontal patients: Prevalence and associated factors. *Int. J. Med. Biomed. Res.* **2013**, *2*, 125–132. [CrossRef]
3. Silva, M.F.; Cademartori, M.G.; Leite, F.R.M.; López, R.; Demarco, F.F.; Nascimento, G.G. Is periodontitis associated with halitosis? A systematic review and meta-regression analysis. *J. Clin. Periodontol.* **2017**, *44*, 1003–1009. [CrossRef] [PubMed]
4. Tonetti, M.S.; Greenwell, H.; Kornman, K.S. Staging and Grading of Periodontitis: Framework and Proposal of a New Classification and Case Definition. *J. Periodontol.* **2018**, *45* (Suppl. S20), S149–S161. [CrossRef] [PubMed]
5. Donaldson, A.C.; Riggio, M.P.; Rolph, H.J.; Bagg, J.; Hodge, P.J. Clinical Examination of Subjects with Halitosis. *Oral Dis.* **2007**, *13*, 63–70. [CrossRef] [PubMed]
6. Winkel, E.G.; Roldan, S.; Van Winkelhoff, A.J.; Herrera, D.; Sanz, M. Clinical effects of a new mouthrinse containing chlorhexidine, cetylpyridinium chloride and zinc-lactate on oral halitosis. *J. Clin. Periodontol.* **2003**, *30*, 300–306. [CrossRef] [PubMed]

medical sciences forum

MDPI

Proceeding Paper

Identification Procedures of Yeast Species Recovered from Portuguese Intensive Care Units †

Teresa Nascimento [1,2,*], João Inácio [3], Daniela Guerreiro [1], Patrícia Patrício [4], Priscila Diaz [5], Cristina Toscano [6], Isabel Faria [6] and Helena Barroso [1]

1 Egas Moniz Center for Interdisciplinary Research (CiiEM), Egas Moniz School of Health & Science, 2829-511 Caparica, Portugal; dguerreiro@egasmoniz.edu.pt (D.G.); mhbarroso@egasmoniz.edu.pt (H.B.)
2 Instituto de Higiene e Medicina Tropical, Universidade Nova de Lisboa, 1099-085 Lisbon, Portugal
3 School of Pharmacy and Biomolecular Sciences, University of Brighton, Brighton BN1 9PH, UK; j.inacio@brighton.ac.uk
4 Hospital Beatriz Ângelo, 2674-514 Loures, Portugal; patriciopatricia@gmail.com
5 Hospital Prof. Doutor Fernando Fonseca, 2720-276 Amadora, Portugal; priscila.diaz@hff.min-saude.pt
6 Centro Hospitalar Lisboa Ocidental Hospital Egas Moniz, 1349-019 Lisboa, Portugal; ctoscano@chlo.min-saude.pt (C.T.); ifaria@chlo.min-saude.pt (I.F.)
* Correspondence: tnascimento@egasmoniz.edu.pt or a2110112@ihmt.unl.pt
† Presented at the 6th International Congress of CiiEM—Immediate and Future Challenges to Foster One Health, Almada, Portugal, 5–7 July 2023.

Abstract: Yeast species other than *Candida albicans* are becoming more clinically relevant in the Intensive Care Unit (ICU) landscape, making it important to identify them correctly. The aim of this study was to evaluate the concordance of the identification of clinical isolates from ICUs, at the species level, by conventional and molecular methods. All isolates (n = 371) underwent identification by cultural, MALDI-TOF MS and PCR. The direct concordance between conventional and molecular identification was 92% (341/371). These results allow us to conclude that culture-based methodologies are still useful to reliably identify the most frequent yeasts, but for rare, uncommon or cryptic species, technologies such as MALDI-TOF MS or PCR are needed.

Keywords: *Candida* spp.; MALDI-TOF MS; PCR; yeasts; identification

Citation: Nascimento, T.; Inácio, J.; Guerreiro, D.; Patrício, P.; Diaz, P.; Toscano, C.; Faria, I.; Barroso, H. Identification Procedures of Yeast Species Recovered from Portuguese Intensive Care Units. *Med. Sci. Forum* **2023**, *22*, 35. https://doi.org/10.3390/msf2023022035

Academic Editors: José Brito, Nuno Taveira and Ana I. Fernandes

Published: 16 August 2023

1. Introduction

Invasive fungal infections are increasing, and *Candida* spp. is the main cause [1]. Yeast species other than *Candida albicans* are becoming more frequent, and some of them may have variable patterns of susceptibility to antifungal agents, making it important to identify them correctly [2]. Molecular amplification methods (polymerase chain reaction (PCR)) and proteomic analysis by Matrix-assisted Laser Desorption Ionization–Time-of-flight Mass Spectrometry (MALDI-TOF MS) techniques have emerged as alternative methods [3,4]. The aim of this study was to evaluate the concordance of the identification of ICUs isolates, at the species level, by culture methods based on standard morphological and biochemical criteria, MALDI-TOF MS and PCR.

2. Materials and Methods

During a two-year multicenter prospective observational study in ICU patients from two tertiary hospitals located in the Lisbon metropolitan area, 988 axillar/inguinal swabs were taken and identified at the species level to unveil the prevalence of *C. auris* at Portuguese ICUs. This investigation has been approved by the Institutional Ethical Board of all institutions enrolled.

All swabs' isolates (n = 371) were plated on Sabouraud Dextrose Agar and chromogenic agar (CHROMagar®). For *C. albicans* suspected colonies, a filamentation test was made.

API® Candida or API® 20 C AUX galleries (bioMérieux, Marcy l'Etoile, France) were used for identification according to the enzyme profile and sugar assimilation pattern. PCR assays were optimized for *C. auris* and *Candida* cryptic species identification [5,6]. MALDI-TOF MS methods analysis with Vitek–MS®(bioMérieux, Marcy l'Etoile, France) system was used for definitive identification.

The following species presumptively identified by phenotypic methods from ICU patient samples were studied: *C. albicans* complex (n = 183), *C. parapsilosis* complex (n = 115), *C. glabrata* complex (n = 39), *C. tropicalis* (n = 15), *C. guilliermondii* (n = 3), *C. famata* (n = 2), *C. kefyr* (n = 1), *Saccharomyces cerevisae* (n = 1), *Rhodotorula* sp. (n = 9) and *Trichosporon* sp. (n = 3). ATCC collection strains *C. parapsilosis* 22019, *C. glabrata* 15126, *C. albicans* 90028 and *C. auris* DSMZ 21087 were included in the study.

3. Results

Identification of the 371 isolates by MALDI-TOF MS included 355 *Candida* species isolates. Namely, *C. albicans* (n = 185), *C. parapsilosis* complex (n = 112) [*C. parapsilosis sensu* stricto (n = 109), *C. orthopsilosis* (n = 2), *C. metapsilosis* (n = 1)], *C. glabrata* (n = 36), *C. tropicalis* (n = 15), *C. lusitaniae* (n = 4) and *C. guilliermondii* (n = 3). Other yeast species included: *Rhodotorula rubra* (n = 9); *Trichosporon inkin* (n = 5); *Trichosporon asahii* (n = 1); and *S. cerevisae* (n = 1). No isolate of *C. auris* was retrieved within this cohort. The direct concordance between the conventional identification method and MALDI-TOF MS was 92% (341/371). Discrepancies were observed with the following species: *C. parapsilosis*; *C. glabrata*; *C. tropicalis*; *C. guilliermondii*; *C. famata*; and *C. kefyr*. In this work, MALDI-TOF MS allowed the correct identification of the yeasts *C. lusitaniae*, *C. guilliermondii*, *S. cerevisae*, *T. inkin* and *T. asahii* erroneously identified by conventional methods such as *C. parapsilosis*. There was a 100% correlation between MALDI-TOF MS and PCR assays for the identification of cryptic species of *C. albicans*, *C. parapsilosis*, *C. glabrata* complexes and *C. auris*.

4. Discussion

The accurate identification of *Candida* and other yeast species is extremely important, contributing to the increase of knowledge about the epidemiology of these microorganisms in the ICU setting. As already published by other authors, a limitation of conventional methods is the inability to identify cryptic species of *C. albicans*, *C. parapsilosis* and *C. glabrata* or new emerging species like *C. auris* [7–9].

These results allow us to conclude that conventional methodologies are still useful to reliably identify the most frequently isolated yeast species from clinical samples, but when dealing with rare, uncommon, or cryptic *Candida* species, it is important to confirm them using technologies such as MALDI-TOF MS. The PCR assays used allowed a reliable species identification and has the potential to be implemented in a cost-effective manner into epidemiological studies to broaden the limited knowledge of cryptic and rare species.

Author Contributions: Conceptualization, T.N. and H.B.; methodology, T.N., H.B., D.G., I.F., P.D., C.T. and P.P.; investigation, T.N., H.B., D.G., I.F. and C.T.; writing—original draft preparation, T.N.; writing—review and editing, T.N., H.B. and J.I.; supervision, H.B. and J.I.; project administration, H.B.; funding acquisition, H.B. All authors have read and agreed to the published version of the manuscript.

Funding: This research was funded by Egas Moniz Cooperativa de Ensino Superior, CRL, grant number EI 19/01.

Institutional Review Board Statement: All procedures performed in studies involving human participants were in accordance with the ethical standards of the institutional and/or national research committee and with the 1964 Helsinki declaration and its later amendments or comparable ethical standards. The study was approved by the Ethical Committee of the Prof. Doutor Fernando Fonseca Hospital on 13/11/2019 (54/2019) and by the Ethical Committee of the Beatriz Ângelo Hospital on 21/07/2021 (3655/2021).

Informed Consent Statement: Informed consent was obtained from all subjects involved in the study.

Data Availability Statement: MDPI Research Data Policies.

Conflicts of Interest: The authors declare no conflict of interest.

References

1. Sharma, M.; Chakrabarti, A. Candidiasis and Other Emerging Yeasts. *Curr. Fungal Infect. Rep.* **2023**, *17*, 15–24. [CrossRef] [PubMed]
2. Koehler, P.; Stecher, M.; Cornely, O.A.; Koehler, D.; Vehreschild, M.J.G.T.; Bohlius, J.; Wisplinghoff, H.; Vehreschild, J.J. Morbidity and mortality of candidaemia in Europe: An epidemiologic meta-analysis. *Clin. Microbiol. Infect.* **2019**, *25*, 1200–1212. [CrossRef] [PubMed]
3. Keighley, C.; Garnham, K.; Harch, S.A.J.; Robertson, M.; Chaw, K.; Teng, J.C.; Chen, S.C. Diagnostic Challenges and Emerging Opportunities for the Clinical Microbiology Laboratory. *Curr. Fungal Infect. Rep.* **2021**, *15*, 116–126. [CrossRef] [PubMed]
4. Maldonado, I.; Cataldi, S.; Garbasz, C.; Relloso, S.; Striebeck, P.; Guelfand, L.; Moral, L.L.; Romeo, A.M.; Schijman, M.; Ponce, G.; et al. Identification of Candida yeasts: Conventional methods and MALDI-TOF MS. *Rev. Iberoam. Micol.* **2018**, *35*, 151–154. [CrossRef] [PubMed]
5. Arastehfar, A.; Fang, W.; Pan, W.; Liao, W.; Yan, L.; Boekhout, T. Identification of nine cryptic species of Candida albicans, *C. glabrata*, and *C. parapsilosis* complexes using one-step multiplex PCR. *BMC Infect. Dis.* **2018**, *18*, 480. [CrossRef] [PubMed]
6. Kordalewska, M.; Zhao, Y.; Lockhart, S.R.; Chowdhary, A.; Berrio, I.; Perlin, D.S. Rapid and Accurate Molecular Identification of the Emerging Multidrug-Resistant Pathogen Candida auris. *J. Clin. Microbiol.* **2017**, *55*, 2445–2452. [CrossRef] [PubMed]
7. Zhao, Y.; Tsang, C.C.; Xiao, M.; Chan, J.F.W.; Lau, S.K.P.; Kong, F.; Xu, Y.; Woo, P.C.Y. Yeast identification by sequencing, biochemical kits, MALDI-TOF MS and rep-PCR DNA fingerprinting. *Med. Mycol.* **2018**, *56*, 816–827. [CrossRef] [PubMed]
8. Lockhart, S.R.; Lyman, M.M.; Sexton, D.J. Tools for Detecting a "Superbug": Updates on Candida auris Testing. *J. Clin. Microbiol.* **2022**, *60*, e0080821. [CrossRef]
9. Rosenvinge, F.S.; Dzajic, E.; Knudsen, E.; Malig, S.; Andersen, L.B.; Løvig, A.; Arendrup, M.C.; Jensen, T.G.; Gahrn-Hansen, B.; Kemp, M. Performance of matrix-assisted laser desorption-time of flight mass spectrometry for identification of clinical yeast isolates. *Mycoses* **2013**, *56*, 229–235. [CrossRef]

medical sciences forum

MDPI

Proceeding Paper

Susceptibility Patterns of *Candida* spp. Collected from Intensive Care Units: A Prospective Study in 2020–2022 †

Teresa Nascimento [1,2,*], João Inácio [3], Daniela Guerreiro [1], Patrícia Patrício [4], Priscila Diaz [5], Cristina Toscano [6], Isabel Faria [6] and Helena Barroso [1]

1 Egas Moniz Center for Interdisciplinary Research (CiiEM), Egas Moniz School of Health & Science, 2829-511 Almada, Portugal; dguerreiro@egasmoniz.edu.pt (D.G.); mhbarroso@egasmoniz.edu.pt (H.B.)
2 Instituto de Higiene e Medicina Tropical, Universidade Nova de Lisboa, 1099-085 Lisbon, Portugal
3 School of Pharmacy and Biomolecular Sciences, University of Brighton, Brighton BN1 9PH, UK; j.inacio@brighton.ac.uk
4 Hospital Beatriz Ângelo, 2674-514 Loures, Portugal; patriciopatricia@gmail.com
5 Hospital Prof. Doutor Fernando Fonseca, 2720-276 Amadora, Portugal; priscila.diaz@hff.min-saude.pt
6 Centro Hospitalar Lisboa Ocidental, Hospital Egas Moniz, 1349-019 Lisboa, Portugal; ctoscano@chlo.min-saude.pt (C.T.); ifaria@chlo.min-saude.pt (I.F.)
* Correspondence: tnascimento@egasmoniz.edu.pt or a2110112@ihmt.unl.pt
† Presented at the 6th International Congress of CiiEM—Immediate and Future Challenges to Foster One Health, Almada, Portugal, 5–7 July 2023.

Abstract: The frequency of *Candida* spp. isolates as a cause of hospital infections has risen, and in vitro antifungal susceptibility testing (AST) plays an increasingly important role in guiding therapeutic decisions. This multicenter study aimed to determine AST pattern of *Candida* spp. isolates from Intensive Care Unit (ICU) patients. In total, 674 patients were enrolled, and axillar/inguinal swabs were collected at admission and during the ICU stay (5th and 8th day). In vitro AST was performed on 355 *Candida* spp. isolates, according to the concentration gradient Etest® strip technique. The overall susceptibility rates were 100%, 99.7%, 98.3%, and 97.7% for amphotericin B, voriconazole, anidulafungin, and fluconazole, respectively. The current study demonstrates that antifungal resistance remains infrequent among *Candida* spp. isolates in Portugal's ICUs.

Keywords: *Candida* spp.; ICUs; Fluconazole; Voriconazole; Amphotericin B; Anidulafungin; resistance

Citation: Nascimento, T.; Inácio, J.; Guerreiro, D.; Patrício, P.; Diaz, P.; Toscano, C.; Faria, I.; Barroso, H. Susceptibility Patterns of *Candida* spp. Collected from Intensive Care Units: A Prospective Study in 2020–2022. *Med. Sci. Forum* **2023**, *22*, 45. https://doi.org/10.3390/msf2023022045

Academic Editors: José Brito, Nuno Taveira and Ana I. Fernandes

Published: 25 August 2023

1. Introduction

The frequency of *Candida* spp. isolates as a cause of hospital infections has risen in recent years, leading to high morbidity and mortality rates [1]. Studies have shown that *Candida* species are the second or the third most common cause of septicemia in Intensive Care Unit (ICU) patients [2], and *Candida* spp. colonization can be a predictor of candidemia among this cohort of patients [3]. Most often empirical antifungals are prescribed for candidemia, mainly based on locally and country-wide antifungal surveillance data, which differ for every geographic region [4]. Therefore, this study aimed to evaluate antifungal susceptibility pattern of *Candida* spp. colonization isolates from ICU patients.

2. Materials and Methods

In this multicenter prospective observational study, 675 patients in ICU were recruited from January 2020 through December 2022. Collection of axillar/inguinal swabs was made at admission and during the ICU stay (5th and 8th day). Patient data were obtained through a form containing epidemiological and clinical information. This investigation has been approved by the Institutional Ethical Board of all institutions enrolled. Isolates were identified by cultural, Matrix Assisted Laser Desorption Ionization-Time of Flight Mass Spectrometry (MALDI-TOF MS), and molecular methods. In vitro antifungals susceptibility tests (AST) were performed for fluconazole, voriconazole, amphotericin B, and anidulafungin, according to the concentration gradient Etest® strip technique, following the manufacturer's instructions. *C. parapsilosis* ATCC 22019 and *C. krusei* ATCC 6258 standard strains were used as quality controls. Results were interpreted based on the clinical breakpoints recommended by the European Committee on Antimicrobial Susceptibility Testing (EUCAST).

3. Results

In total, 355 *Candida* species isolates were recovered from 988 samples. E-test determination was performed for the isolates, namely, *C. albicans* (n = 185), the *C. parapsilosis* complex (n = 112) (*C. parapsilosis sensu stricto* (n = 109), *C. orthopsilosis* (n = 2), and *C. metapsilosis* (n = 1)), *C. glabrata* (n = 36), *C. tropicalis* (n = 15), *C. lusitaniae* (n = 4), and *C. guilliermondii* (n = 3).

Most *Candida* species exhibited susceptibility to antifungals. Amphotericin B, voriconazole, and anidulafungin were the drugs for which all *Candida* species showed more susceptibility, which were, respectively, 100%, 99.7%, and 97.5%. The overall rate of resistance to fluconazole was 2.3%. For fluconazole, NAC isolates were more resistant (1.4%) than *C. albicans* (0.8%). The rates of susceptibility to fluconazole were 96.8, 95.5, and 100% in *C. albicans, C. parapsilosis,* and *C. tropicalis*, respectively. For the other triazole tested, voriconazole, resistance was only observed for *C. albicans* (1.6%) isolates, but 3.2% (6/185) and 2.7% (3/112) of *C. albicans* and *C. parapsilosis*, respectively, presented MICs within the intermediate category. Resistance to drugs within the same class was detected, with two strains of *C. albicans* and one strain of *C. parapsilosis* showing resistance to the two azoles tested. Resistance to anidulafungin was observed for three species: *C. tropicalis* (6.7%), *C. albicans* (3.8%), and *C. parapsilosis* (2.7%).

4. Discussion

The current study demonstrates the prevalence and antifungal susceptibility pattern of *Candida* species in Portugal's ICUs. Within this cohort, antifungal resistance is infrequent among *Candida* isolates. Similarly to published data, in our isolates, azole resistance was noted in *C. albicans, C. parapsilosis,* and *C. glabrata*, whereas echinocandin resistance was noted in *C. albicans, C. parapsilosis,* and *C. tropicalis* [5–8].

The overall rates of resistance were 2.3 and 0.3% for fluconazole and voriconazole, respectively, which were in line with previous reports, particularly from other European countries [2]. Studies on antifungal susceptibility profiles in Portugal are not recent and used small samples, making it impossible to draw conclusions about antifungal resistance [9,10]. Still, despite the differences in practical details and origin of the isolates, Faria-Ramos et al. [10] reported higher resistance rates: 5 and 12% for fluconazole and voriconazole, respectively.

In this study, three *Candida* spp. exhibited resistance to anidulafungin, namely, *C. tropicalis* (6.7%, 1/15), *C. albicans* (1.6%, 3/185), and *C. parapsilosis* (0.9%, 1/112). *C. glabrata* and other *Candida* spp. did not show any resistance to anidulafungin. These results were in line with the resistance values reported in the literature [6,11,12].

In conclusion, these observations emphasized the importance of knowing the local epidemiology and resistance patterns for *Candida* spp. within institutions to guide therapeutic decisions.

Author Contributions: Conceptualization, T.N. and H.B.; methodology, T.N., H.B., D.G., I.F., P.D., C.T. and P.P.; investigation, T.N., H.B., D.G., I.F. and C.T.; writing—original draft preparation, T.N.; writing—review and editing, T.N., H.B. and J.I.; supervision, H.B. and J.I.; project administration, H.B.; funding acquisition, H.B. All authors have read and agreed to the published version of the manuscript.

Funding: This research was funded by Egas Moniz Cooperativa de Ensino Superior, CRL, grant number EI 19/01.

Institutional Review Board Statement: All procedures performed in studies involving human participants were in accordance with the ethical standards of the institutional and/or national research committee and with the 1964 Helsinki declaration and its later amendments or comparable ethical standards. The study was approved by the Ethical Committee of the Prof. Doutor Fernando Fonseca Hospital on 13/11/2019 (54/2019) and by the Ethical Committee of the Beatriz Ângelo Hospital on 21/07/2021 (3655/2021).

Informed Consent Statement: Written informed consent was obtained from all subjects involved in the study.

Data Availability Statement: Not applicable.

Conflicts of Interest: The authors declare no conflict of interest.

References

1. Poissy, J.; Rouzé, A.; Cornu, M.; Nseir, S.; Sendid, B. The Changing Landscape of Invasive Fungal Infections in ICUs: A Need for Risk Stratification to Better Target Antifungal Drugs and the Threat of Resistance. *J. Fungi* **2022**, *8*, 946. [CrossRef] [PubMed]
2. Bassetti, M.; Giacobbe, D.R.; Vena, A.; Trucchi, C.; Ansaldi, F.; Antonelli, M.; Adamkova, V.; Alicino, C.; Almyroudi, M.P.; Atchade, E.; et al. Incidence and outcome of invasive candidiasis in intensive care units (ICUs) in Europe: Results of the EUCANDICU project. *Crit. Care* **2019**, *23*, 219. [CrossRef] [PubMed]
3. Alenazy, H.; Alghamdi, A.; Pinto, R.; Daneman, N. *Candida* colonization as a predictor of invasive candidiasis in non-neutropenic ICU patients with sepsis: A systematic review and meta-analysis. *Int. J. Infect. Dis.* **2021**, *102*, 357–362. [CrossRef] [PubMed]
4. Galia, L.; Pezzani, M.D.; Compri, M.; Callegari, A.; Rajendran, N.B.; Carrara, E.; Tacconelli, E. The Combacte Magnet Epi-Net Network. Surveillance of Antifungal Resistance in Candidemia Fails to Inform Antifungal Stewardship in European Countries. *J. Fungi* **2022**, *8*, 249. [CrossRef] [PubMed]
5. Castanheira, M.; Deshpande, L.M.; Messer, S.A.; Rhomberg, P.R.; Pfaller, M.A. Analysis of global antifungal surveillance results reveals predominance of Erg11 Y132F alteration among azole-resistant *Candida parapsilosis* and *Candida tropicalis* and country-specific isolate dissemination. *Int. J. Antimicrob. Agents* **2020**, *55*, 105799. [CrossRef] [PubMed]
6. Vannini, M.; Emery, S.; Lieutier-Colas, F.; Legueult, K.; Mondain, V.; Retur, N.; Gastaud, L.; Pomares, C.; Hasseine, L. Epidemiology of candidemia in NICE area, France: A five-year study of antifungal susceptibility and mortality. *J. Med. Mycol.* **2022**, *32*, 101210. [CrossRef] [PubMed]
7. Rogers, T.R.; Verweij, P.E.; Castanheira, M.; Dannaoui, E.; White, P.L.; Arendrup, M.C.; Subcommittee on Antifungal Susceptibility Testing (AFST) of the ESCMID European Committee for Antimicrobial Susceptibility Testing (EUCAST). Molecular mechanisms of acquired antifungal drug resistance in principal fungal pathogens and EUCAST guidance for their laboratory detection and clinical implications. *J. Antimicrob. Chemother.* **2022**, *77*, 2053–2073. [CrossRef] [PubMed]
8. Dabas, Y.; Xess, I.; Pandey, M.; Ahmed, J.; Sachdev, J.; Iram, A.; Singh, G.; Mahapatra, M.; Seth, R.; Bakhshi, S.; et al. Epidemiology and Antifungal Susceptibility Patterns of Invasive Fungal Infections (IFIs) in India: A Prospective Observational Study. *J. Fungi* **2021**, *8*, 33. [CrossRef] [PubMed]
9. Costa-de-Oliveira, S.; Pina-Vaz, C.; Mendonça, D.; Gonçalves Rodrigues, A. A first Portuguese epidemiological survey of fungaemia in a university hospital. *Eur. J. Clin. Microbiol. Infect. Dis.* **2008**, *27*, 365–374. [CrossRef] [PubMed]
10. Faria-Ramos, I.; Neves-Maia, J.; Ricardo, E.; Santos-Antunes, J.; Silva, A.T.; Costa-de-Oliveira, S.; Cantón, E.; Rodrigues, A.G.; Pina-Vaz, C. Species distribution and in vitro antifungal susceptibility profiles of yeast isolates from invasive infections during a Portuguese multicenter survey. *Eur. J. Clin. Microbiol. Infect. Dis.* **2014**, *33*, 2241–2247. [CrossRef] [PubMed]
11. Lindberg, E.; Hammarström, H.; Ataollahy, N.; Kondori, N. Species distribution and antifungal drug susceptibilities of yeasts isolated from the blood samples of patients with candidemia. *Sci. Rep.* **2019**, *9*, 3838. [CrossRef] [PubMed]
12. Díaz-García, J.; Gómez, A.; Machado, M.; Alcalá, L.; Reigadas, E.; Sánchez-Carrillo, C.; Pérez-Ayala, A.; Gómez-García De La Pedrosa, E.; González-Romo, F.; Cuétara, M.S.; et al. Blood and intra-abdominal *Candida* spp. from a multicentre study conducted in Madrid using EUCAST: Emergence of fluconazole resistance in *Candida parapsilosis*, low echinocandin resistance and absence of *Candida auris*. *J. Antimicrob. Chemother.* **2022**, *77*, 3102–3109. [CrossRef] [PubMed]

medical sciences forum

MDPI

Proceeding Paper

Sexual Dimorphism in the Prevalence of Musculoskeletal Disorders among Dental Students †

Manuel Barbosa de Almeida * and Marion Moreira

Egas Moniz Center for Interdisciplinary Research (CiiEM); Egas Moniz School of Health & Science, Campus Universitário, Quinta da Granja, Caparica, 2829-511 Almada, Portugal; marionmoreira614@gmail.com
* Correspondence: mbalmeida@egasmoniz.edu.pt
† Presented at the 6th International Congress of CiiEM—Immediate and Future Challenges to Foster One Health, Almada, Portugal, 5–7 July 2023.

Abstract: Work-related musculoskeletal disorders (MSDs) are concerning among dental students. The study aims to compare the prevalence of MSDs between male and female dental students. A total of 63 dental students completed the Nordic Musculoskeletal Questionnaire. The results showed that female students had a higher prevalence of symptoms in the neck and lower back compared to those of the male students. In the weekly assessment, male students had a higher occurrence of symptoms in the hands and wrists. These findings emphasize the need to address musculoskeletal issues among dental students, targeting their different etiologies to prevent disabilities and early retirements associated with MSDs.

Keywords: musculoskeletal disorders; dental students; sexual dimorphism; ergonomy

Citation: Almeida, M.B.d.; Moreira, M. Sexual Dimorphism in the Prevalence of Musculoskeletal Disorders among Dental Students. *Med. Sci. Forum* **2023**, *22*, 36. https://doi.org/10.3390/msf2023022036

Academic Editors: José Brito, Nuno Taveira and Ana I. Fernandes

Published: 16 August 2023

1. Introduction

In recent years, work-related musculoskeletal disorders (MSD) have become one of the most important health issues among health care workers [1]. Up to 95.8% of dentists could develop MSDs in their lifetime and 92% reported musculoskeletal pain in the last year [1–3]. This issue began early in dentist's careers, with a prevalence among students between 44% and 93% [2,4,5], with an increasing number of preclinical dental students voicing concerns about MSDs [3]. MSDs contribute considerably to sick leave, reduced the students' productivity and terminating their clinical practice earlier than they desired [1,3].

MSDs are defined by the World Health Organization (WHO) as a disorder of the muscles, ligaments, tendons, joints nerves, and bones not directly resulting from an acute or instantaneous event (e.g., slips or falls). MSDs are work-related when the work environment promotes their development [1]. Dental professionals have higher risk of developing musculoskeletal disorders or symptoms compared to that of the general population [1,2]. Although increasing evidence is suggesting that psychosocial factors may be associated with the prevalence of MSD, the physical burden of clinical work associated with incorrect postures or poor body mechanics have been described as the major factors associated with MSDs referred by dental health workers [1,3,4]. The characteristics of clinical work and prevalence results demonstrate that female students are more susceptible to develop MSDs [6–9].

Therefore, the aim of this study is to compare the prevalence of musculoskeletal disorders between male and female dental students.

2. Materials and Methods

The participants of this study were students from the fourth and fifth years of their integrated master's degree in dentistry at Egas Moniz School of Health & Science enrolled in clinical practice. We provided an intuitive and easy online form of the Nordic Musculoskeletal Questionnaire (NMQ) to collect the data, which has shown good reliability and moderate validity for use in epidemiological studies [10].

3. Results

A total of 117 students were contacted to participate, and 85 answered the questionnaire, achieving a response rate of 72.7%. Ten were excluded due to being older than 30 years, and 12 were excluded for having been diagnosed with ongoing musculoskeletal injuries, leaving a total of 63 participants. The total was composed of 81% women and 19% men with the following characteristics: 22.54 ± 1.51 years, 1.67 ± 0.08 m, and 59.95 ± 11.11 kg. Descriptive statistics by sex are described in Table 1.

Table 1. Sociodemographic data by sex.

	Students	Age (years)	Height (m)	Weight (kg)
	n (%)	Mean ± SD [range]	Mean ± SD	Mean ± SD
Female	51 (81%)	22.57 ± 1.54 [20–28]	1.64 ± 0.06	56.67 ± 8.53
Male	12 (19%)	22.42 ± 1.44 [20–25]	1.78 ± 0.09	73.92 ± 10.12

Abbreviations. SD = Standard deviation; kg = kilogram; m = meter.

The occurrence rate by sex was assessed within the last year and week, and they are represented in Figure 1. In a 12-month span, the most frequently reported symptoms among women were in the neck (68.6%) and lower back (66.7%), while among the men, they were in the lower back (58.3%) and shoulders (41.7%). A total 92.2% of the female students reported symptoms in at least one body region over the past year, while among the men, the total prevalence was 91.6%.

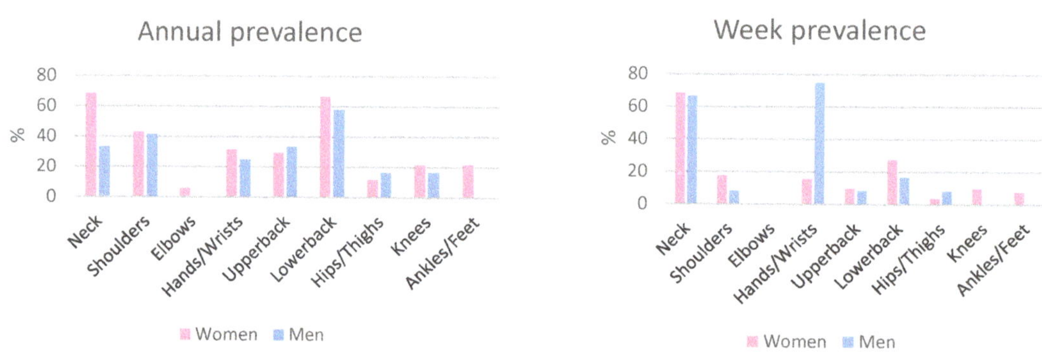

Figure 1. Annual and weekly prevalence of musculoskeletal symptoms among dental students.

The weekly higher occurrence rates were in the neck (68.6%) and lower back (27.5%) among the women and wrists and hands (75%) and neck (66.7%) among the men. For a week-long period, 58.8% of the female students reported symptoms in one or more body regions, while among the men, this rate was 50%.

4. Discussion

The oral cavity represents a very small working area [2]. Dentists' work position necessitates them leaning their head towards the patients with their arms distant from their body and continuously rotating their trunk, while maintaining high levels of attention and concentration for long periods. The repeated use of these positions results in excessive and continued pressure placed on the musculoskeletal structures in the neck, shoulders, trunk, and waist, exacerbating and highlighting the MSD impact among people in this profession and ultimately leading to a reduced working efficiency and prematurely disabled dentists [2,4,5,9]. Dental procedures, such as the filling of a cavity or the preparation of a root canal, require static postures [11], which are defined as positions that are maintained for more than four seconds [12]. The kinematic analysis of the work-related musculoskeletal loading of trunks determined that the static positions of a dentist's head and trunk are generally retained for 27.4% and 23.6% of the treatment time, respectively [2].

The present study aimed to monitor and compare the development of musculoskeletal complaints between male and female dental students.

The neck and lower back are the most frequently reported sites of pain, with occurrence rates being generally higher in comparison to those of previous studies [1,6–9]. However, it is important to note that comparisons across studies should be made cautiously, considering factors such as the study design and sample sizes.

Ergonomics and sexual dimorphism could play significant roles in understanding the occurrence and development of musculoskeletal disorders in dentistry students. The occurrence of neck pain is more common higher among women within the general population [13], and this is explained by the existence of different physiological mechanisms for pain perception between sexes. Despite the asymmetry in the male and female participants that could influence the weight of each change between the groups, our results reflect a larger difference between the sexes when comparing to the general population, with women (68.6%) having an annual rate that is twice as large as that of men (33.3%). Since dental work often involves prolonged periods of head tilting, which can lead to the overloading of neck muscles and cervical spine joints, an association between this and the described sexual dimorphism for pain perception could present a possible explanation for this higher incidence. Conversely, symptoms in the lower back are commonly associated with incorrect sitting postures, forward bending, and a lack of hip tilting during work. Although the prevalence of neck and lower back symptoms is high, a disparity between sexes was only observed for the neck, which could suggest that the physical burden associated with clinical practice is potentially greater on the neck than it is on the lower back. The total annual prevalence of MSDs for all body regions was similar between the male (91.6%) and female (92.2%) students and is in line with reported outcomes in previous studies on dental students [2,4,5] and dentists [1–3].

In a weekly report, the symptom rates between men and women were similar, except in the hands and wrists, with men (75%) having a more than five times larger rate than women (15.7%). Improper techniques with the sustained contraction of wrist muscles and inappropriate ergonomic support were previously observed as the causes of these symptoms [14] and could explain these differences. The overall rate of MSDs within a one-week period is slightly higher among the female students (58.8%) compared to that of the male students (50%). However, it is noteworthy that this means that one out of every two students experienced some form of MSD in the previous week. These findings raise concerns regarding the need to urgently develop and implement preventive strategies tailored to the specific needs of each student group, both in the short and long terms. Education on work-related MSDs, preventive exercises, posture correction, and ergonomic aid implementation could be important measures to reduce the differences between the sexes and the global prevalence of MSDs among dental students.

5. Conclusions

Our findings highlight the importance of addressing musculoskeletal issues in dental professionals, even among students, to prevent potential disabilities and early retirement associated with these conditions. Addressing these issues with an intervention targeting their different etiologies and students' sexual differences could promote better outcomes and reduce the general MSD prevalence rates among dental students.

Author Contributions: Conceptualization, M.B.d.A.; methodology, M.B.d.A.; formal analysis, M.B.d.A.; investigation, M.B.d.A. and M.M.; resources, M.B.d.A.; data curation, M.B.d.A.; writing—original draft preparation, M.B.d.A.; writing—review and editing, M.B.d.A. and M.M.; supervision, M.B.d.A.; project administration, M.B.d.A. All authors have read and agreed to the published version of the manuscript.

Funding: This research received no external funding.

Institutional Review Board Statement: Egas Moniz Ethics Commission approval: CE-1121.

Informed Consent Statement: Informed consent was obtained from all participants in the study.

Data Availability Statement: Mendeley Data, V1, doi: 10.17632/z5mtnn9xh3.1, available online: https://data.mendeley.com/datasets/z5mtnn9xh3/1.

Acknowledgments: To André Júdice for his invaluable support throughout the course of this study and to the dental students who participated in this study.

Conflicts of Interest: The authors declare no conflict of interest.

References

1. Hayes, M.; Cockrell, D.; Smith, D. A systematic review of musculoskeletal disorders among dental professionals. *Int. J. Dent. Hyg.* **2009**, *7*, 159–165. [CrossRef] [PubMed]
2. Ohlendorf, D.; Naser, A.; Haas, Y.; Haenel, J.; Fraeulin, L.; Holzgreve, F.; Erbe, C.; Betz, W.; Wanke, E.M.; Brueggmann, D. Prevalence of musculoskeletal disorders among dentists and dental students in Germany. *Int. J. Environ. Res. Public Health* **2020**, *17*, 8740. [CrossRef] [PubMed]
3. Santucci, N.M.; Jellin, J.; Davenport, T.E. Dental and physical therapy faculty collaborate in assessing and educating dental students on musculoskeletal disorders. *J. Dent. Educ.* **2021**, *85*, 53–59. [CrossRef] [PubMed]
4. Kapitán, M.; Pilbauerová, N.; Vavřičková, L.; Šustová, Z.; Machač, S. Prevalence of musculoskeletal disorders symptoms among Czech dental students. Part 1: A questionnaire survey. *Acta Medica* **2019**, *61*, 131–136. [CrossRef] [PubMed]
5. Aboalshamat, K.T. Nordic assessment of occupational disorders among dental students and dentists in Saudi Arabia. *J. Int. Soc. Prev. Community Dent.* **2020**, *10*, 561. [CrossRef] [PubMed]
6. Felemban, R.A.; Sofi, R.A.; Alhebshi, S.A.; Alharbi, S.G.; Farsi, N.J.; Abduljabbar, F.H.; Farsi, J.M. Prevalence and predictors of musculoskeletal pain among undergraduate students at a dental school in Saudi Arabia. *Clin. Cosmet. Investig. Dent.* **2021**, *13*, 39–46. [CrossRef] [PubMed]
7. Lestari, A.I.; Palupi, R. Better Early Prevention: Dental Student? s Awareness of Musculoskeletal Disorders. *Sys. Rev. Pharm.* **2020**, *11*, 941–945.
8. Sulimany, A.M. Cervical and Lumbar Pain among Dental Interns in Saudi Arabia: A National Cross-sectional Study. *J. Contemp. Dent. Pract.* **2021**, *22*, 860–866. [PubMed]
9. Zafar, H.; Almosa, N. Prevalence of work-related musculoskeletal disorders among dental students of King Saud University, Riyadh, Kingdom of Saudi Arabia. *J. Contemp. Dent. Pract.* **2019**, *20*, 449–453. [PubMed]
10. Mesquita, C.C.; Ribeiro, J.C.; Moreira, P. Portuguese version of the standardized Nordic musculoskeletal questionnaire: Cross cultural and reliability. *J. Public Health* **2010**, *18*, 461–466. [CrossRef]
11. Nowak, J.; Erbe, C.; Hauck, I.; Groneberg, D.A.; Hermanns, I.; Ellegast, R.; Ditchen, D.; Ohlendorf, D. Motion analysis in the field of dentistry: A kinematic comparison of dentists and orthodontists. *BMJ Open* **2016**, *6*, e011559. [CrossRef] [PubMed]
12. Ohlendorf, D.; Erbe, C.; Hauck, I.; Nowak, J.; Hermanns, I.; Ditchen, D.; Ellegast, R.; Groneberg, D.A. Kinematic analysis of work-related musculoskeletal loading of trunk among dentists in Germany. *BMC Musculoskelet Disord.* **2016**, *17*, 427. [CrossRef] [PubMed]
13. Fejer, R.; Kyvik, K.O.; Hartvigsen, J. The prevalence of neck pain in the world population: A systematic critical review of the literature. *Eur. Spine J.* **2006**, *15*, 834–848. [CrossRef] [PubMed]
14. De Sio, S.; Traversini, V.; Rinaldo, F.; Colasanti, V.; Buomprisco, G.; Perri, R.; Mormone, F.; La Torre, G.; Guerra, F. Ergonomic risk and preventive measures of musculoskeletal disorders in the dentistry environment: An umbrella review. *PeerJ* **2018**, *6*, e4154. [CrossRef] [PubMed]

medical sciences forum

MDPI

Proceeding Paper

Relationship between Positive and Adverse Childhood Experiences and Resilience in Adulthood †

Ana Matos [1,*], Diana Pais [1], Inês Canha [1], Maria Duarte [1], Rúben Pereira [1], Jorge Cardoso [1,2,3] and Telma C. Almeida [1,2,3]

1 Egas Moniz School of Health & Science, Caparica, 2829-511 Almada, Portugal; didimcp01@gmail.com (D.P.); canha.ines@gmail.com (I.C.); mariajoaoduarte2002@gmail.com (M.D.); rubenpereira.5@hotmail.com (R.P.); jcardoso@egasmoniz.edu.pt (J.C.); talmeida@egasmoniz.edu.pt (T.C.A.)
2 Egas Moniz Center for Interdisciplinary Research (CiiEM), Egas Moniz School of Health & Science, Caparica, 2829-511 Almada, Portugal
3 LabPSI—Laboratório de Psicologia Egas Moniz, Caparica, 2829-511 Almada, Portugal
* Correspondence: anafranciscamms@outlook.com
† Presented at the 6th International Congress of CiiEM—Immediate and Future Challenges to Foster One Health, Almada, Portugal, 5–7 July 2023.

Abstract: Childhood experiences have a significant impact on the development of resilience, shaping an individual's capacity to cope with stress and adversity throughout their lifespan. This study aimed to analyze the relationship between positive childhood experiences (PCEs), adverse childhood experiences (ACEs), and resilience in adulthood. The results revealed that emotional and physical abuse and emotional and physical neglect are negatively correlated with PCEs and resilience. PCEs protect individuals against ACEs and promote better health and resilience in adulthood. The findings emphasize the importance of reducing ACEs and increasing PCEs in childhood, preventing long-term impacts and mental health outcomes, and increasing resilience levels.

Keywords: childhood experiences; resilience; adults

Citation: Matos, A.; Pais, D.; Canha, I.; Duarte, M.; Pereira, R.; Cardoso, J.; Almeida, T.C. Relationship between Positive and Adverse Childhood Experiences and Resilience in Adulthood. *Med. Sci. Forum* **2023**, *22*, 48. https://doi.org/10.3390/msf2023022048

Academic Editors: José Brito, Nuno Taveira and Ana I. Fernandes

Published: 4 September 2023

1. Introduction

Childhood experiences are important for an individual's development and well-being throughout their life [1,2]. While adverse childhood experiences (ACEs), such as abuse or neglect, can interfere with the development of resilience, decreasing one's levels of it [3,4], positive childhood experiences (PCEs), like having a stable routine, can promote it [5], enhancing mental and physical health in adulthood [6–10].

Resilience plays an important role during a child's development and corresponds to the ability to adapt to adversity and overcome difficulties [6,11]. ACEs can affect resilience through difficulties in establishing emotional bonds, lower self-esteem, and psychopathological symptoms [4], leading to an impaired capacity for developing resilience [12]. On the other hand, PCEs can contribute to healthy relationships and future social experiences [5], having a positive effect against adversity through an improvement in resilience and avoidance of adversity reoccurrence later in life [7,11]. Research suggests that the quality of parental care and support received during childhood can impact the development of resilience and an individual's ability to cope effectively with stressors later in life [13,14]. The present study aims to analyze the relationship between childhood experiences and resilience in adulthood.

2. Materials and Methods

Our sample comprised 76 Portuguese adults, 54 women (71.1%) and 22 men (28.9%), with ages between 18 and 70 ($M = 35.59$, $SD = 15.30$). Participants answered questions from sociodemographic questionnaires online, the Benevolent Childhood Experiences Scale [9], the Childhood Adverse Experiences Questionnaire [15], and the Resilience Scale [16]. All participants granted their consent to participate in this study. The study followed the ethical principles outlined in the Declaration of Helsinki [17] and was approved by the Institutional Review Board of Egas Moniz School of Health and Science.

3. Results and Discussion

The results showed that emotional abuse is positively correlated with physical abuse ($r = 0.696$, $p < 0.001$) and with emotional ($r = 0.546$, $p < 0.001$) and physical neglect ($r = 0.332$, $p = 0.003$). Physical abuse is positively correlated with both emotional ($r = 0.416$, $p < 0.001$) and physical neglect ($r = 0.256$, $p = 0.026$). There are also positive correlations between emotional and physical neglect ($r = 0.432$, $p < 0.001$). Negative correlations were found between emotional neglect and self-determination ($r = -0.343$, $p = 0.002$), adaptability ($r = -0.281$, $p = 0.014$), resilience ($r = -0.333$, $p = 0.003$), and PCEs ($r = -0.600$, $p < 0.001$). Both physical ($r = -0.251$, $p = 0.029$) and emotional abuse ($r = -0.252$, $p = 0.028$), as well as physical neglect ($r = -0.384$, $p < 0.001$), are negatively correlated with PCEs.

These findings are consistent with the existing literature, showing that multiple ACEs (e.g., emotional neglect) are susceptible to co-occurring [18] due to risk factors, such as dysfunctional family dynamics [19], and are associated with lower levels of PCEs (e.g., supportive family environment) [20,21] and resilience [21]. Research suggests that the tendency to experience several ACEs [19] can generate cumulative problems in an individual's life [22]. This can interfere with emotional support and a nurturing environment [23], contributing to fewer PCEs [20,21] and impairments in the development of resilience [21]. Later in life, exposure to ACEs can lead to impairments in the psychological and social functioning of an individual [4]. On the other hand, studies have also reported that the presence of at least six types of PCEs could mitigate the effects of ACEs [24,25], leading to better mental health [10] and the development of resilience through an adequate response to adversity [25].

The results also show that women experience more emotional negligence, $t(74) = 11.252$, $p = 0.001$, sexual abuse, $t(74) = 11.594$, $p < 0.001$, physical abuse, $t(74) = 15.810$, $p < 0.001$, and emotional abuse, $t(74) = 4.366$, $p = 0.004$, than men. Our results are in line with those of other research. The literature argues that women are more likely to experience ACEs due to gender-based violence (e.g., sexual abuse) [26].

Assessments of ACEs, resilience, and positive factors (PCEs) will allow a better understanding of an individual's needs and a better adjustment of the support needed for positive functioning and well-being [12]. The present study may be constrained by limitations, such as a non-representative sample, due to its insufficient size and lack of heterogeneity in gender representation, restricting the ability to accurately capture the diverse characteristics of the larger population. Our results show the importance of implementing preventive actions to promote childhood protection and foster close relationships for stable and healthy mental development.

Author Contributions: Conceptualization, T.C.A.; methodology, T.C.A., A.M., D.P., I.C., M.D. and R.P.; validation, T.C.A.; formal analysis, A.M., D.P., I.C., M.D., R.P. and T.C.A.; investigation, A.M., D.P., I.C., M.D. and R.P.; data curation, T.C.A., A.M., D.P., I.C., M.D. and R.P.; writing—original draft preparation, A.M., D.P., I.C., M.D. and R.P.; writing—review and editing, T.C.A. and J.C.; supervision, T.C.A.; project administration, T.C.A. All authors have read and agreed to the published version of the manuscript.

Funding: This research received no external funding.

Institutional Review Board Statement: The study was conducted following the Declaration of Helsinki and was approved by the Institutional Review Board of Egas Moniz School of Health and Science.

Informed Consent Statement: Informed consent was obtained from all of the subjects involved in the study.

Data Availability Statement: The data presented in this study are available upon request from the corresponding author. The data are not publicly available due to ethical reasons.

Acknowledgments: The authors would like to express the most profound gratitude to all the participants who voluntarily provided information to this study.

Conflicts of Interest: The authors declare no conflict of interest.

References

1. Lee, H.; Schafer, M. Are Positive Childhood Experiences Linked to Better Cognitive Functioning in Later Life?: Examining the Role of Life Course Pathways. *J. Aging Health* **2021**, *33*, 217–226. [CrossRef]
2. Webster, E.M. The Impact of Adverse Childhood Experiences on Health and Development in Young Children. *Glob. Pediatr. Health* **2022**, *9*, 2333794X221078708. [CrossRef]
3. Morgado, A. O Impacto na Saúde das Experiências Adversas na Infância (Tese de Mestrado). Faculdade de Medicina Universidade do Porto. 2020. Available online: https://sigarra.up.pt/fep/en/pub_geral.show_file?pi_doc_id=297573 (accessed on 18 May 2023).
4. Rawlins, B.; Brooks, M.; Khan, R. Posttraumatic stress symptoms mediate the relationship between adverse childhood experiences, avoidant personality traits and resilience. *Anxiety Stress Coping* **2020**, *33*, 590–601. [CrossRef] [PubMed]
5. Narayan, A.J.; Rivera, L.M.; Bernstein, R.E.; Harris, W.W.; Lieberman, A.F. Positive childhood experiences predict less psychopathology and stress in pregnant women with childhood adversity: A pilot study of the benevolent childhood experiences (BCEs) scale. *Child Abus. Negl.* **2018**, *78*, 19–30. [CrossRef] [PubMed]
6. Skodol, A.E.; Bender, D.S.; Pagano, M.E.; Shea, M.T.; Yen, S.; Sanislow, C.A.; Grilo, C.M.; Daversa, M.T.; Stout, R.L.; Zanarini, M.C.; et al. Positive Childhood Experiences. *J. Clin. Psychiatry* **2007**, *68*, 1102–1108. [CrossRef] [PubMed]
7. Crandall, A.; Miller, J.R.; Cheung, A.; Novilla, L.K.; Glade, R.; Novilla, M.L.B.; Magnusson, B.M.; Leavitt, B.L.; Barnes, M.D.; Hanson, C.L. ACEs and counter-ACEs: How positive and negative childhood experiences influence adult health. *Child Abus. Negl.* **2019**, *96*, 104089. [CrossRef] [PubMed]
8. Daines, C.L.; Hansen, D.; Novilla, M.L.B.; Crandall, A. Effects of positive and negative childhood experiences on adult family health. *BMC Public Health* **2021**, *21*, 651. [CrossRef]
9. Almeida, T.C.; Guarda, R.; Cunha, O. Positive childhood experiences and adverse experiences: Psychometric properties of the Benevolent Childhood Experiences Scale (BCEs) among the Portuguese population. *Child Abus. Negl.* **2021**, *120*, 105179. [CrossRef]
10. Doom, J.R.; Seok, D.; Narayan, A.J.; Fox, K.R. Adverse and Benevolent Childhood Experiences Predict Mental Health During the COVID-19 Pandemic. *Advers. Resil. Sci.* **2021**, *2*, 193–204. [CrossRef]
11. Babić, R.; Babić, M.; Rastović, P.; Ćurlin, M.; Šimić, J.; Mandić, K.; Pavlović, K. Resilience in Health and Illness. *Psychiatr. Danub.* **2020**, *32*, 226–232.
12. Bethell, C.D.; Carle, A.; Hudziak, J.; Gombojav, N.; Powers, K.; Wade, R.; Braveman, P. Methods to Assess Adverse Childhood Experiences of Children and Families: Toward Approaches to Promote Child Well-being in Policy and Practice. *Acad. Pediatr.* **2017**, *17*, S51–S69. [CrossRef] [PubMed]
13. Wigley, I.L.C.M.; Mascheroni, E.; Bulletti, F.; Bonichini, S. COPEWithME: The Role of Parental Ability to Support and Promote Child Resilient Behaviors during the COVID-19 Emergency. *Front. Psychol.* **2021**, *12*, 732745. [CrossRef] [PubMed]
14. Urban, T.H.; Friedman, D.; Kaskas, M.M.; Caruso, A.J.; Canenguez, K.M.; Rotter, N.; Wozniak, J.; Basu, A. Brief report of protective factors associated with family and parental well-being during the COVID-19 pandemic in an outpatient child and adolescent psychiatric clinic. *Front. Psychol.* **2022**, *13*, 883955. [CrossRef] [PubMed]
15. Silva, S.; Maia, Â. *Versão Portuguesa do Family ACE Questionnaire (Questionário da História de Adversidade na Infância)*; Psiquilíbrios Edições: Braga, Portugal, 2008.
16. Pinheiro, M.; Matos, A.; Pestana, C.; Oliveira, S.; Costa, J. The Resilience Scale: A study in a Portuguese adult sample. *Eur. Proc. Soc. Behav. Sci.* **2015**, *6*, 1–220. [CrossRef]
17. World Medical Association. World Medical Association Declaration of Helsinki: Ethical principles for medical research involving human subjects. *JAMA* **2013**, *310*, 2191–2194. [CrossRef]
18. Almeida, T.C.; Ramos, C.; Brito, J.; Cardoso, J. The juvenile victimization questionnaire: Psychometric properties and polyvictimization among Portuguese youth. *Child. Youth Serv. Rev.* **2020**, *113*, 105001. [CrossRef]
19. Dong, M.; Anda, R.F.; Felitti, V.J.; Dube, S.R.; Williamson, D.F.; Thompson, T.J.; Loo, C.M.; Giles, W.H. The interrelatedness of multiple forms of childhood abuse, neglect, and household dysfunction. *Child Abus. Negl.* **2004**, *28*, 771–784. [CrossRef]
20. Afifi, T.; Taillieu, T.; Cheung, K.; Katz, L.; Tonmyr, L.; Sareen, J.; Turner, S. Adverse childhood experiences and positive childhood experiences are independently associated with adult health outcomes in a representative Canadian sample. *J. Public Health* **2020**, *42*, 588–596.
21. Goldenson, J.; Kitollari, I.; Lehman, F. The Relationship Between ACEs, Trauma-Related Psychopathology and Resilience in Vulnerable Youth: Implications for Screening and Treatment. *J. Child Adolesc. Trauma* **2020**, *14*, 151–160. [CrossRef]
22. Hamby, S.; Elm, J.H.L.; Howell, K.H.; Merrick, M.T. Recognizing the cumulative burden of childhood adversities transforms science and practice for trauma and resilience. *Am. Psychol.* **2021**, *76*, 230–242. [CrossRef]

23. Rebicova, M.L.; Veselska, Z.D.; Husarova, D.; Klein, D.; Geckova, A.M.; Van Dijk, J.P.; Reijneveld, S. Does family communication moderate the association between adverse childhood experiences and emotional and behavioural problems? *BMC Public Health* **2020**, *20*, 1264. [CrossRef] [PubMed]
24. Baglivio, M.T.; Wolff, K.T. Positive Childhood Experiences (PCE): Cumulative Resiliency in the Face of Adverse Childhood Experiences. *Youth Violence Juv. Justice* **2020**, *19*, 139–162. [CrossRef]
25. Crandall, A.; Broadbent, E.; Stanfill, M.; Magnusson, B.M.; Novilla, M.L.B.; Hanson, C.L.; Barnes, M.D. The influence of adverse and advantageous childhood experiences during adolescence on young adult health. *Child Abus. Negl.* **2020**, *108*, 104644. [CrossRef] [PubMed]
26. Haahr-Pedersen, I.; Perera, C.; Hyland, P.; Vallières, F.; Murphy, D.; Hansen, M.; Spitz, P.; Hansen, P.; Cloitre, M. Females have more complex patterns of childhood adversity: Implications for mental, social, and emotional outcomes in adulthood. *Eur. J. Psychotraumatol.* **2020**, *11*, 1708618. [CrossRef]

medical sciences forum

MDPI

Proceeding Paper

Employability and Dental Caries Experience †

Eduardo Guerreiro *, Ana Cristina Manso and José João Mendes

Egas Moniz Center for Interdisciplinary Research (CiiEM), Egas Moniz School of Health & Science, 2829-511 Caparica, Portugal; cmanso@egasmoniz.edu.pt (A.C.M.); jmendes@egasmoniz.edu.pt (J.J.M.)
* Correspondence: eguerreiro@egasmoniz.edu.pt
† Presented at the 6th International Congress of CiiEM—Immediate and Future Challenges to Foster One Health, Almada, Portugal, 5–7 July 2023.

Abstract: One of the most common oral diseases and a public health concern is dental caries. Risk assessment and health factor identification are necessary for prevention. Our research focuses on evaluating how dental caries experience is affected by employability. Radiological, oral, sociodemographic, and behavioral questionnaires were included in the investigation. Individuals who are unemployed display an increased occurrence of tooth decay. Regarding employability, the results showed a significantly lower experience in employed participants when compared to unemployed participants. Employability should be considered as a relevant health determinant of dental health.

Keywords: dental caries; public health; risk factor; employability

1. Introduction

Dental caries, affecting approximately 2.3 billion permanent dentitions worldwide, stands out as one of the most prevalent diseases. This condition is marked by a dysbiosis of the oral biofilm, primarily caused by fermentable carbohydrates [1]. As well as a lower perceived quality of life, dental caries are associated with considerable economic burdens [2]. Dental caries can result in eating difficulties, tooth loss, and toothache and may also lead to absence from school and work if not effectively managed [3,4]. Several groups within the population are at greater risk of dental caries, resulting in an unequal distribution of caries among the population [1]. The risk is increased by a number of factors, including cariogenic bacteria, cariogenic diets, oral hygiene habits, dental anxiety, socioeconomic status, education level, smoking habits, and others [5]. Gaining a thorough comprehension of these factors at the population level is imperative for devising effective oral health promotion strategies and policies [6]. Providing aggregated data on caries experience and factors related to it [7,8], a retrospective analysis was performed of a cohort of initial dental visits to a renowned Portuguese university hospital. Our primary objective was to assess caries experience and employability within the study population.

2. Materials and Methods

2.1. Sample Size, Study Design, and Setting

A secondary analysis of initial patient visits to a university dental clinic, specifically the Egas Moniz Dental Clinic in Almada, Portugal, is presented in this study. Data were collected for this study from an ongoing database of first-time patients that began on 1 January 2016 and was completed on 13 March 2020 using a nonprobability sampling approach. The initial appointments encompassed a thorough process, incorporating a self-reported health questionnaire, a comprehensive clinical examination of the entire mouth, and radiographic evaluations (including panoramic X-rays and/or bitewings). In addition, this study collected data on age, sex, education level, employment status, general medical history, medications used, smoking habits, and oral hygiene practices via self-reported questionnaire. A dental student observed the participants during the examination, and

Citation: Guerreiro, E.; Manso, A.C.; Mendes, J.J. Employability and Dental Caries Experience. *Med. Sci. Forum* **2023**, *22*, 44. https://doi.org/10.3390/msf2023022044

Academic Editors: José Brito, Nuno Taveira and Ana I. Fernandes

Published: 23 August 2023

a qualified clinical assistant verified the final diagnosis. Patients were informed about their condition and treatment plan. This study required patients to give written consent, to be willing to participate, and to be 18 or older to be eligible. Patients with incomplete data or edentulous teeth were excluded. Due to the possibility of dental caries causing edentulism, it was excluded from this study. A higher prevalence of periodontitis has been reported in the observed population [9], which could result in an overestimation of dental caries, especially in the missing teeth component. This study did not include patients with disabilities or special needs since they are managed by the Egas Moniz Dental Clinic's Special Needs Department.

2.2. Dependent Variables

We employed a dichotomous variable (present or absent) to assess caries experience. Additionally, we utilized the DMF index as a continuous variable since it is the most used index for dental caries.

2.3. Independent Variables

Self-reported questionnaires were used to collect sociodemographic and behavioral information. Age, sex, education level, and occupation are important independent variables for determining health determinants and sociodemographic factors. Caries are predicted using several of these variables [10]. Males and females were divided into two groups by gender. The subjects were divided into four age groups based on their age: 18–24, 25–44, 45–64, and 65+. Regarding occupation, subjects were categorized into four categories: student, employed, unemployed, and retired.

2.4. Statistical Analysis

The data analysis was conducted using IBM SPSS Statistics version 28.0 for Windows (IBM Corp., Armonk, NY, USA). An analysis of the collected data was conducted using the Kruskal–Wallis test, a non-parametric statistical test, followed by an analysis of the pairwise differences using the Bonferroni test as a post hoc analysis.

3. Results

Participant Inclusion and Characteristics

The 9860 incoming patients met the eligibility criteria in 9349 (94.8%) cases, while 511 were excluded. In total, 36 (59.9%) of those excluded were younger than 18, 204 (39.9%) were edentulous, and 1 (0.2%) had an incomplete triage questionnaire. Among the 9349 participants, 59.8% were females, and 64.3% were aged 25 to 64, according to Table 1. In terms of occupation, the majority, 53.3%, were employed, with a smaller proportion of 11.6% unemployed, while 35.2% were either students or retired.

Table 1. Participant sociodemographic characterization (n = 9349).

Variable	Sub-Variable	*n* (%)
Gender	Female	5592 (59.8%)
	Male	3757 (40.2%)
Group by age (years)	18 to 24	1867 (20.0%)
	25 to 44	2907 (31.1%)
	45 to 64	3101 (33.2%)
	≥65	1474 (15.8%)
Occupation	Student	1616 (17.3%)
	Employed	4980 (53.3%)
	Unemployed	1083 (11.6%)
	Retired	1670 (17.9%)

This study involved a total of 9349 individuals, out of which 8521 (91.1%) reported dental caries experience. Table 2 reveals that unemployed individuals exhibited a statistically significant increase in both decayed teeth and missing teeth ($p < 0.001$) in comparison to their employed counterparts. The study findings indicate that the mean number of filled teeth was significantly higher among employed individuals compared to their unemployed counterparts.

Table 2. Dental caries and sociodemographic, health, and behavioral factors (n = 8521).

Variable	Sub-Variable	n (%)	Decayed	Missing	Filled	DMFT Mean
Gender	Female	5090 (59.7%)	5.8 (4.3) [a]	6.6 (7.4) [a]	3.3 (3.6) [a]	15.7 (8.2) [a]
	Male	3431 (40.3%)	6.3 (4.8) [b]	6.5 (7.4) [a]	2.7 (3.3) [b]	15.5 (8.2) [a]
Group by age (in years)	18 to 24	1496 (17.6%)	4.6 (4.5) [a]	0.7 (1.5) [a]	1.9 (2.6) [a]	7.3 (6.3) [a]
	25 to 44	2702 (31.7%)	6.9 (4.9) [b]	3.5 (4.5) [b]	3.6 (3.7) [b]	14.1 (7.0) [b]
	45 to 64	2932 (34.4%)	6.2 (4.1) [c]	9.4 (7.2) [c]	3.7 (3.8) [c]	19.2 (6.7) [c]
	≥65	1391 (16.3%)	5.5 (4.1) [d]	13.9 (7.9) [d]	2.1 (2.8) [a]	21.5 (7.0) [d]
Occupation	Student	1255 (14.7%)	4.1 (4.2) [a]	0.9 (2.2) [a]	2.2 (2.8) [a]	7.2 (5.5) [a]
	Employed	4661 (54.7%)	6.4 (4.5) [b]	5.7 (6.2) [b]	3.7 (3.8) [b]	15.8 (7.3) [b]
	Unemployed	1024 (12.0%)	7.4 (5.1) [c]	8.3 (7.9) [c]	2.7 (3.4) [b]	18.4 (7.5) [c]
	Retired	1581 (18.6%)	5.6 (4.1) [d]	13.4 (8.0) [d]	2.2 (2.8) [a]	21.2 (7.1) [d]

Data are mean (standard deviation). Different letters indicate statistically different mean values (Bonferroni, $p < 0.05$). Abbreviations: DMFT: Decayed, Missing, Filled Teeth index; n—number of participants. Statistical analysis for a significance level $p < 0.05$.

4. Discussion

This study highlights a high incidence of dental caries in the population. Among the participants, 91.1% had at least one tooth affected by dental caries, emphasizing its widespread prevalence globally [1,11]. The results might concern both the population in the Portuguese capital and those in less developed rural areas. Further epidemiological studies can aid the government in devising improved oral health policies [12].

Occupation is a significant variable. Unemployed participants showed higher dental caries rates, consistent with previous studies suggesting poorer oral health among the unemployed. Occupational settings can influence oral health, and socioeconomic status based on occupation is a common variable in epidemiological studies. Workplace stress, healthcare policies, and health insurance can affect oral health in older adults [13].

There are several potential explanations for how unemployment can affect oral health. One possibility is that dental treatment costs can be high even for those who are employed, and in certain countries, access to public dental care is limited or practically nonexistent [13,14].

Dental caries is a dangerous disease, but due to its risks and dynamic nature, more epidemiological research is needed to understand how it affects Portugal's population.

5. Conclusions

The incidence of caries in this population was quite high, and the existence of this condition was found to be significantly linked with occupation. In understanding dental caries, variables and determinants must be understood; however, behavioral and cognitive biases must be considered as well. These data could be utilized to prepare comprehensive epidemiological research and public health interventions in the future.

Author Contributions: E.G. designed, performed, and analyzed the experiments, provided reagents/materials/analysis tools, prepared figures and/or tables, drafted or reviewed drafts of the paper, and approved the final version. The final draft of this paper was approved by A.C.M. and J.J.M., who analyzed the data, provided reagents and materials, and authored or reviewed drafts of this paper. All authors have read and agreed to the published version of the manuscript.

Funding: This research received no external funding.

Institutional Review Board Statement: In compliance with the Declaration of Helsinki, this study received approval from the Ethics Committee of Instituto Universitario Egas Moniz (ID no. 898).

Informed Consent Statement: This study was conducted with informed consent provided by all subjects.

Data Availability Statement: All data generated or analyzed in this study are incorporated within this article. For additional inquiries, please contact the corresponding author.

Acknowledgments: Egas Moniz School of Health and Science.

Conflicts of Interest: The authors have no conflict of interest to declare.

References

1. Fontana, M.; Gonzalez-Cabezas, C. Evidence-Based Dentistry Caries Risk Assessment and Disease Management. *Dent. Clin. N. Am.* **2019**, *63*, 119–128. [CrossRef] [PubMed]
2. Peres, M.A.; Macpherson, L.M.; Weyant, R.J.; Daly, B.; Venturelli, R.; Mathur, M.R.; Listl, S.; Celeste, R.K.; Guarnizo-Herreño, C.C.; Kearns, C.; et al. Oral diseases: A global public health challenge. *Lancet* **2019**, *394*, 249–260. [CrossRef]
3. Bogale, B.; Engida, F.; Hanlon, C.; Prince, M.J.; Gallagher, J.E. Dental caries experience and associated factors in adults: A cross-sectional community survey within Ethiopia. *BMC Public Health* **2021**, *21*, 180. [CrossRef]
4. Frencken, J.E.; Sharma, P.; Stenhouse, L.; Green, D.; Laverty, D.; Dietrich, T. Global epidemiology of dental caries and severe periodontitis—A comprehensive review. *J. Clin. Periodontol.* **2017**, *44* (Suppl. 18), S94–S105. [CrossRef]
5. Smith, J.; Griffiths, K.; Judd, J.; Crawford, G.; D'Antoine, H.; Fisher, M.; Bainbridge, R.; Harris, P. Ten years on from the World Health Organization Commission of Social Determinants of Health: Progress or procrastination? *Health Promot. J. Aust.* **2018**, *29*, 3–7. [CrossRef]
6. Hujoel, P.P.; Hujoel, M.L.A.; Kotsakis, G. Personal oral hygiene and dental caries: A systematic review of randomised controlled trials. *Gerodontology* **2018**, *35*, 282–289. [CrossRef]
7. Zewdu, T.; Abu, D.; Agajie, M.; Sahilu, T. Dental caries and associated factors in Ethiopia: Systematic review and meta-analysis. *Environ. Health Prev. Med.* **2021**, *26*, 21. [CrossRef]
8. Fraihat, N.; Madae'en, S.; Bencze, Z.; Herczeg, A.; Varga, O. Clinical Effectiveness and Cost-Effectiveness of Oral-Health Promotion in Dental Caries Prevention among Children: Systematic Review and Meta-Analysis. *Int. J. Environ. Res. Public Health* **2019**, *16*, 2668. [CrossRef] [PubMed]
9. Botelho, J.; Machado, V.; Proença, L.; Alves, R.; Cavacas, M.A.; Amaro, L.; Mendes, J.J. Study of Periodontal Health in Almada-Seixal (SoPHiAS): A cross-sectional study in the Lisbon Metropolitan Area. *Sci. Rep.* **2019**, *9*, 15538. [CrossRef] [PubMed]
10. Yassin, S.; Tikare, S.; AlKahtani, Z.; AlFaifi, F.; AlFaifi, W.S.; AlFaifi, E.; Omair, A.; Ravi, K. Caries preventive practices and dental caries among boys aged 6–15 in Saudi Arabia. *Sci. Rep.* **2020**, *21*, 97–102. [CrossRef]
11. Lukacs, J.R.; Largaespada, L.L. Explaining sex differences in dental caries prevalence: Saliva, hormones, and "life-history" etiologies. *Am. J. Hum. Biol.* **2006**, *18*, 540–555. [CrossRef]
12. Guerreiro, E.; Botelho, J.; Machado, V.; Proença, L.; Mendes, J.J.; Manso, A.C. Caries Experience and Risk Indicators in a Portuguese Population: A Cross-Sectional Study. *Int. J. Environ. Res. Public Health* **2023**, *20*, 2511. [CrossRef]
13. Nobre, M.A.d.A.; Sezinando, A.M.; Fernandes, I.C.; Araújo, A.C. Influence of Smoking Habits on the Prevalence of Dental Caries: A Register-Based Cohort Study. *Eur. J. Dent.* **2021**, *15*, 714–719. [CrossRef]
14. Fattah, M.A.A.; Barghouth, M.H.; Wassel, M.O.; Deraz, O.H.; Khalil, A.E.; Sarsik, H.M.; Mohsen, A.M.A.; Qenawy, A.S.; El Fadl, R.K.A. Epidemiology of dental caries in permanent dentition: Evidence from a population-based survey in Egypt. *BMC Public Health* **2022**, *22*, 2438. [CrossRef]

medical sciences forum

MDPI

Proceeding Paper

X-ray Vision: Self-Prevention of Oral Health Knowledge among Adolescents Undergoing Orthodontic Treatment †

Aucéane Karramkan *, José Grillo Evangelista, Valter Pedroso Alves, Ana Sintra Delgado and Maria do Rosário Dias

Egas Moniz School of Health & Science, Campus Universitário, Quinta da Granja, 2829-511 Caparica, Portugal
* Correspondence: aakarramkan@gmail.com; Tel.: +351-930660137
† Presented at the 6th International Congress of CiiEM—Immediate and Future Challenges to Foster One Health, Almada, Portugal, 5–7 July 2023.

Abstract: Oral health is particularly important in adolescents undergoing orthodontic treatment. However, studies have shown that, in this period, behaviors that are favorable toward oral health appear to decrease. Measures such as self-prevention and self-promotion do not seem to be as effective as they should be. This study has assessed dental hygiene habits and knowledge regarding oral health using a questionnaire that was given to the participants. The sample consisted of 200 subjects aged between 10 and 24 years old. The results suggest that there is a lack of knowledge surrounding oral-self-care and prove the need to invest in oral health education.

Keywords: oral health education; adolescents; self-prevention; health promotion

1. Introduction

Oral health is crucial, particularly in adolescent patients because of the physical and mental transformations that happen in this phase. Therefore, the habits acquired in this phase tend to have future repercussions on the self-esteem and health behavior of the individual. However, studies have shown that, in this period, behaviors that are favorable toward oral health appear to decrease since oral health is considered to be less important from their perspective. Measures such as self-prevention and self-promotion do not seem to be as effective as they should be [1–4]. This line of study aims to identify the mental representation of a part of the body, namely, the inner oral cavity. In the present study, we seek the perception of adolescents on this topic with the mere objective of contributing to the creation of innovative oral health strategies and health instruments.

Citation: Karramkan, A.; Evangelista, J.G.; Alves, V.P.; Delgado, A.S.; Dias, M.d.R. X-ray Vision: Self-Prevention of Oral Health Knowledge among Adolescents Undergoing Orthodontic Treatment. *Med. Sci. Forum* **2023**, *22*, 5. https://doi.org/10.3390/msf2023022005

Academic Editors: José Brito, Nuno Taveira and Ana I. Fernandes

Published: 8 August 2023

2. Materials and Methods

This study assessed the dental hygiene habits and knowledge regarding oral health using a questionnaire that was given to the participants. The sample consisted of 200 subjects aged between 10 and 24 years old in consultations at the Egas Moniz University Clinic [5]. They were first asked to draw the interior of their oral cavity before and after undergoing orthodontic treatment. Then, they were invited to answer a questionnaire concerning oral health behaviors and knowledge. Finally, the patients were given information to help them maintain and improve their self-care. The data were then analyzed using various tools; among them was a content analysis grid, created especially for this study, that was made of analytical categories and subcategories [6]. The data obtained were also analyzed using the statistical software IBM SPSS 28.

3. Results

While adolescents are at risk of developing dental health problems due to their oral health habits but also their knowledge and motivation, the results showed that there is a significant lack of knowledge in adolescents, especially in terms of frequency of brushing, duration, and technique of brushing. Only 67 patients (33.5%) revealed that they brush the inside surface of the tooth, and most of the sample (62%) admitted brushing their teeth for approximately 2 min. Although 78.5% of the adolescents, through the orthodontic consultations, reported that they had access to information regarding their health, the results showed that they did not use it properly. Regarding the results obtained from the drawings, they indicate that there is a low area of knowledge surrounding the interior of the oral cavity. Although the subjects were asked to draw the interior of the oral cavity, 55.5% of drawings represented a *broad smile* in M1 (before) and 60.5% represented that in M2 (after the orthodontic treatment), supported by an *extra-oral view*.

4. Discussion

Firstly, the results showed that most of the subjects demonstrated a lack of oral health knowledge. When asked *how do you brush your teeth?*, most of the patients admitted brushing the occlusal and facial face, whereas 67 subjects (33.5%) stated that they brushed the lingual face. Then, when asked *are you using any additional tools to clean your teeth?*, 85 patients (42.5%) responded that they were not. Nevertheless, not only this, but in most cases they revealed to have already received some information about oral hygiene, and they also disclosed that they perceived this information well (Tables 1 and 2).

Table 1. Descriptive analysis regarding the information about oral hygiene received by the patients.

Have You Already Received Some Information about Oral Hygiene since You Started Orthodontic Treatment?	N	%
Yes	157	78.5
More or less	40	20
No	3	1
Total	200	100

Table 2. Descriptive analysis regarding the perception of the received information.

If So, Did You Understand the Received Information Well?	N	%
Yes	196	99.5
No	1	0.5
Yes, but I still have some doubts	0	0.0
Total	197	100

Furthermore, it is surprising that most of the adolescents in the study self-evaluated their oral health as being good (64%) or reasonable (25.5%) when asked.

In relation to the inner oral cavity knowledge, the patients were asked if they knew the number of teeth in their mouth. The results showed that only 26.5% were assumed to know it, and 58.5% responded incorrectly (Table 3).

Table 3. Descriptive analysis regarding the self-perception of the number of teeth.

Do You Know How Many Teeth You Have in Your Mouth?	N	%
Yes	53	26.5
More or less	107	53.5
No	40	20
Total	200	100

Secondly, concerning the results of the content analysis of the drawings obtained, the data indicate that the self-perception of the inner oral cavity was perceived as a *smile* by the subjects. In fact, the adolescents' drawings represented a *broad smile* from an *extra-oral view*, which was in contradiction with what they were assigned to do—draw the interior of your oral cavity (Table 4, Figure 1).

Table 4. Descriptive analysis regarding the *view* category.

View Category	M1		M2	
	N	%	N	%
Extra-oral view	111	55.5	122	61
Intra-oral view	89	44.5	77	38.5
Total	200	100	200	100

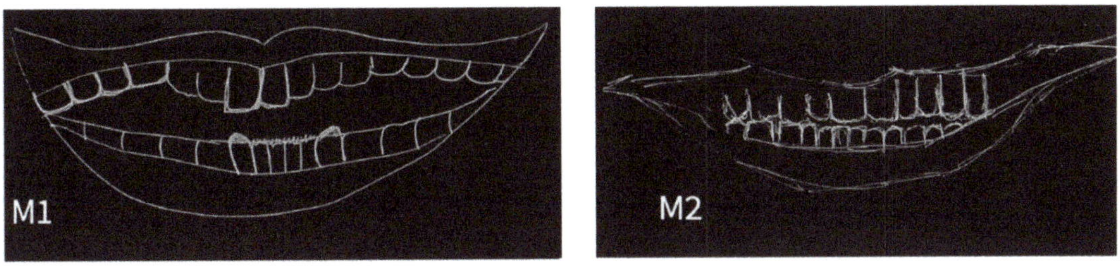

Figure 1. Drawings from a patient representing a *broad smile* from an *extra-oral view*.

5. Conclusions

To conclude, adolescents need to receive better education on oral hygiene care in order to enable self-prevention and self-promotion and hence facilitate this. Information should be given regularly as it showed efficacy in the short term and adolescents should be continually reminded of it. To this end, the shortage of knowledge proves the need to invest in oral health education.

Author Contributions: Conceptualization, M.d.R.D. and A.K.; methodology, M.d.R.D., A.K. and J.G.E.; software, A.K.; formal analysis, A.K.; investigation, A.K. and V.P.A.; data curation, A.K. writing, A.K. supervision, M.d.R.D., J.G.E., V.P.A. and A.S.D. All authors have read and agreed to the published version of the manuscript.

Funding: This research received no external funding.

Institutional Review Board Statement: The study was conducted in accordance with the Declaration of Helsinki and approved by the Ethics Committee of Egas Moniz, School of Health and Science (protocol code PT-173/22; date: 26 January 2023), for studies involving humans.

Informed Consent Statement: Informed consent was obtained from all subjects involved in the study.

Data Availability Statement: Data are available upon reasonable request from the author.

Conflicts of Interest: The authors declare no conflict of interest.

References

1. Awartani, F.; Atassi, F. Oral Hygiene Status among Orthodontic Patients. *J. Contemp. Dent. Pract.* **2010**, *11*, 25–32. [CrossRef]
2. Inquimbert, C.; Clement, C.; Couatarmanach, A.; Tramini, P.; Bourgeois, D.; Carrouel, F. Oral Hygiene Practices and Knowledge among Adolescents Aged between 15 and 17 Years Old during Fixed Orthodontic Treatment: Multicentre Study Conducted in France. *Int. J. Environ. Res. Public Health* **2022**, *19*, 2316. [CrossRef] [PubMed]
3. Jorge, S.M. Impacto do Reforço da Motivação num Programa de Promoção de Saúde Oral. Master's Thesis, Universidade do Porto, Porto, Portugal, 2013.
4. Martignon, S.; Ekstrand, K.R.; Lemos, M.I.; Lozano, M.P.; Higuera, C. Plaque, Caries Level and Oral Hygiene Habits in Young Patients Receiving Orthodontic Treatment. *Community Dent. Health* **2010**, *27*, 133–138. [CrossRef] [PubMed]

5. Sawyer, S.M.; Azzopardi, P.S.; Wickremarathne, D.; Patton, G.C. The Age of Adolescence. *Lancet Child Adolesc. Health* **2018**, *2*, 223–238. [CrossRef] [PubMed]
6. Karramkan, A.; Do Rosário Dias, M.; Grillo Evangelista, J.; Dias Dos Santos, H.; Pedroso Alves, V.; Sintra Delgado, A. X-ray Vision: Representação Mental da Cavidade Oral dos Adolescentes em Tratamento Ortodôntico. In Proceedings of the III JEM Jornadas Cientificas Egas Moniz, Caparica, Portugal, 30 January–3 February 2023.

medical sciences forum

MDPI

Proceeding Paper

Oral Health in Nursing Home Residents—Preliminary Results of an Exploratory Cross-Sectional Pilot Study †

Joana Pombo Lopes [1,2,*], Diogo Sousa-Catita [1,3,4], Paulo Mascarenhas [1,3], Jorge Fonseca [1,3] and José Grillo Evangelista [1,2]

1 Egas Moniz Center for Interdisciplinary Research (CiiEM), Egas Moniz School of Health & Science, Campus Universitário, Quinta da Granja, 2829-511 Almada, Portugal; diogo.rsc2@gmail.com (D.S.-C.); pmascarenhas@egasmoniz.edu.pt (P.M.); jorgedafonseca@gmail.com (J.F.); josegrillo87@gmail.com (J.G.E.)
2 Morfology Department, Egas Moniz School of Health & Science, Campus Universitário, Quinta da Granja, 2829-511 Almada, Portugal
3 PaMNEC-Grupo de Patologia Médica, Nutrição e Estudos Clínicos, Egas Moniz School of Health & Science, Campus Universitário, Quinta da Granja, 2829-511 Almada, Portugal
4 Residências Montepio–Serviços de Saúde, SA, Sede–Rua Julieta Ferrão nº 10–5º, 1600-131 Lisboa, Portugal
* Correspondence: joanampmlopes@gmail.com
† Presented at the 6th International Congress of CiiEM—Immediate and Future Challenges to Foster One Health, Almada, Portugal, 5–7 July 2023.

Abstract: This study was conducted to assess the oral status of nursing home residents diagnosed with dementia in the Lisbon region, Portugal. In this cross-sectional observational pilot study, the oral and dental status were evaluated by determining the Decayed-Missing-Filled-Tooth (DMFT) index, frequency of oral hygiene, presence of erosion lesions or mucosal lesions, presence of dentures, and Shorted Xerostomia Inventory (SXI-5). This study aims to be a starting point for a broad analysis of this population and a further determination of potential associations between oral status and anamnestic factors as well as dementia data.

Keywords: oral health; nursing home; dementia

Citation: Lopes, J.P.; Sousa-Catita, D.; Mascarenhas, P.; Fonseca, J.; Evangelista, J.G. Oral Health in Nursing Home Residents—Preliminary Results of an Exploratory Cross-Sectional Pilot Study. *Med. Sci. Forum* **2023**, *22*, 8. https://doi.org/10.3390/msf2023022008

Academic Editors: José Brito, Nuno Taveira and Ana I. Fernandes

Published: 9 August 2023

1. Introduction

According to the World Health Organization (WHO), dementia is a group of disorders that affect memory, thinking, and the ability to carry out daily activities [1]. It is the leading cause of disability, dependence, and mortality, particularly among the elderly. Few studies have attempted to establish a link between oral health and brain health [2–4]; however, evidence suggests that people with dementia are more likely to have poor oral health, such as increased dental plaque, poorer periodontal condition, more coronal and root caries, higher Decayed-Missing-Filled-Tooth (DMFT) scores, a smaller number of remaining teeth, and fewer dentures to replace lost teeth [2,5–7]. In addition, some studies have linked oral disease to the subsequent risk of developing cognitive impairment and dementia [8]. The prevalence of oral health problems increases in cognitively impaired people, particularly those with dementia, due to their declining cognitive abilities, memory loss, learning disabilities, attention deficits, and deteriorating motor skills [3]. This vulnerability is even greater in care home residents, where oral health is often overlooked. In this sense, this study aimed to evaluate the oral status of nursing home residents diagnosed with dementia in the Lisbon region, to further determine possible associations between oral health status and anamnestic factors as well as dementia data.

2. Materials and Methods

This was an exploratory cross-sectional pilot study among elderly people living in nursing homes in the Lisbon region, Portugal, conducted between March and May 2023.

The study was reviewed and approved by the Ethics Committee of the Egas Moniz School of Health and Science, Portugal (1182, 26 January 2023), and Montepio Residences. Residents or their legal guardian were informed about the study verbally and in writing and gave written informed consent. A total of nine elderly people living in three different nursing homes in Lisbon (Residência Montepio da Parede, Parque das Nações, and Montepio) were included in this study. The inclusion criteria included newly admitted residents in Montepio Residences, written informed consent, and a diagnosis of dementia by the assistant clinician. Participants who did not meet these criteria were excluded from the study. Oral examinations were performed by a single dentist (J.P.L.) using a headlamp, mouth mirror, and exploratory and periodontal probes, mostly in the resident's room, but occasionally in other designated private rooms for observation. Residents were examined in a sitting or reclining position. The examination was recorded, transcribed into the patient's medical record, and entered into the study database. The examinations included oral health questionnaires and a full-mouth examination. The DMFT index was used to assess caries experience [9]. DMFT is a numerical expression of caries prevalence obtained by summing the number of decayed (D), missing (M), and filled (F) teeth (T). The sum of these numbers gives the DMFT index value, which can range from 0 to 32, including wisdom teeth. In addition, oral hygiene frequency and the assessment of the presence of erosive lesions, mucosal lesions, and dentures were completed. The Short Xerostomia Inventory (SXI) is a sum rating scale comprising five domains, and it assigns a continuous scale score between 5 and 15, indicating the severity of chronic xerostomia [10]. The data collected were used to determine the urgency of intervention required. Data collection followed the recommendations of the World Health Organization (WHO) in Oral Health Surveys–Basic Methods, 3rd edition [9]. Cognitive impairment was assessed using the Mini Mental Status Examination (MMSE). This scale allows patients to be scored on a scale from 0 to 30 points. A person is considered to have cognitive impairment if they score 14 or less and are illiterate. If the person has 1 to 11 years of schooling, a score of 22 or less indicates cognitive impairment. However, if the person has more than 11 years of schooling, a score of 27 or less indicates cognitive impairment [11].

3. Results

A total of nine participants (77.8% female) were evaluated. The mean MMSE score was 13.1. The mean age of the participants was 83.4 years, with a range of 74–92 years. All had completed primary education, only. The mean number of comorbidities was 4.4, and the participants had a mean of 6.1 for permanent medications. The data on population characteristics are shown in Table 1.

Table 1. Characteristics of the total study population.

MMSE, mean (range)	13.1 (2–22)
Gender, n (%)	
Female	7 (77.8%)
Male	2 (22.2%)
Age in years, mean (range)	83.4 (74–92)
Education level, (n)	Basic education (9)
Number of medical conditions, mean (range)	4.4 (3–6)
Number of medications, mean (range)	6.1 (1–10)

n = number of cases.

Information regarding the oral health evaluation is presented in Table 2. Two patients were edentulous (22.2%) and, therefore, did not participate in the dental examinations. The mean number of remaining teeth was 15.3, ranging from 4 to 25 teeth still present. The mean DMFT index was 22.1, with a total of 32 decayed teeth, 111 missing teeth, and 12 filled teeth accounted for. Regarding oral hygiene, most patients reported brushing their teeth at least once a day (44.4%). Mucosal lesions were found in five patients and consisted of ulceration (n = 1), prosthetic stomatitis (n = 1), hemangioma (n = 1), fibroma (n = 2), and a white lesion on the dorsum of the tongue referred for biopsy and differential diagnosis. Removable dentures were used by only two participants (22.2%), and both were acrylic removable dentures. SXI scale score was, on average, 9 (5–13). Regarding the urgency of intervention after the oral health study, all of the individuals are in need of treatment, and one of the patients was referred for further evaluation.

Table 2. Oral health status observed in dementia patients.

Prevalence of edentulism (%)	22.2
Number of remaining teeth among dentate patients	
1–9 (%)	22.2
10–19 (%)	22.2
20 or more (%)	33.3
DMFT index, mean (range)	22.1 (16–32)
No. Decayed Teeth (DT), n	32
No. Missing Teeth (MT), n	111
No. Filled Teeth (FT), n	12
Oral hygiene frequency, n (%)	
Never	0 (0)
Once a day	4 (44.4)
More than once a day	3 (33.3)
Presence of erosion lesions, n (%)	
No signs of erosion	2 (22.2)
Enamel lesion	2 (22.2)
Dentinal lesion	2 (22.2)
Pulp involvement	0 (0)
Presence of oral mucosal alterations, n (%)	5 (55.6)
Wearing a removable dental prosthesis (%)	
Yes	22.2
No	77.8
SXI-5, mean (range)	9 (5–13)
Intervention urgency	
No treatment needed	0 (0)
Preventive or routine treatment needed	3 (33.3)
Prompt treatment including scaling needed	3 (33.3)
Immediate (urgent) treatment needed	2 (22.2)
Referred for comprehensive evaluation	1 (11.1)

4. Discussion

The results are consistent with the literature and confirm poor oral health in people with dementia; a high prevalence of edentulism, no dentures to replace lost teeth, more coronal and root caries, higher Decayed-Missing-Filled-Tooth (DMFT) scores, and a need for intervention in all individuals were observed. The frequency of oral hygiene assessed was not consistent with the oral hygiene observed, which was generally very poor. A complete periodontal assessment is of paramount importance, as well as the determination of plaque and gingival index to provide a quantitative and comparable analysis of oral hygiene and gingival inflammation. Further studies should focus on the relationship between the patients' oral health, the type of dementia, and the severity of cognitive impairment. Moreover, exploring the carers' oral health knowledge, practices, and barriers they face when performing oral hygiene will provide a better understanding of how to improve the oral health of people with dementia.

5. Conclusions

These preliminary results are in line with the literature and can serve as a basis for further research into the oral health of people with dementia. The findings of this study should encourage patients, carers, health professionals, and dentists to ensure regular preventive dental examinations and treatments for individuals with dementia.

Author Contributions: Conceptualization, J.P.L., J.F. and J.G.E.; methodology, J.P.L. and J.G.E.; formal analysis, J.P.L.; investigation, J.P.L. and D.S.-C.; data curation, J.P.L. and D.S.-C.; writing—original draft preparation, J.P.L.; writing—review and editing, P.M., J.F. and J.G.E.; visualization, P.M., J.F., D.S.-C. and J.G.E.; supervision, J.F. and J.G.E. All authors have read and agreed to the published version of the manuscript.

Funding: This research received no external funding.

Institutional Review Board Statement: The study was conducted in accordance with the Declaration of Helsinki and approved by the Ethics Committee of Egas Moniz School of Health and Science (protocol code 1182, dated 26 January 2023).

Informed Consent Statement: Informed consent was obtained from all subjects or their guardian involved in the study.

Data Availability Statement: All data will be provided to any interested part upon its requesting.

Conflicts of Interest: The authors declare no conflict of interest.

References

1. WHO. Dementia: Fact Sheet. Available online: https://www.who.int/news-room/fact-sheets/detail/dementia (accessed on 18 May 2021).
2. Tran, T.D.; Krausch-Hofmann, S.; Duyck, J.; de Almeida Mello, J.; De Lepeleire, J.; Declerck, D.; Declercq, A.; Lesaffre, E. Association between Oral Health and General Health Indicators in Older Adults. *Sci. Rep.* **2018**, *8*, 8871. [CrossRef] [PubMed]
3. Foley, N.C.; Affoo, R.H.; Siqueira, W.L.; Martin, R.E. A Systematic Review Examining the Oral Health Status of Persons with Dementia. *JDR Clin. Transl. Res.* **2017**, *2*, 330–342. [CrossRef] [PubMed]
4. Zeng, L.; Zong, Q.; Xu, S.; An, F.; Ungvari, G.S.; Bressington, D.T.; Cheung, T.; Qin, M.; Chen, L.; Xiang, Y. Oral Health in Patients with Dementia: A Meta-analysis of Comparative and Observational Studies. *Int. J. Geriatr. Psychiatry* **2021**, *36*, 467–478. [CrossRef] [PubMed]
5. Delwel, S.; Scherder, E.J.A.; Baat, C.; Binnekade, T.T.; Wouden, J.C.; Hertogh, C.M.P.M.; Maier, A.B.; Perez, R.S.G.M.; Lobbezoo, F. Orofacial Pain and Its Potential Oral Causes in Older People with Mild Cognitive Impairment or Dementia. *J. Oral Rehabil.* **2019**, *46*, 23–32. [CrossRef] [PubMed]
6. Hamza, S.; Asif, S.; Bokhari, S.H. Oral Health of Individuals with Dementia and Alzheimer's Disease: A Review. *J. Indian Soc. Periodontol.* **2021**, *25*, 96. [CrossRef] [PubMed]
7. Lauritano, D.; Moreo, G.; Della Vella, F.; Di Stasio, D.; Carinci, F.; Lucchese, A.; Petruzzi, M. Oral Health Status and Need for Oral Care in an Aging Population: A Systematic Review. *Int. J. Environ. Res. Public Health* **2019**, *16*, 4558. [CrossRef] [PubMed]
8. Orr, M.E.; Reveles, K.R.; Yeh, C.; Young, E.H.; Han, X. Can Oral Health and Oral-derived Biospecimens Predict Progression of Dementia? *Oral Dis.* **2020**, *26*, 249–258. [CrossRef] [PubMed]
9. Petersen, P.E.; Baez, R.J.; World Health Organization. *Oral Health Surveys: Basic Methods*, 5th ed.; World Health Organization: Geneva, Switzerland, 2013; ISBN 978-92-4-154864-9.
10. Thomson, W.M.; Chalmers, J.M.; Spencer, A.J.; Williams, S.M. The Xerostomia Inventory: A Multi-Item Approach to Measuring Dry Mouth. *Community Dent. Health* **1999**, *16*, 12–17. [PubMed]
11. Guerreiro, M.; Silva, A.P.; Botelho, A.; Leitão, O.; Castro-caldas, A.; Garcia, C. Adaptação à População Portuguesa Da Tradução Do Mini Mental State Examination (MMSE). *Rev. Port. De Neurol.* **1994**, *1*, 9–10.

medical sciences forum

MDPI

Proceeding Paper

Evaluation of Glycemia, Glycosylated Hemoglobin and Body Mass Index in Adolescents †

Sérgio Valério [1,*], Ricardo Carregosa [2], Frederico Marx [2], Inês de Jesus [2], Ana Beatriz Piães [2], Catarina Bernardes [1], Catarina Godinho [1], Jorge Fonseca [1], Maria João Hilário [1] and Sara Alves [1]

1 Egas Moniz Center for Interdisciplinary Research (CiiEM), Egas Moniz School of Health & and Science, 2829-511 Almada, Portugal
2 Egas Moniz School of Health & and Science, 2829-511 Almada, Portugal
* Correspondence: sfcvalerio@gmail.com
† Presented at the 6th International Congress of CiiEM—Immediate and Future Challenges to Foster One Health, Almada, Portugal, 5–7 July 2023.

Abstract: Early screening of diabetes is crucial. This study's objectives are to evaluate the levels of glycemia, HbA1c and body mass index (BMI) in students of secondary schools, to describe and frame the levels evaluated and detect potential cases of pre-diabetes or unknown diabetes. A Level I descriptive study with a sample of 234 students allows us to discover that 89% had unaltered blood levels of glycemia. Furthermore, 5% had altered glycemia levels and 69% of female and 73% of male students had a normal BMI value. Of the students, 21% and 16% of females and males, respectively, were overweight, and 9% of males were "Obese".

Keywords: blood glucose; glycosylated hemoglobin; teenagers; BMI; diabetes

1. Introduction

The assessment of blood glucose, glycosylated hemoglobin and BMI are increasingly important to prevent diseases [1–3]. Measuring these at schools plays a key role in education for healthy lifestyle habits. Some measurements have already been taken, namely the removal of some carbohydrate-rich foods at bars and training at curricular units in the health area [4]. Diabetes is characterized by an increase in the level of glucose in the blood, i.e., hyperglycemia, and is associated with several complications, including retinopathy, nephropathy, neuropathy and an increased risk of cardiovascular disease. Type 2 diabetes in children [5] and young people is clearly related to the increase in childhood obesity, as a result of the currently lived sedentary lifestyle and a predominantly hypercaloric diet [6]. In this research paper, we evaluated the levels of glycemia, glycosylated hemoglobin, body mass index (BMI) and measurement of weight-for-height/length [2] in students of secondary schools. As a secondary outcome, we described and framed the levels assessed, detected potential cases of prediabetes or detected unknown positive cases or severe acute malnutrition cases [6], clarifying and referring to medical consultation to confirm the pathology. An informed consent form was drawn up and delivered to secondary schools, so that parents would be aware of the contents of screening and they could consent or refuse the participation of their child.

Citation: Valério, S.; Carregosa, R.; Marx, F.; de Jesus, I.; Piães, A.B.; Bernardes, C.; Godinho, C.; Fonseca, J.; Hilário, M.J.; Alves, S. Evaluation of Glycemia, Glycosylated Hemoglobin and Body Mass Index in Adolescents. *Med. Sci. Forum* **2023**, *22*, 31. https://doi.org/10.3390/msf2023022031

Academic Editors: José Brito, Nuno Taveira and Ana I. Fernandes

Published: 15 August 2023

2. Materials and Methods

The sample consisted of 234 students with ages between 14 and 19 years from two secondary schools (Moita e Ramada) that were selected using the convenience method. Of the students, 145 were from the E.S. (secondary school) of Moita (62%) and 89 are students from E.S. da Ramada (38%). A mixed methodology (quantitative and qualitative) and a Level I descriptive study were employed.

Informed consents were given to those responsible for the education of the students. Only after we had obtained their consent we proceeded to the collection and assessment of the aforementioned parameters.

Cases with values of glycemia under 70 mg/dL resulting from prolonged periods of fasting were excluded.

In order to measure glycemia, we used glucometers and the respective glucose strips of this device. In the case of BMI measurement, a tape measure placed on the wall was used to determine height and a calibrated scale to determine weight. In addition to this, we used another tape to measure the abdominal circumference of young people. For capillary blood sampling, we used cotton wool and alcohol for disinfection and lancets to prick one of the fingers, in order to obtain a drop of capillary blood to analyze. After pricking we put a band-aid on it to stop the bleeding.

For the measurement of glycated hemoglobin, we use the HbA1c measurement device provided by Siemens Healthcare Diagnostics Portugal, as well as HbA1c, which consists of a cartridge with a reagent and capillary titrator. All instruments were calibrated and controlled before use.

3. Results and Discussion

The average height of the students in the sample was 1.68 m and the average weight of the pupils in the sample was 64 kg. The density of pupils decreased as higher weight values were considered.

The BMI values of the pupils in the sample ranged from 15.1 to 34.9 kg/m^2 (when rounded to the nearest unit), having an average value of 22.5 kg/m^2. For females, the mean value was 22.8 kg/m^2. For males, the average was 22.2 kg/m^2.

In this investigation, we encountered a great majority of students (89%) with unaltered blood levels of glycemia and 5% (11 cases) with altered glycemia levels.

From the 11 elevated cases of glycemia, 4 were tested with high values of HbA1c ($\geq$6). Of those four students, three were "Overweight" based on their BMI. These results, although not expressive, may be indicative of a positive relationship between overweight and glycemic overweight, and the level of glycemia of an individual. In fact, being overweight may be one of the causes of possible changes in blood glucose levels, which may eventually lead to diabetes.

Both female and male students presented a very similar distribution. The majority of female and male students (69% female students and 73% male students) had a normal BMI value (values from 18.5 to <25 that fall within the healthy weight range). There was also a considerable proportion of overweight students (21% females and 16% males had BMI values within 25.0 to <30), and "Obese" (9% males had BMI values of 30 or higher). At the opposite end, no pupils suffered from "Severe Acute Malnutrition" (weight-for-height/length < -3SD or mid-upper arm circumference < 115 mm), and only a very residual percentage of students suffered from "Moderate Acute Malnutrition" (1% female and 2% male students). These results, although not very significant in number, may be indicative of a positive relationship between overweight and positive relationship between overweight and an individual's glycemic level. In fact, overweight weight may be one of the causes of possible alterations in glycemia levels, with the possibility of eventually developing diabetes. All red-flag cases were reported to those responsible for students' education in a closed envelope.

In future, it is important to carry out similar actions in a timely manner, to raise awareness among teachers and students of the importance of this screening for their health and promote greater adherence, motivating parents and guardians to encourage their children to participate.

With this screening work in schools, we have also been able to demonstrate how important it is for higher-education establishments, due to their capabilities and involving students and teachers, to provide a service of social responsibility to the community.

The two screening actions in Moita and Ramada secondary schools explained that there should be a joint effort in the educational establishments with the respective curricular units, students and teachers, in order to create greater proximity with the community, promoting an active attitude of solidarity, and thus contributing to their contribution to the promotion of citizenship.

In this study, there are some questions that should be deepened, and which, due to the results obtained, become urgent. It is necessary to have a screening with a larger sample, taking into consideration the lifestyles in urban and rural areas, cross-reference results obtained using different geographical regions and with different geographic regions with different lifestyles, eating habits and social conditions. Thus more studies should be conducted to assess the gaps in health literacy among the younger age groups, in order to promote more training and awareness on diabetes.

The percentage of young people with pre-diabetes in this study is worrying and there is a need for in-depth research using a larger, if not massive, sample of Portuguese schools, as there is a challenge ahead which must be considered with seriousness. To this end, faculties must play a fundamental role in researching and deepening the various issues mentioned in this document. Additionally, creating multidisciplinary teams to promote health literacy and prevention in the fight against diabetes is a social responsibility in the context of academic classes.

Thus, we believe that our work leaves an open door for further investigations on diabetes in adolescents, which, although may represent an enormous effort within institutions, should be seen as a necessary and fundamental challenge for the preservation of a healthy life among the community.

Author Contributions: Conceptualization, R.C., F.M., I.d.J. and A.B.P.; methodology, S.V.; software, S.V.; validation, C.B., C.G. and J.F.; formal analysis, M.J.H.; investigation, R.C., F.M. and A.B.P.; resources, R.C.; data curation, F.M. and I.d.J.; writing—S.V.; writing edition—S.V. and M.J.H.; visualization, M.J.H.; supervision, S.V. and S.A.; project administration, S.V. and S.A. All authors have read and agreed to the published version of the manuscript.

Funding: This research received no external funding.

Institutional Review Board Statement: The study was conducted according to the guidelines of the Declaration of Helsinki, and approved by the Institutional Review Board of Instituto Universitario Egas Moniz (date of approval: 8 July 2021).

Informed Consent Statement: Not applicable.

Data Availability Statement: Not applicable.

Conflicts of Interest: The authors declare no conflict of interest.

References

1. Simões, H.S.; Duarte, F.S. Diabetes Tipo 2 na Infância e Adolescência-Novos Doentes, Novos Desafios. *Rev. Port. Diabetes* **2015**, *10*, 23–25.
2. Serra, P.; Botet, J.; Roux, J.A. Clinica e Investigacion en Arteriosclerosis. In *Más Allá del Control Glucémico en el Tratamiento de la Diabetes Mellitus Tipo 2*, 2nd ed.; Elsevier: Barcelona, Spain, 2015; pp. 205–211.
3. Neu, A.; Bürger-Büsing, J.; Danne, T.; Dost, A.; Holder, M.; Holl, R.W.; Holterhus, P.M.; Kapellen, T.; Karges, B.; Kordonouri, O.; et al. Diagnosis, Therapy and Follow-Up of Diabetes Mellitus in Children and Adolescents. *Exp. Clin. Endocrinol. Diabetes* **2019**, *127*, S39–S72. [CrossRef] [PubMed]
4. Ziegler, R.; Neu, A. Diabetes in Childhood and Adolescence. *Dtsch. Arztebl. Int.* **2018**, *115*, 146–156. [CrossRef] [PubMed]
5. Bertakis, K.D.; Azari, R.; Helms, L.J.; Callahan, E.J.; Robbins, J.A. Gender differences in the utilization of health care services. *J. Fam. Pract.* **2000**, *49*, 239–244.
6. Koren, D.; Levitsky, L.L. Type 2 Diabetes Mellitus in Childhood and Adolescence. *Pediatr. Rev.* **2021**, *42*, 167–179. [CrossRef] [PubMed]

medical sciences forum

MDPI

Proceeding Paper

Oral Health in Psychotropic-Medicated Outpatients of the Lisbon Psychiatric Hospital Centre (CHPL) †

Cecília Rozan *, Luís Proença and Cristina Manso

Egas Moniz Center for Interdisciplinary Research (CiiEM), Egas Moniz School of Health & Science, 2829-511 Almada, Portugal; lproenca@egasmoniz.edu.pt (L.P.); cmanso@egasmoniz.edu.pt (C.M.)

* Correspondence: ceciliarozan@gmail.com; Tel.: +351-962300218

† Presented at the 6th International Congress of CiiEM—Immediate and Future Challenges to Foster One Health, Almada, Portugal, 5–7 July 2023.

Abstract: We conducted a cross-sectional study to assess the oral health status in psychotropic-medicated outpatients of the Lisbon Psychiatric Hospital Centre (CHPL). Sixty outpatients of the General Psychiatry consultation consented to participate in this study. An oral examination and a structured questionnaire were conducted. Among the 60 subjects, 61.70% were males and the mean age was 51.53 ± 11.15 years. The prevalence of dental caries was 98.3%, and the decayed, missing, filled index (DMFT) mean score was 18.03 ± 9.39. Considering the tooth brushing frequency, 30% and 40% brushed their teeth once and twice daily, respectively.

Keywords: psychotic disorders; oral health; dental caries

Citation: Rozan, C.; Proença, L.; Manso, C. Oral Health in Psychotropic-Medicated Outpatients of the Lisbon Psychiatric Hospital Centre (CHPL). *Med. Sci. Forum* **2023**, 22, 38. https://doi.org/10.3390/msf2023022038

Academic Editors: José Brito, Nuno Taveira and Ana I. Fernandes

Published: 17 August 2023

1. Introduction

In Portugal, 22.9% of the total population have experienced a mental health disorder at some stage in their life [1]. There is evidence that people with a serious mental illness experience worse oral health outcomes than the general population [2]. Oral health is an important aspect of general health and well-being because of its effects on pain, eating, speech and quality of life [3]. People who have experienced a mental health disorder are a heterogeneous group, who commonly exhibit many factors which may contribute to poor oral health, including xerostomia caused by psychiatric medication, lack of motivation for self-care, poor oral hygiene and difficulties to access dental care [4]. The aim of this study is to assess the prevalence of dental caries and self-care behaviour in psychotropic-medicated outpatients of the Lisbon Psychiatric Hospital Centre (CHPL).

2. Materials and Methods

A cross-sectional study was conducted over a period of 4 months (March–June 2022) at the outpatients department of the General Psychiatry and Non-Medical Nursing consultation (injectable) at the CHPL. The inclusion criteria for the patients were as follows: (1) had a psychiatric diagnosis according the International Classification of Diseases-10 [5]; (2) have had the psychiatric condition for at least 1 year; (3) have taken antipsychotic medication for at least 1 year; (4) being treated as an outpatient; and (5) over 18 years old. A sample of 60 outpatients was selected randomly from the psychiatry department of CHPL of both genders, aged between 27 and 72 years who consented to participate in this study. Approved by the Ethics Committee of the Lisbon Psychiatric Hospital Centre and the Instituto Universitário Egas Moniz. The prevalence of dental caries was assessed using the decayed, missing, and filled teeth index (DMFT). Subsequently, a questionnaire was applied regarding sociodemographic variables, psychiatric pathology, and oral hygiene habits. Data were submitted to descriptive analysis using IBM SPSS Statistics®v.28 software.

3. Results

Amongst 60 outpatients, 61.70% (37) were males and the mean age of the study population was 51.53 ± 11.15. The majority of the population, 53.34% (32), had primary or less education level; 46.7% (28) had less than 1 monthly family income of minimum wages, and 46.7% (28) were retired (Table 1). The frequency distribution of psychiatric patients was 66.70% (40) schizophrenia, schizotypal, and delusional disorders, 31.70% (19) epilepsy, and 1.7% (1) neurotic, stress-related, and somatoform disorder (Table 2). The prevalence of dental caries was 98.3% and the DMFT score was 18.03 ± 9.39 among our population. Overall DMFT scores ranged from 0 to 32, whereas the mean of decayed (D) score was 3.63 ± 3.9, the mean of missing (M) score was 12.81 ± 11.03, and the mean of filled(F) score was 1.56 ± 2.95 (Table 3). According the tooth brushing frequency, 40.0% of the population brushed the teeth twice daily (Table 4). The results demonstrate a statistically significant correlation between the prevalence of dental caries and subjects who never brush their teeth ($p = 0.018$) (Table 5). A total of 63.30% of the population had their last dental visit over a year ago (Table 6).

Table 1. Distribution of participants according to sociodemographic characteristics.

Variables		Absolute Frequency	Relative Frequency
Gender	Male	37	61.70%
	Female	23	38.30%
Educational Level	Primary or less	32	53.34%
	Secondary or higher	15	25.00%
	Graduate student or higher	13	21.70%
Employment status	Employed	15	25.00%
	Unemployed	17	28.40%
	Retired	28	46.70%
Income	Not known/No response	4	6.70%
	Less than 1 monthly family income wage	28	46.70%
	1–2 monthly income	23	38.30%
	2–4 monthly income	5	8.30%

Table 2. Distribution of participants according to their psychiatric diagnosis.

Psychiatric Diagnosis	Absolute Frequency	Relative Frequency
Schizophrenia, schizotypal, and delusional disorders	40	66.70%
Epilepsy	19	31.70%
Neurotic, stress-related, and somatoform disorder	1	1.7%

Table 3. Distribution of mean scores of DMFT.

	Mean	SD	Minimum	Maximum
DMFT	18.03	9.39	3	32
D	3.63	3.9	0	18
M	12.81	11.03	0	32
F	1.56	2.95	0	16

Table 4. Distribution of tooth brushing frequency.

Tooth Brushing Frequency	Absolute Frequency	Relative Frequency
Not known/No response	5	8.30%
Never	8	13.30%
Sometimes	5	8.30%
Once daily	18	30.00%
Twice daily	24	40.00%

Table 5. Distribution of DMFT mean scores among tooth brushing frequency.

Tooth Brushing Frequency	DMFT	p-Value [1]
Not known/No response	32.0 ± 0.0	-
Never	23.75 ± 10.64	0.018
Sometimes	16.2 ± 6.61	0.381
Once daily	16.0 ± 8.58	0.356
Twice daily	15.13 ± 7.93	0.084

[1] A significance level of 0.05 was considered statistically significant.

Table 6. Frequency of last dental visit.

Last Dental Visit	Absolute Frequency	Relative Frequency
Not known/No response	1	1.70%
Over a year	38	63.30%
Unless a year	21	35.00%

4. Discussion and Conclusions

The DMFT index is classified in five levels: 0.0–4.9 very low, 5.0–8.9 low, 9.0–13.9 moderate, 14.0–17.9 high, and more than 18 very high [6]. The most recent study of the prevalence of caries in the general population is from the World Health Organization from the year 2000. The prevalence of caries in permanent teeth is between 25% and 50% [7], lower than for the data obtained in our study. Goud et al. (2021) determined that dental caries was prevalent in 87.3% of psychiatric outpatients in India, the mean of DMFT scores was 4.06, and the mean of decayed (D) score was 3.4 [8]. Deepali et al. (2021) conducted a study on oral health in patients of schizophrenia in Haryana, and 72% and 3% of the subjects brushed their teeth once and twice daily, respectively [9]. An epidemiological study conducted by Nielsen J et al. (2011) indicated that 43% of participants had visited the dentist within one year. Despite the prevalence of suboptimal oral health among mental health disorder patients compared with the general population, dental services are often underutilized by this heterogeneous group due to the stigma, helplessness, low self-esteem, and low income [10,11]. The poor oral health of people with psychological disorders remains a largely forgotten problem. The findings of this study suggest that the oral health concerns of this heterogeneous group are unrecognized, especially with regard to dental caries, and they lack the responsibility to take care of their oral health. The increased focus on the physical health of people with psychiatric illness should include consideration of oral health, and a closer collaboration between dental and mental health professionals is also important.

Author Contributions: Conceptualization, C.R. and C.M.; methodology, C.R. and C.M.; validation, C.R.; formal analysis, C.R. and C.M.; investigation, C.R.; resources, C.R.; data curation, L.P.; writing—original draft preparation, C.R.; writing—review and editing, C.R. and C.M.; supervision, C.R. and C.M.; project administration, C.R. and C.M. All authors have read and agreed to the published version of the manuscript.

Funding: This work was financed by national funds through the FCT—Foundation for Science and Technology, I.P., under the project UIDB/04585/2020.

Institutional Review Board Statement: The study was conducted in accordance with the Declaration of Helsinki and approved by the Ethics Committee of Egas Moniz School of Health and Science (protocol code 1013, approved in 27 January 2022).

Informed Consent Statement: Informed consent was obtained from all subjects involved in the study.

Data Availability Statement: The data presented in this study are available on request from the corresponding author. The data are not publicly available, as they are part of an ongoing study.

Acknowledgments: The authors thank the outpatients who participated in the study. We are indebted to the occupational therapy and nursing staff at the CHPL.

Conflicts of Interest: The authors declare no conflict of interest.

References

1. Almeida, J.; Xavier, M.; Cardoso, G.; Pereira, A.; Gusmao, R.; Corrêa, B.; Silva, J. *Estudo Epidemiológico Nacional de Saúde Mental. 1° Relatório;* World Mental Health Surveys Initiative: Lisboa, Portugal, 2013.
2. Palmier-Claus, J.E.; Shiers, D.; French, P.; Harris, R.; Laverty, L. Oral health in psychosis: An unmet need. *Schizophr. Res.* **2019**, *204*, 442. [CrossRef] [PubMed]
3. Sheiham, A. Oral health, general health and quality of life. *Bull. World Health Organ.* **2005**, *83*, 644. [PubMed]
4. Wey, M.C.; Loh, S.; Doss, J.G.; Abu Bakar, A.K.; Kisely, S. The oral health of people with chronic schizophrenia: A neglected public health burden. *Aust. N. Z. J. Psychiatry* **2015**, *50*, 685–694. [CrossRef] [PubMed]
5. World Health Organization. *The ICD-10 Classification of Mental and Behavioural Disorders: Clinical Descriptions and Diagnostic Guidelines;* World Health Organization: Switzerland, Geneva, 1992.
6. World Health Organization. *Global Data on Dental Caries Prevalence (DMFT) in Adults Aged 35–44 Years Global Oral Data Bank, Oral Health/Area Profile Programme;* World Health Organization: Geneva, Switzerland, 2000.
7. Kassebaum, N.J.; Smith, A.G.C.; Bernabé, E.; Fleming, T.D.; Reynolds, A.E.; Vos, T.; Murray, C.J.L.; Marcenes, W.; GBD 2015 Oral Health Collaborators. Global, Regional, and National Prevalence, Incidence, and Disability-Adjusted Life Years for Oral Conditions for 195 Countries, 1990–2015: A Systematic Analysis for the Global Burden of Diseases, Injuries, and Risk Factors. *J. Dent. Res.* **2017**, *96*, 380–387. [CrossRef] [PubMed]
8. Goud, V.; Kannaiyan, K.; Rao, B.V.; Abidullah, M.; Dharani, V.; Nayak, M. Oral Health Status and Treatment Needs of Psychiatric Outpatients Aged 18–64 Years in District Civil Hospital, Raichur, Karnataka: A Cross-Sectional Study. *J. Pharm. Bioallied Sci.* **2021**, *13* (Suppl. S1), S598–S601. [CrossRef] [PubMed]
9. Agarwal, D.; Kumar, A.; BC, M.; Kumar, V.; Sethi, S. Oral health perception and plight of patients of schizophrenia. *Int. J. Dent. Hyg.* **2021**, *19*, 121–126. [CrossRef] [PubMed]
10. Nielsen, J.; Munk-Jorgensen, P.; Skadhede, S.; Correll, C.U. Determinants of poor dental care in patients with schizophrenia: A historical, prospective database study. *J. Clin. Psychiatry* **2011**, *72*, 140–143. [CrossRef] [PubMed]
11. Sogi, G.M.; Khan, S.A.; Bathla, M.; Sudan, J. Oral health status, self-perceived dental needs, and barriers to utilization of dental services among people with psychiatric disorders reporting to a tertiary care center in Haryana. *Dent. Res. J.* **2020**, *17*, 360–365. [CrossRef]

medical sciences forum

MDPI

Proceeding Paper

Assessment of Xerostomia in Outpatients of the Lisbon Psychiatric Hospital Centre (CHPL) †

Cecilia Rozan *, **Luís Proença** and **Cristina Manso**

Egas Moniz Center for Interdisciplinary Research (CiiEM), Egas Moniz School of Health & Science, 2829-511 Almada, Portugal; lproenca@egasmoniz.edu.pt (L.P.); cmanso@egasmoniz.edu.pt (C.M.)
* Correspondence: crozan@egasmoniz.edu.pt; Tel.: +351-962300218
† Presented at the 6th International Congress of CiiEM—Immediate and Future Challenges to Foster One Health, Almada, Portugal, 5–7 July 2023.

Abstract: This study aimed to assess the prevalence of xerostomia in psychotropics-medicated outpatients of the Lisbon Psychiatric Hospital Centre (CHPL). For this cross-sectional study, 60 outpatients who underwent a general psychiatry consultation signed a consent form and answered a questionnaire assessing Summated Xerostomia Inventory (SXI-PL), and sialometry was performed. Amongst the 60 subjects, 61.70% were male, and the mean age was 51.53 ± 11.15 years. Among the population, the prevalence of hyposialia was 16.7%, and that of xerostomia was 23.3%. Higher SXI-PL scores (7.60 ± 2.26) were reported in patients with normal saliva flow. The symptom that showed a higher severity of xerostomia was "My mouth feels dry" (1.78 ± 0.81).

Keywords: psychotic disorders; hyposalivation; xerostomia

1. Introduction

Saliva is a biochemically complex fluid consisting of a mixture of water, proteins, glycoproteins and ions that collaborate to carry out several functions for the homeostasis of oral health, including lubrication of oral tissues. It also plays a major role in the process of remineralization and antimicrobial activity and acts as a physical barrier due its numerous immune and nonimmune defence components. Salivary secretion dysfunction is mainly attributed to adverse drug effects, leading the symptom of mouth dryness (xerostomia) because of hyposialia [1]. Hyposialia has been documented upon exposure to antipsychotics and other medications (first-generation antipsychotics, second-generation antipsychotics and anticholinergics) which often disturb saliva secretion [2]. The aim of this study was to assess the prevalence of hyposialia and xerostomia in psychotropics-medicated outpatients of the Lisbon Psychiatric Hospital Centre (CHPL).

2. Materials and Methods

A cross-sectional study was conducted from March to June 2022 for a period of 4 months at the outpatient department of the General Psychiatry and Non-Medical Nursing Consultation Department at the CHPL. The inclusion criteria for the patients were as follows: (1) has a psychiatric diagnosis according to the International Classification of Diseases-10 [3]; (2) has had a psychiatric condition for at least 1 year; (3) has taken antipsychotic medication for at least 1 year; (4) is being treated as an outpatient; and (5) is over 18 years old. A sample of 60 outpatients were selected randomly from the psychiatry department of the CHPL, consisting of individuals of both genders aged between 27 and 72 years who consented to participate in this study. This study was approved by the Ethics Committee of the Lisbon Psychiatric Hospital Centre and the Egas Moniz Ethics Committee with the approval number 1126. Subsequently, a questionnaire was applied regarding sociodemographic variables and SXI-PL. Sialometry was performed for unstimulated (USFR) and stimulated (SSFR) salivary flow rates. Hyposialia was considered when

Citation: Rozan, C.; Proença, L.; Manso, C. Assessment of Xerostomia in Outpatients of the Lisbon Psychiatric Hospital Centre (CHPL). *Med. Sci. Forum* **2023**, *22*, 37. https://doi.org/10.3390/msf2023022037

Academic Editors: José Brito, Nuno Taveira and Ana I. Fernandes

Published: 17 August 2023

USFR < 0.1 mL/min and/or SSFR < 0.7 mL/min [4]. Data were submitted to descriptive analysis using IBM SPSS Statistics® v.28 software.

3. Results

Amongst the 60 outpatients, the majority were male, at 61.70% (37), and the mean age of the study population was 51.53 ± 11.15. The majority of the population, 53 34% (32), had completed primary school or had a lower education level; 46.7% (28) had fewer than 1 monthly family income at minimum wage and 46.7% (28) were retired (Table 1). The frequency distribution of psychiatric patients was as follows: schizophrenia, schizotypal and delusional disorders, 66.70% (40); epilepsy, 31.70% (19); and neurotic, stress-related and somatoform disorder, 1 (1.7%) (Table 2).

Table 1. Distribution of participants according to sociodemographic characteristics.

Variables		Absolute Frequency	Relative Frequency
Gender	Male	37	61.70%
	Female	23	38.30%
Educational Level	Primary or less	32	53.34%
	Secondary or higher	15	25.00%
	Graduate student or higher	13	21.70%
Employment status	Employed	15	25.00%
	Unemployed	17	28.40%
	Retired	28	46.70%
Income	Not known/no response	4	6.70%
	Fewer than 1 monthly family income at minimum wage	28	46.70%
	1–2 monthly incomes	23	38.30%
	2–4 monthly incomes	5	8.30%

Table 2. Distribution of participants according to their psychiatric diagnosis.

Psychiatric Diagnosis	Absolute Frequency	Relative Frequency
Schizophrenia, schizotypal and delusional disorders	40	66.70%
Epilepsy	19	31.70%
Neurotic, stress-related, and somatoform disorder	1	1.7%

The prevalence of hyposialia was 16.7% and xerostomia was 23.3% among our population (Table 3). Overall SXI-PL scores ranged from 5 to 15 with a mean of (7.41 ± 2.28) and the severe symptom that showed higher perception of xerostomia was "My mouth feels dry" (1.78 ± 0.81) (Table 4). The mean SXI-PL scores were higher in patients with normal saliva flow (7.60 ± 2.26). The results demonstrate no statistically significant correlation between SXI-PL scores and hyposalivation ($p > 0.05$) (Table 5).

Table 3. Distribution of the frequency of salivary flow conditions.

	Absolute Frequency (n)	Absolute Frequency (n)
Normal	50	83.30%
Hyposialia	10	16.7%

Table 4. Distribution of the mean scores of Summated Xerostomia Inventory (SXI-PL).

SXI-PL	Mean	SD
My mouth feels dry when eating a meal	1.38	0.66
My mouth feels dry	1.76	0.81
I have difficulty in eating dry foods	1.26	0.54
I have difficulty swallowing certain foods	1.33	0.57
My lips feel dry	1.68	0.81
Total	7.41	2.28

Table 5. Distribution of SXI-PL mean scores among salivary flow conditions.

	Hyposialia	Normal	Total	*p*-Value [1]
SXI-PL	6.5 ± 2.27	7.60 ± 2.26	7.85 ± 2.37	>0.05

[1] A significance level of 0.05 was considered statistically significant.

4. Discussion

The SXI-PL score was higher in patients with normal saliva flow conditions among our population. Patients with mental disorders often have behavioural pattern impairment and an inability to distinguish symptoms of a concurrent physical illness and do not often consider their oral status [5,6]. Thomson et al.'s (2000) and Hopcraft et al.'s (2010) presented evidence for oral reactions being drug-induced is variable; both studies demonstrated that hospitalized patients who endured dry mouth symptoms more likely had been exposed to more than two kinds of drugs (cardiovascular, psychiatric and allergy drugs) than those who did not complain of dry mouth. [7,8]. Putten et al. (2011) concluded that "My mouth feels dry" (1.8 points) was the most severe symptom of xerostomia reported among a Dutch population [9]. The phenomenon of poor oral health among people with psychological disorders remains a largely neglected problem, and concerns have arisen regarding the iatrogenic effects of antipsychotics and xerostomia-inducing drugs on oral health. Greater awareness of these repercussions could help to protect against the poor outcomes seen in chronic psychosis and relieve the burden of the implications for public health [10].

5. Conclusions

The major limitations of this study were the small size of our sample population and the fact that it was conducted for a single institution, which did not ensure causal relationships and did not give a general perspective of the whole population suffering from mental disorders. Future studies should consider a broader sample size among different institutions. Closer collaboration between mental health clinicians and dentists should also be considered to remove the barriers to care and improve the oral health measures for this neglected population.

Author Contributions: Conceptualization, C.R. and C.M.; methodology, C.R. and C.M.; validation, C.R.; formal analysis, C.R. and C.M.; investigation, C.R.; resources, C.R.; data curation, L.P.; writing—original draft preparation, C.R.; writing—review and editing, C.R. and C.M.; supervision, C.R. and C.M.; project administration, C.R. and C.M. All authors have read and agreed to the published version of the manuscript.

Funding: This work was financed by national funds through the FCT—Foundation for Science and Technology, I.P.—under the project no. UIDB/04585/2020.

Institutional Review Board Statement: The study was conducted in accordance with the Declaration of Helsinki and approved by the Ethics Committee of Egas Moniz School of Health and Science (protocol code 1013, approved on 27 January 2022).

Informed Consent Statement: Informed consent was obtained from all subjects involved in the study.

Data Availability Statement: The data presented in this study are available on request from the corresponding author. The data are not publicly available as they are part of an ongoing study.

Acknowledgments: The authors thank the outpatients who participated in the study. We are indebted to the occupational therapy and nursing staff at the CHPL.

Conflicts of Interest: The authors declare no conflict of interest.

References

1. Niklander, S.; Veas, L.; Barrera, C.; Fuentes, F.; Chiappini, G.; Marshall, M. Risk factors, hyposalivation and impact of xerostomia on oral health-related quality of life. *Braz. Oral Res.* **2017**, *31*, e14. [CrossRef] [PubMed]
2. Wey, M.C.; Loh, S.; Doss, J.G.; Abu Bakar, A.K.; Kisely, S. The oral health of people with chronic schizophrenia: A neglected public health burden. *Aust. N. Z. J. Psychiatry* **2015**, *50*, 685–694. [CrossRef] [PubMed]
3. World Health Organization. *The ICD-10 Classification of Mental and Behavioural Disorders: Clinical Descriptions and Diagnostic Guidelines;* World Health Organization: Switzerland, Geneva, 1992.
4. Navazesh, M.; Christensen, C.M.; Brightman, V.J. Clinical criteria for the diagnosis of salivary gland hypofunction. *J. Dent. Res.* **1992**, *71*, 1363–1369. [CrossRef] [PubMed]
5. Slack-Smith, L.; Hearn, L.; Scrine, C.; Durey, A. Barriers and enablers for oral health care for people affected by mental health disorders. *Aust. Dent. J.* **2017**, *62*, 6–13. [CrossRef] [PubMed]
6. Goud, V.; Kannaiyan, K.; Rao, B.V.; Abidullah, M.; Dharani, V.; Nayak, M. Oral Health Status and Treatment Needs of Psychiatric Outpatients Aged 18–64 Years in District Civil Hospital, Raichur, Karnataka: A Cross-Sectional Study. *J. Pharm. Bioallied Sci.* **2021**, *13* (Suppl. S1), S598–S601. [CrossRef] [PubMed]
7. Thomson, W.M.; Chalmers, J.M.; Spencer, A.J.; Slade, G.D. Medicationand dry mouth: Findings from a cohort study of older people. *J. Public Health Dent.* **2000**, *60*, 12–20. [CrossRef] [PubMed]
8. Hopcraft, M.; Tan, C. Xerostomia: An update for clinicians. *Aust. Dent. J.* **2010**, *55*, 238–244. [CrossRef] [PubMed]
9. Putten, G.J.; Brand, H.S.; Schols, J.M.; Baat, C. The diagnostic suitability of a xerostomia questionnaire and the association between xerostomia, hyposalivation and medication use in a group of nursing home residents. *Clin. Oral Investig.* **2011**, *15*, 185–192. [CrossRef] [PubMed]
10. Choi, J.; Price, J.; Ryder, S.; Siskind, D.; Solmi, M.; Kisely, S. Prevalence of dental disorders among people with mental illness: An umbrella review. *Aust. N. Z. J. Psychiatry* **2022**, *56*, 949–963. [CrossRef] [PubMed]

medical sciences forum

MDPI

Proceeding Paper

Risk Factors Associated with Mini-Implant Failure: A Retrospective Study †

Maria de Fátima Martins [1], Pedro Mariano Pereira [1,2], Hélder Nunes Costa [1,2] and Iman Bugaighis [1,3,*]

1 Department of Orthodontics, Egas Moniz School of Health and Science, Caparica, 2829-511 Almada, Portugal; maria.afonso.martins@gmail.com (M.d.F.M.); pmarianop@sapo.pt (P.M.P.); hnc@sapo.pt (H.N.C.)
2 Egas Moniz Center of Interdisciplinary Research (CiiEM), Egas Moniz School of Health and Science, Caparica, 2829-511 Almada, Portugal
3 The Libyan Authority for Scientific Research, Tripoli P.O. Box 80045, Libya
* Correspondence: isbugaighis@yahoo.com
† Presented at the 6th International Congress of CiiEM—Immediate and Future Challenges to Foster One Health, Almada, Portugal, 5–7 July 2023.

Abstract: This study aimed to evaluate the failure rate of mini-implants used at the orthodontic external clinic, Egas Moniz School of Health and Science. A retrospective cross-sectional investigation was performed. This included the insertion of 232 TADs into 125 consecutive patients and was always performed with an immediate loading protocol. The examined variables were: gender, age, Angle's classification, presence of pathologies, medication, smoking, receiving jaw, placement side, and insertion site. Descriptive statistics were used and inferential analysis was performed, revealing the six-month failure rate of the used TADs to be 25%. A significant association was found between being a smoker, the failure rate ($p = 0.036$), and the placement site ($p = 0.003$).

Keywords: miniscrew; orthodontic; mini-implant; skeletal anchorage; temporary anchorage device

Citation: Martins, M.d.F.; Mariano Pereira, P.; Costa, H.N.; Bugaighis, I. Risk Factors Associated with Mini-Implant Failure: A Retrospective Study. *Med. Sci. Forum* **2023**, 22, 30. https://doi.org/10.3390/msf2023022030

Academic Editors: José Brito, Nuno Taveira and Ana I. Fernandes

Published: 14 August 2023

1. Introduction

Anchorage management is an essential component of planning a successful orthodontic treatment. A reliable anchorage system can ensure predictable tooth movement accordingly [1]. Over the past century, numerous intra-oral and extra-oral anchorage techniques have been utilized to secure adequate anchorage. Traditionally, maximum anchorage control was mainly obtained using extraoral devices. This was carried out with or without intraoral auxiliary appliances, such as elastics and transpalatal arches [1–3]. A paradigm shift occurred when new bone-anchoring devices were developed for orthodontic purposes. Mini-implant (MI) evolution in the past two decades has enabled clinicians to perform challenging orthodontic treatment procedures that were previously considered inconceivable when relying on conventional anchorage devices. Generally, MI stability for more than six months is considered a success. However, a wide range of success rates were reported in the literature, ranging from 55.6% to 100% [4]. This wide discrepancy might be due to the multiple contributing factors that may cause MI failure, such as factors related to the implant, related to the patient, related to location, and related to orthodontics and implant maintenance. Identifying these factors is crucial to avoiding MI loss during treatment, which might compromise the planned orthodontic treatment outcome. Numerous studies have examined the factors that cause failures in MIs, as evidenced by previous research [5–8]. A recent systematic review and meta-analysis found that success rates varied across studies due to inconsistent factors. Another review established a correlation between the failure rate of MIs and specific factors, while a separate investigation restricted the assessment of success or failure to the placement area of the MI. This underscores the scarcity of studies that comprehensively analyze all contributing factors collectively, indicating a need for more research. Therefore, the objective of this study was to investigate the failure rate of MIs and all associated factors that contribute to such failures.

2. Materials and Methods

The present study was a retrospective investigation that evaluated the rate of success of MI when integrated with fixed orthodontic treatment between January 2017–December 2019. The investigation was undertaken at the orthodontic clinic, Egas Moniz School of Health and Science University clinic. Ethical approval was obtained from the Ethics Committee of the same school, and the participants granted their informed consent. The sample comprised 232 Mis. These were inserted into 125 consecutive orthodontic patients (mean age 23.4, SD = 11.7 years) by the same operator. A titanium Ti-6AI-4V MI was positioned at all the insertion sites (Osstem Implant Biofisa Ø1.8xH10xG/H1.5): infra-zygomatic, buccal shelf, anterior and posterior inter-radicular space, retromolar area, and the tuberosity zone. Insertion was always performed using an immediate loading protocol. Failure referred to the detection of an expected loss or the presence of mobility requiring the replacement of the MI in the used prior six months. The IBM Package for Social Science (SPSS, V26) was used for the analysis. The overall MI failure rate and the failure rate of each variable were determined. The chi-square test was used to evaluate a possible association between each variable and the MI failure level. The Mantel–Haenzel odd ratio (OR) and 95% confidence interval (CI) were calculated at a significance level of 5%.

3. Results

The six-month failure rate of the used MI was 25%. A significant association existed between being a smoker and a failure rate at $p = 0.036$ (Figure 1). Furthermore, the placement site significantly determined the stability of the MI, where the interradicular region had the highest success rate at 84.4% ($p = 0.003$). In contrast, there was a similar association ($p > 0.05$) between the following variables and the level of MI failure: gender, presence or absence of pathologies, whether the patient was medicated or non-medicated, type of malocclusion, recipient jaw, and direction of movement aided by the MI.

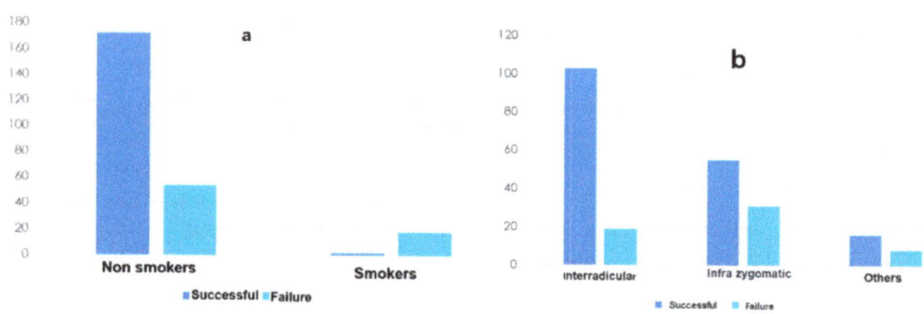

		Failure n (%)		p	OR (95% CI)
		Yes	No		
Smoker	No	54 (23.9)	172 (76.1)	0.035	1
	Yes	4 (66.7)	2 (33.3)		6.37 (1.13-35.74)

		Failure n (%)		p	OR (95% CI)
		Yes	No		
Position	Interradicular	19 (15.6)	103 (84.4)	0.003	1
	Infrazygomatic	31 (36.0)	55 (64.0)		3.06 (1.58-5.90)
	Others	8 (33.3)	16 (66.7)		2.71 (1.02-7.22)

Figure 1. (**a**) The number and the percentage of succeeded and failed mini-implants (MI) associated with smoking habit (**a**) and position of insertion (**b**) with the confidence interval and the p value.

4. Discussion

This was a retrospective observational study that evaluated the failure rate and the factors contributing to the failure of MI after its integration into fixed orthodontic treatment. Similar studies categorized the causes of MI failure to host-related and implant-related factors [4–8]. The MI used in this study had the same characteristics and were placed by the same clinician. Therefore, the following discussion includes only the host-related factors.

This study revealed a significant association between MI failure and smoking. Numerous investigations reported the same findings [9–12]. Several complications that might compromise the MI survival time are associated with smoking. This includes such issues as peri-implantitis and mucositis, leading to MI loosening [9]. Bayat and Bauss [10] reported a deleterious effect for the heavy smoking, which is the consumption of of >10 cigarettes/day, on the success rate of MI. However, they did not notice significant discrepancies in the success rate between light smokers and non-smokers. In our study, the patients were categorized only into two groups according to their self-reporting; smokers and non-smokers, without any discrimination made being between the smoking intensity. It is recommended that researchers undertake prospective studies to refine the methodology and categorization of the investigation.

5. Conclusions

Based on the results obtained in the present study, it is possible to conclude that the failure rate of MI, used as orthodontic anchorage in the orthodontic external clinic at the Egas Moniz School of Health and Science, was 25%. MI that had been inserted into the infra-zygomatic region failed the most (36%), and those inserted into the interdental region failed the least (15.6%). Gender, the existence or non-existence of pathologies, use or lack of use of medication and MI-supported tooth movement were the least influential variables. Conversely, the position of MI placement and smoking were the most influential factors contributing to the high MI failure rate.

Author Contributions: M.d.F.M.—the principal investigator, collecting study data, taking direct measurements, using the software, validating the measurements, collecting references and writing up the proceeding draft; P.M.P.—Conceptualization, Supervision, co-editing and co-writing up, Revising the final draft of the Proceeding; H.N.C.—Conceptualization revision, co-editing and co-writing up, Revising the final draft of the Proceeding; I.B.—second supervisor, Co-editing and co-writing up the proceeding. All authors have read and agreed to the published version of the manuscript.

Funding: This research received no external funding.

Institutional Review Board Statement: The study was conducted in accordance with the Declaration of Helsinki, and approved by the Institutional Review Board of Egas Moniz School of Health and Science, (protocol code 962 in May 2021).

Informed Consent Statement: Informed consent was obtained from all subjects involved in the study.

Data Availability Statement: Data is available upon request.

Conflicts of Interest: The authors declare no conflict of interest.

References

1. Jaradat, M.; Al Omari, S. Orthodontic Mini Implants, an Update. *EC Dent. Sci.* **2021**, *20*, 107–116.
2. Park, H.S.; Jeong, S.H.; Kwon, O.W. factors affecting the clinical success of screw implants used as orthodontic anchorage. *Am. J. Orthod. Dentofac. Orthop.* **2006**, *130*, 18–25. [CrossRef]
3. Alharbi, F.; Almuzian, M.; Bearn, D. Anchorage effectiveness of orthodontic miniscrews compared to headgear and transpalatal arches: A systematic review and meta-analysis. *Acta Odontol. Scand.* **2019**, *77*, 88–98. [CrossRef] [PubMed]
4. Ramirez-Ossa, D.M.; Escobar-Correa, N.; Ramírez-Bustamante, M.A.; Agudelo-Suarez, A.A. An umbrella review of the effectiveness of temporary anchorage devices and the factors that contribute to their success or failure. *J. Evid. Based Dent. Pract.* **2020**, *20*, 101402. [CrossRef] [PubMed]
5. Schatzle, M.; Mannchen, R.; Zwahlen, M.; Lang, N.P. Survival and failure rates of orthodontic temporary anchorage devices: A systematic review. *Clin. Oral Implant. Res.* **2009**, *20*, 1351–1359. [CrossRef] [PubMed]
6. Dalessandri, D.; Salgarello, S.; Dalessandri, M.; Lazzaroni, E.; Piancino, M.; Paganelli, C.; Maiorana, C.; Santoro, F. Determinants for success rates of temporary anchorage devices in orthodontics: A metaanalysis. *Eur. J. Orthod.* **2014**, *36*, 303–313. [CrossRef] [PubMed]
7. Rodriguez, J.C.; Suarez, F.; Chan, H.L.; Padial-Molina, M.; Wang, H.L. Implants for orthodontic anchorage: Success rates and reasons of failures. *Implant Dent.* **2014**, *23*, 155–161. [CrossRef] [PubMed]
8. Papageorgiou, S.N.; Zogakis, I.P.; Papadopoulos, M.A. Failure rates and associated risk factors of orthodontic miniscrew implants: A meta-analysis. *Am. J. Orthod. Dentofac. Orthop.* **2012**, *142*, 577–595.e7. [CrossRef] [PubMed]
9. Turri, A.; Rosseti, P.H.; Canullo, L.; Grusovin, M.G.; Dahlin, C. Prevalence of peri-implantitis in medically compromised patients and smokers: A Systematic Review. *Int. J. Oral Maxillofac. Implant.* **2016**, *31*, 1. [CrossRef] [PubMed]
10. Bayat, E.; Bauss, O. Effect of smoking on the failure rates of orthodontic miniscrews. *J. Orofac. Orthop.* **2010**, *71*, 117–124. [CrossRef] [PubMed]

11. Jaramillo-Bedoya, D.; Villegas-Giraldo, G.; Agudelo-Suárez, A.A.; Ramírez-Ossa, D.M. A scoping review about the characteristics and success-failure rates of temporary anchorage devices in orthodontics. *Dent. J.* **2022**, *10*, 1–17. [CrossRef] [PubMed]
12. Jambi, S.; Walsh, T.; Sandler, J.; Benson, P.E.; Skeggs, R.M.; O'Brien, K.D. Reinforcement of anchorage during orthodontic brace treatment with implants or other surgical methods. *Cochrane Database Syst. Rev.* **2014**, *2014*, CD005098. [CrossRef] [PubMed]

medical sciences forum

MDPI

Proceeding Paper

Emotional Intelligence and Life Satisfaction in a Sample of Adult Victims and Non-Victims of Violence †

Catarina Silva [1], Joana Menino [1], Laura Afonso [1], Maria Madaleno [1], Raquel Costa [1], Jorge Cardoso [1,2,3] and Telma C. Almeida [1,2,3,*]

1 Egas Moniz School of Health & Science, Caparica, 2829-511 Almada, Portugal; catagsilva.egasmoniz@gmail.com (C.S.); joanamenino16@gmail.com (J.M.); laurafilipaoafonso@gmail.com (L.A.); madalenomaria@hotmail.com (M.M.); raquelpbcosta01@hotmail.com (R.C.); jcardoso@egasmoniz.edu.pt (J.C.)
2 Egas Moniz Center for Interdisciplinary Research (CiiEM), Egas Moniz School of Health & Science, Caparica, 2829-511 Almada, Portugal
3 LabPSI–Laboratório de Psicologia Egas Moniz, Caparica, 2829-511 Almada, Portugal
* Correspondence: talmeida@egasmoniz.edu.pt
† Presented at the 6th International Congress of CiiEM—Immediate and Future Challenges to Foster One Health, Almada, Portugal, 5–7 July 2023.

Abstract: Individuals who have higher emotional intelligence show higher levels of life satisfaction. This study analyzed the impact of emotional intelligence on life satisfaction in a sample of 144 individuals aged between 18 and 77 years (M = 36.97, SD = 15.87), comparing victims and non-victims of violence. Participants answered an online protocol comprising a sociodemographic questionnaire, the Wong and Law Emotional Intelligence Scale (WLEIS), and the Satisfaction with Life Scale (SLWS). Age and emotional intelligence impact life satisfaction, and victims showed low emotional intelligence and satisfaction scores. As emotional intelligence is an essential skill in emotional evaluation and regulation, its relationship with life satisfaction should also affect quality of life.

Keywords: emotional intelligence; life satisfaction; victims; non-victims

Citation: Silva, C.; Menino, J.; Afonso, L.; Madaleno, M.; Costa, R.; Cardoso, J.; Almeida, T.C. Emotional Intelligence and Life Satisfaction in a Sample of Adult Victims and Non-Victims of Violence. *Med. Sci. Forum* **2023**, *22*, 47. https://doi.org/10.3390/msf2023022047

Academic Editors: José Brito, Nuno Taveira and Ana I. Fernandes

Published: 31 August 2023

1. Introduction

Emotional intelligence is the ability to understand, control, and regulate emotions in oneself and others [1]. It can be explained by the influence of subjective well-being on the subject's life [2]. On the other hand, life satisfaction is a person's assessment of their overall quality of life [3]. Positive mental health outcomes and longevity are essential for good life satisfaction [4]. Emotional intelligence focuses on emotional problem-solving, is related to life satisfaction, and can be a protective factor for mental health [5]. Therefore, emotional intelligence is responsible for individual differences in life satisfaction [6], and emotions are the source of emotional intelligence [7].

Individuals who suffered victimization experiences show more vulnerability and have lower emotional intelligence. They present difficulties in evaluating, regulating, and using emotions (they experience more negative emotions than positive ones), increasing stress and dissatisfaction with life [4]. Therefore, the main objective of this study is as follows: to identify the relationship between emotional intelligence and life satisfaction; to compare a sample of victims and non-victims of violence in adulthood and a sample of men and women regarding emotional intelligence and life satisfaction; to analyze the variables that explain the variance of satisfaction with life in a sample of the Portuguese population.

2. Materials and Methods

This study comprised a sample of 144 individuals aged between 18 and 77 years (M = 36.97, SD = 15.87). Of the participants, 41 were men (28.47%), and 103 (71.53%) were women. Nearly half of the sample (n = 66, 45.8%) reported being victims of violence in adulthood. The study design is cross-sectional with a non-probabilistic sample. Participant recruitment was conducted by disseminating the study through personal and social contact networks. Participants answered online the sociodemographic questionnaire, the Wong and Law Emotional Intelligence Scale (WLEIS) [8], and the Satisfaction with Life Scale (SLWS) [9]. The WLEIS [8] measures emotional intelligence in four areas: self-emotion appraisal, others' emotion appraisal, use of emotion, regulation of emotion. The SLWS [9] evaluates several components of subjective well-being, assessing overall satisfaction with life. All participants granted their consent to participate in this study. The study followed the ethical principles outlined in the Declaration of Helsinki [10] and was approved by the Egas Moniz Ethics Committee.

3. Results

3.1. Correlation Analyses

The results show statistically significant positive correlations between satisfaction with life and emotional intelligence ($r = 0.541, p < 0.001$), self-emotion appraisal ($r = 0.428$, $p < 0.001$), others' emotion appraisal ($r = 0.363, p < 0.001$), regulation of emotion ($r = 0.364$, $p < 0.001$) and use of emotion ($r = 0.511, p < 0.001$).

3.2. Comparison Analyses

Victims of violence in adulthood show lower scores than non-victims on self-emotion appraisal ($M = 14.167$, $SD = 3.317$) [$F(1,142) = 4.668$, $p = 0.032$], use of emotion ($M = 13.803$, $SD = 3.892$) [$F(1,142) = 5.837$, $p = 0.017$], regulation of emotion ($M = 13.542$, $SD = 3.633$) [$F(1,142) = 6.855$, $p = 0.010$], emotional intelligence ($M = 56.727$, $SD = 10.051$) [$F(1,142) = 5.204$, $p = 0.024$], and satisfaction with life ($M = 17.182$, $SD = 4.543$) [$F(1,142) = 5.046$, $p = 0.026$]. Women show higher scores on the emotional intelligence assessment of others' emotions compared to men ($M = 16.204$, $SD = 2.374$) [$F(1,142) = 4.152$, $p = 0.043$].

3.3. Regression Analyses

The explanatory model of satisfaction with life, created using a Multiple Linear Regression, showed that the model is significant. Durbin–Watson was 2.17, and VIF was <3. Sex, regulation of emotion, and others' emotion appraisal are not significant. We performed a new model only with significant paths. The model is significant ($F(3,140) = 22.89$, $p < 0.001$) and explains 33% of the variance in satisfaction with life. Age ($\beta = -0.19$, $p = 0.01$), self-emotion appraisal ($\beta = 0.28$, $p = 0.001$), and use of emotion ($\beta = 0.39$, $p < 0.001$) are significant predictors of satisfaction with life.

4. Discussion

The results show that emotional intelligence is related to life satisfaction. When emotional intelligence increases, life satisfaction also increases. Those outcomes align with other studies [6,11], which showed that individuals with higher emotional intelligence are more satisfied with their lives and are more able to understand and pay attention to their emotions, leading to greater personal fulfillment and achievement. Thus, emotional intelligence is necessary to control emotions and determine life satisfaction [12].

The regression analysis verified that age, self-emotion appraisal, and use of emotion are predictors of satisfaction with life. Increasing age predicts a decreased perception of satisfaction with life. The results corroborate Garcia et al.'s [13] study, which found a negative association between life satisfaction and age. On the other hand, an increase in self-emotion appraisal and use of emotion explains the higher scores for satisfaction with life. That is, individuals with high levels of self-emotion appraisal can understand and express their emotions, and individuals with higher levels of use of emotion can use their emotions in productive activities and personal performance, leading to higher levels of satisfaction with life. Our results align with other studies indicating that emotional intelligence skills impact satisfaction with life [14]. The literature reports that subjects with high emotional intelligence competencies demonstrate a good adaptive capacity, as they understand and regulate their emotions and those of others, which allows them to solve social problems beneficially [15]. Previous studies showed that emotional intelligence influences social skills and relationships, e.g., [16]. Therefore, subjects with high emotional intelligence, who can associate others' emotions along with their own, have good satisfaction with life, a consequence of each individual's assessment of their life according to their past and present [17]. Other variables not studied in this research, such as job satisfaction, financial situation, compassion fatigue, and education, should also explain life satisfaction [18].

The results heightened the significant differences between victims and non-victims of violence in adulthood regarding satisfaction with life and emotional intelligence. In the study of Mannarini et al. [19], victims also reported lower life satisfaction. High emotional intelligence makes individuals more motivated, allowing them to solve problems and achieve goals using their emotions. Thus, victims with higher emotional intelligence may be able to get through abusive situations and feel less anxiety and fear. Victims with higher emotional intelligence exhibit less negative humor when coping with traumatic experiences [20].

Furthermore, differences were found between men and women when evaluating other people's emotions, with women showing higher scores. This is corroborated by studies that reported that women show better emotional intelligence skills [18], can better recognize their own and others' emotions, and show greater levels of perception and empathy than men [21].

In short, the results reinforce previous studies in which emotional intelligence is directly related to life satisfaction; this is associated with individuals' ability to understand, access, and regulate their emotions, allowing them to adapt to different contexts.

Author Contributions: Conceptualization, T.C.A.; methodology, T.C.A., C.S., J.M., L.A., M.M. and R.C.; validation, T.C.A.; formal analysis, C.S., J.M., L.A., M.M., R.C. and T.C.A.; investigation, C.S., J.M., L.A., M.M. and R.C.; data curation, T.C.A., C.S., J.M., L.A., M.M. and R.C.; writing—original draft preparation, C.S., J.M., L.A., M.M., R.C. and T.C.A.; writing—review and editing, T.C.A. and J.C.; supervision, T.C.A.; project administration, T.C.A. All authors have read and agreed to the published version of the manuscript.

Funding: This research received no external funding.

Institutional Review Board Statement: The study was conducted according to the Declaration of Helsinki, and approved by the Ethics Committee of the Egas Moniz School of Health and Science.

Informed Consent Statement: Informed consent was obtained from all the participants.

Data Availability Statement: The data presented in this study are available on request from the corresponding author. The data are not publicly available due to ethical, local reasons.

Conflicts of Interest: The authors declare no conflict of interest.

References

1. Afolabi, O.A.; Balogun, A.G. Impacts of psychological security, emotional intelligence and self-efficacy on undergraduates' life satisfaction. *Psychol. Thought* **2017**, *10*, 247–261. [CrossRef]
2. Olasupo, M.O.; Idemudia, E.S.; Kareem, D.B. Moderated mediation roles of social connectedness and optimism on emotional intelligence and life satisfaction. *Heliyon* **2021**, *7*, e07029. [CrossRef] [PubMed]
3. Veenhoven, R. The study of life satisfaction. In *A Comparative Study of Satisfaction with Life in Europe*; Saris, W.E., Veenhoven, R., Scherpenzeel, A.C., Bunting, B., Eds.; Eötvös University Press: Budapest, Hungary, 1996; pp. 11–18.
4. Xiang, Y.; Yuan, R.; Zhao, J. Childhood maltreatment and life satisfaction in adulthood: The mediating effect of emotional intelligence, positive affect and negative affect. *J. Health Psychol.* **2021**, *26*, 2460–2469. [CrossRef] [PubMed]
5. Prior, A.I.S. *Inteligência Emocional, Bem-Estar e Saúde Mental: Um Estudo Com Técnicos de Apoio à Vítima (Tese de Mestrado)*; Universidade Fernando Pessoa: Porto, Portugal, 2017.
6. Palmer, B.; Donaldson, C.; Stough, C. Emotional intelligence and life satisfaction. *Personal. Individ. Differ.* **2022**, *33*, 1091–1100. [CrossRef]
7. Helmi, A.F. Meta-analysis of the correlation between emotional intelligence and life satisfaction. *Anatol. J. Educ.* **2021**, *6*, 63–74. [CrossRef]

8. Rodrigues, N.; Rebelo, T.; Coelho, J.V. Adaptação da escala de inteligência emocional de Wong e Law (WLEIS) e análise da sua estrutura factorial e fiabilidade numa amostra portuguesa. *Psychologica* **2011**, *55*, 189–207. [CrossRef] [PubMed]
9. Simões, A. Ulterior validação de uma escala de satisfação com a vida [Further validation of a life satisfaction scale] (SLWS). *Rev. Port. Pedagog.* **1992**, *3*, 503–515.
10. World Medical Association. World medical association declaration of Helsinki: Ethical principles for medical research involving human subjects. *JAMA* **2013**, *310*, 2191–2194. [CrossRef] [PubMed]
11. Teixeira, P.S.S. *Inteligência Emocional e Satisfação Com a Vida*; Universidade Lusófona de Humanidades e Tecnologias: Lisboa, Portugal, 2008.
12. Delhoma, I.; Satorresb, E.; Meléndezb, J.C. Can we improve emotional skills in older adults? Emotional intelligence, life satisfaction, and resilience. *Psychosoc. Interv.* **2020**, *29*, 133–139. [CrossRef]
13. Garcia, R.C.; Lima, M.P.; Carona, C. The relation between helicopter parenting and satisfaction with life in emerging adults living with their parents: The moderating role of gender and age group. *Cent. Eur. J. Paediatr.* **2022**, *18*, 150–160. [CrossRef]
14. Quintana-Orts, C.; Mérida-López, S.; Rey, L.; Extremera, N. A closer look at the emotional intelligence construct: How do emotional intelligence facets relate to life satisfaction in students involved in bullying and cyberbullying? *Eur. J. Investig. Health Psychol. Educ.* **2021**, *11*, 711–725. [CrossRef] [PubMed]
15. Gonzaga, L.; Sarmento, A.; Caldeira, C. Inteligência emocional e otimismo em sujeitos adultos. *Rev. UI_IPSantarém* **2016**, *4*, 313–328.
16. Ain, N.U.; Munir, M.; Suneel, I. Role of emotional intelligence and grit in life satisfaction. *Heliyon* **2021**, *7*, e06829. [CrossRef] [PubMed]
17. Schoeps, K.; Montoya-Castilla, I.; Raufelder, D. Does stress mediate the association between emotional intelligence and life satisfaction during adolescence? *J. Sch. Health* **2019**, *89*, 354–364. [CrossRef] [PubMed]
18. Milutinović, D.; Marcinowicz, L.; Jovanović, N.B.; Dragnić, N. Impact of compassion satisfaction and compassion fatigue on satisfaction with life in Serbian and Polish nurses: A cross-sectional study. *Int. Nurs. Rev.* **2023**, *70*, 194–203. [CrossRef] [PubMed]
19. Mannarini, S.; Taccini, F.; Rossi, A.A. Women and violence: Alexithymia, relational competence and styles, and satisfaction with life: A comparative profile analysis. *Behav. Sci.* **2021**, *11*, 147. [CrossRef] [PubMed]
20. Cabras, C.; Mondo, M.; Diana, A.; Sechi, C. Relationships between trait emotional intelligence, mood states, and future orientation among female italian victims of intimate partner violence. *Heliyon* **2020**, *6*, e05538. [CrossRef] [PubMed]
21. Sanchez-Núñez, M.T.; Fernández-Berrocal, P.; Montañés, J.; Latorre, J.M. Does emotional intelligence depend on gender? The socialization of emotional competencies in men and women and its implications. *Electron. J. Res. Educ. Psychol.* **2008**, *15*, 455–474.

medical sciences forum

MDPI

Proceeding Paper

Performance of ID NOW Influenza A&B 2 †

Susana Bandarra [1], Célia Neto [1], Lurdes Monteiro [1,*] and Laura Brum [2]

1 Molecular Pathology Laboratory, SYNLAB Lisboa, Av. Columbano Bordalo Pinheiro, 75A, 1070-061 Lisbon, Portugal; susana.bandarra@synlab.pt (S.B.); celia.neto@synlab.pt (C.N.)

2 Clinical Laboratories SYNLAB Portugal, Av. Columbano Bordalo Pinheiro, 75A, 1070-061 Lisbon, Portugal; laura.brum@synlab.pt

* Correspondence: lurdes.monteiro@synlab.pt; Tel.: +351-217-216-060

† Presented at the 6th International Congress of CiiEM—Immediate and Future Challenges to Foster One Health, Almada, Portugal, 5–7 July 2023.

Abstract: ID NOW™ INFLUENZA A&B 2 is a point-of-care assay for rapid molecular diagnosis of Influenza A and B. The present study aims to evaluate the performance of ID NOW™ INFLUENZA A&B 2 compared to a reference RT-PCR assay. A total of 67 nasopharyngeal swabs from 67 patients for screening of Influenza A and B by RT-PCR (Allplex™ RP1) were also tested with ID NOW assay. Of the seventeen positive Influenza A and five positive Influenza B, fifteen (88%) and five (100%), respectively, were also detected by the ID NOW assay. The overall agreement with the reference test was 95.5%. The sensitivity was 88.2% for Influenza type A and 100% for Influenza type B, and the specificity was equal to or higher than 96% for both.

Keywords: Influenza A; Influenza B; point-of-care; molecular detection

1. Introduction

The Influenza viruses can cause significant seasonal morbidity and mortality, particularly in young children and patients with underlying chronic diseases. Influenza diagnosis is mainly based on clinical symptoms; however, a robust microbiological diagnosis is required for hospitalized patients or those at risk of developing complications, such as pneumonia. In these cases, Influenza diagnosis based on clinical observation alone can be difficult, as a variety of other respiratory viruses with similar symptoms are often circulating at the same time [1]. Rapid and early identification of the viral cause of acute respiratory infection has a critical role in clinical decision making, particularly for patients at risk for severe complications of Influenza who should be treated early in the course of infection. The implications for patient management include hospital admission or discharge, reduction in diagnostic investigations, and administration of antibiotics and antiviral therapy [2]. This not only improves patient monitoring and care and infection control, but also reduces healthcare costs.

Currently, there are several methods available to test for Influenza, divided into antigen detection-based assays, viral culture, and molecular detection. Real-time polymerase chain reaction (RT-PCR) is the gold standard for identifying Influenza viruses but requires specialized technology and skilled molecular technicians.

The ID NOW™ INFLUENZA A&B 2 assay, performed on the ID NOW instrument (Abbott Molecular Inc.), is a point-of-care detection system for Influenza viruses. ID NOW uses isothermal nucleic acid amplification technology, using primers and fluorescent probes specific for the amplification of RNA targets without the need for a thermal cycler. Like other platforms, ID NOW uses disposable test units containing target-specific reagents where extraction, amplification, and detection take place. The ID NOW™ INFLUENZA A&B 2 assay has been described as an easy-to-use device that provides robust and accurate results within 15 min for the detection of Influenza A/B [3,4]. In the present study, we aim

Citation: Bandarra, S.; Neto, C.; Monteiro, L.; Brum, L. Performance of ID NOW Influenza A&B 2. *Med. Sci. Forum* **2023**, *22*, 28. https://doi.org/10.3390/msf2023022028

Academic Editors: José Brito, Nuno Taveira and Ana I. Fernandes

Published: 14 August 2023

to evaluate the performance of the ID NOW™ INFLUENZA A&B 2 assay compared to a reference RT-PCR commonly used in our routine diagnosis of Influenza.

2. Materials and Methods

Nasopharyngeal swabs (NPS) from 67 individuals from hospital centers were tested between January and February 2020. Samples were processed simultaneously with ID NOW™ INFLUENZA A&B 2 assay (Abbott) and the reference method Allplex™ Respiratory Panel (RP) 1 (Seegene) within 24 h of reception at the laboratory. Previously to amplification with Allplex™ RP1, RNA was extracted from 200 µL of a sample using QIACube-QIAamp viral RNA mini kit (Qiagen). The real-time RT-PCR was conducted on the CFX96 (Bio-Rad), and subsequently interpreted by Seegene's Viewer software.

ID Now platform is an automated assay that utilizes molecular isothermal nucleic acid amplification technology and uses as targets polymerase basic gene 2 for Influenza A virus and polymerase acidic gene for Influenza B virus detection. Specificity, sensitivity, negative and positive predictive values (NPV and PPV), and agreement between both tests were analyzed with SPSS software, v.28.

3. Results

A total of 67 NPS from subjects aged between 3 weeks and 96 years (median age = 65 years; 25.4% < 1–18 years, 50.7% > 65 years, and 23.9% [18–64 years]) were tested, including 26 females (39%) and 41 males (61%). Using the reference method, a total of 21 (31.3%) NPS were positive, while 46 (68.7%) samples were negative for both viruses. Using the molecular platform ID NOW, a total of 22 (32.8%) NPS were positive, while 45 (67.2%) were negative for both viruses. The results obtained are detailed in Table 1.

Table 1. Agreement of ID NOW™ INFLUENZA A&B 2 with the reference Allplex™ RP1 method.

	Allplex RP1		Total	Percent Agreement (Influenza A/B)	95% CI [1]
	Detected	Not Detected			
ID NOW					
Positive	20 *	2	22	95.2%	80.7–99.7
Negative	1	44	45	95.7%	87.2–99.3
Total	21	46	67	95.5%	

[1] CI—Confidence interval; * one sample was only positive for Influenza B.

Detailed evaluation of the results showed that among the positive results by the reference method, sixteen samples were positive for Influenza A, four were positive for Influenza B, and one was positive for both Influenza A and Influenza B. By the ID NOW platform, fifteen (from the sixteen RT-PCR positive) were positive for Influenza A and four (from the four RT-PCR positive) for Influenza B, showing a false-negative for Influenza A. Regarding the positive sample for both viruses, only the Influenza B was detected by ID NOW, resulting in an Influenza A false-negative sample, but being a positive sample in global (Table 1). Of the 46 negative samples for both viruses by the reference method, only forty-four samples were negative using ID NOW, two samples positive for Influenza A representing two false-positive results. All negative Influenza B samples on RT-PCR were also negative with ID NOW.

The results comparison between ID NOW and the reference method showed a negative agreement of 96.0% for Influenza A, 100% for B (Table 2), and 95.7% for both viruses (Table 1) and a positive agreement of 88.2% for Influenza A, 100% for B (Table 2), and 95.2% for both viruses (Table 1). The overall agreement of the ID NOW and the reference multiplex RT-PCR was 94.0%, 100%, and 95.5% for Influenza A, Influenza B, and altogether (Influenza A and B), respectively.

Table 2. Estimated diagnostic performance of ID NOW™ INFLUENZA A&B 2.

Measurement	Influenza A	Influenza B
Sensitivity	88.2% (15/17); CI: 67.9–97.9	100% (5/5)
Specificity	96.0% (48/50); CI: 88.2–99.3	100% (62/62)
PPV	88.2% (15/17); CI: 67.9–97.9	100% (5/5)
NPV	96.0% (48/50); CI: 88.2–99.3	100% (62/62)

CI—95% confidence interval; PPV—positive predictive value; NPV—negative predictive value.

The sensitivity for the ID NOW assay was 88.2% (95% CI: 67.9–97.9) for Influenza type A, 100% for Influenza type B, and 95.2% (95% CI: 80.7–99.7) for both. The specificity was equal to or higher than 96% (95% CI: 88.2–99.3) for both. Overall, the positive predictive value (PPV) was 90.9 (95% CI: 74.5–98.4), and the negative predictive value (NPV) was 97.8 (95% CI: 90.6–99.9). In Table 2, all clinical performance parameters are detailed per type of Influenza. The Kappa value was 0.866 (86.6%).

4. Discussion

The main advantage of ID NOW™ INFLUENZA A&B 2 is a shorter turnaround time compared to the RT-PCR Allplex™ RP1 assay or any other currently available molecular assay for Influenza A/B detection. To evaluate the capacity of detecting the Influenza virus in this point-of-care system, this study was conducted by mimicking the conditions of routine diagnostic testing, and the results indicate that the ID NOW™ Influenza A&B 2 has good overall performance and a high degree of concordance with the reference method.

The published results for ID NOW™ INFLUENZA A&B 2 show sensitivities and specificities between 90 and 100% when compared to various RT-PCR assays [3–6]. These high levels of sensitivity and specificity were also found in this study, with a sensitivity > 88% and 100% specificity in the detection of Influenza A and B viruses when compared with RT-PCR Allplex™ RP1 assay. When comparing performance by Influenza type, as in other studies, Influenza B showed the highest sensitivity. However, the number of positive cases for Influenza B was always lower than for Influenza A.

In the case of Influenza A, the two false negatives obtained using the ID NOW platform had a Ct > 37.0 using the reference method (median Ct value of 38.7 $\pm$ 1.8). Thus, since the Ct was close to the limit of detection of the reference method, this indicated that negative results obtained with ID NOW occurred probably in samples with low viral load. Other studies have also reported false negative results for samples with Ct > 37.0 (using the same reference RT-PCR assay) [3] or lower Ct > 31.0 (CDC FLU A/B PCR assay) [4]. In addition, it was verified that positive samples for other respiratory viruses than Influenza A or B were correctly detected as negative for Influenza with ID NOW assay, indicating the absence of cross-reactions with other respiratory viruses.

The rate of invalid results, which can have a negative impact on the cost of the test, the time-to-result, and consequently on patient care, has previously been reported to be relatively low, ranging from 0 to 7.5% [5–7]. In accordance with this, no invalid results were obtained in this study.

In conclusion, the results of this study suggest that the ID NOW™ INFLUENZA A&B 2 assay can be used as a test for rapid and accurate diagnosis of Influenza in clinical practice. ID NOW assay combines high speed, sensitivity, and accuracy. However, it should be emphasized that point-of-care systems cannot replace reference methodology in a transversal way.

Author Contributions: Methodology, C.N.; investigation, data curation, and writing—original draft preparation, S.B.; conceptualization, writing—review and editing, and supervision, L.M.; validation, L.M. and L.B.; project administration, L.M. All authors have read and agreed to the published version of the manuscript.

Funding: This research received no external funding.

Institutional Review Board Statement: Ethical review and approval were waived for this study since this study was carried out on surplus samples that were sent to the laboratory as part of the molecular diagnosis of Influenza A/B, and under no circumstances shall the use of samples for this study jeopardize the diagnosis.

Informed Consent Statement: Patient consent was not required as surplus samples that arrived at the laboratory with a clinician's request to detect Influenza A/B were used.

Data Availability Statement: Not applicable.

Acknowledgments: We thank the Lisbon Molecular Pathology Laboratory team for assistance with testing. The authors gratefully thank Abbott Diagnostics for providing reagents for the comparison assay.

Conflicts of Interest: The authors declare no conflict of interest. The manufacturer had no role in the design of this study; in the collection, analysis, or interpretation of data; in the writing of the manuscript; or in the decision to publish the results.

References

1. Young, S.; Phillips, J.; Griego-Fullbright, C.; Wagner, A.; Jim, P.; Chaudhuri, S.; Tang, S.; Sickler, J. Molecular point-of-care testing for influenza A/B and respiratory syncytial virus: Comparison of workflow parameters for the ID Now and cobas Liat systems. *J. Clin. Pathol.* **2020**, *73*, 328–334. [CrossRef] [PubMed]
2. Landry, M.L. Diagnostic tests for influenza infection. *Curr. Opin. Pediatr.* **2011**, *23*, 91–97. [CrossRef] [PubMed]
3. Farfour, E.; Roux, A.; Ballester, M.; Gagneur, L.; Renaux, C.; Jolly, E.; Vasse, M. Improved performances of the second generation of the ID NOW influenza A&B 2® and comparison with the GeneXpert®. *Eur. J. Clin. Microbiol. Infect. Dis.* **2020**, *28*, 1681–1686. [CrossRef]
4. Kanwar, N.; Michael, J.; Doran, K.; Montgomery, E.; Selvarangan, R. Comparison of the ID Now Influenza A & B 2, Cobas Influenza A/B, and Xpert Xpress Flu Point-of-Care Nucleic Acid Amplification Tests for Influenza A/B Virus Detection in Children. *J. Clin. Microbiol.* **2020**, *58*, e01611-19. [CrossRef] [PubMed]
5. Nguyen Van, J.C.; Caméléna, F.; Dahoun, M.; Pilmis, B.; Mizrahi, A.; Lourtet, J.; Behillil, S.; Enouf, V.; Le Monnier, A. Prospective evaluation of the Alere i Influenza A&B nucleic acid amplification versus Xpert Flu/RSV. *Diagn. Microbiol. Infect. Dis.* **2016**, *85*, 19–22. [CrossRef] [PubMed]
6. Farfour, E.; Yung, T.; Baudoin, R.; Vasse, M. Evaluation of Four Fully Integrated Molecular Assays for the Detection of Respiratory Viruses during the Co-Circulation of SARS-CoV-2, Influenza and RSV. *J. Clin. Med.* **2022**, *11*, 3942. [CrossRef] [PubMed]
7. Chen, J.H.; Lam, H.Y.; Yip, C.C.; Cheng, V.C.; Chan, J.F.; Leung, T.H.; Sridhar, S.; Chan, K.H.; Tang, B.S.; Yuen, K.Y. Evaluation of the molecular Xpert Xpress Flu/RSV assay vs. Alere i influenza A&B assay for rapid detection of influenza viruses. *Diagn. Microbiol. Infect. Dis.* **2018**, *90*, 177–180. [CrossRef] [PubMed]

medical sciences forum

MDPI

Proceeding Paper

Bond Strength Properties of a Dental Adhesive with a Novel Dendrimer—G-IEMA †

Miguel Alves [1], Filipa Branco [1], António H. S. Delgado [1,2], José Brito [1], Luísa Gonçalves [1], Ana Mano Azul [1], Joana Vasconcelos e Cruz [1] and Mário Polido [1,*]

1 Egas Moniz Center for Interdisciplinary Research (CiiEM), Egas Moniz School of Health & Science, 2829-511 Almada, Portugal; malves@egasmoniz.edu.pt (M.A.); filipabs.97@gmail.com (F.B.); ant.delgado94@gmail.com (A.H.S.D.); britojaa@hotmail.com (J.B.); lmlgoncalves@gmail.com (L.G.); anazul2@gmail.com (A.M.A.); joanavcruz@gmail.com (J.V.e.C.)

2 Division of Biomaterials and Tissue Engineering, UCL Eastman Dental Institute, London WC1E 6BT, UK

* Correspondence: mpolido@egasmoniz.edu.pt

† Presented at the 6th International Congress of CiiEM—Immediate and Future Challenges to Foster One Health, Almada, Portugal, 5–7 July 2023.

Abstract: The objectives of this study were to characterize the microtensile bond strength to enamel of two experimental adhesive systems, one containing a novel monomer and the other having the same composition as commercial adhesive systems, and comparing them to commercial materials. Two experimental adhesive systems were developed in the lab, one with Bisphenol A diglycidyl methacrylate (Bis-GMA) and the other with G(2)-isocyanatoethyl methacrylate (G-IEMA) as a substitute for Bis-GMA. Twenty healthy human permanent molars were cut into halves and randomly divided into eight groups based on the application mode. The experimental universal adhesive system without Bis-GMA demonstrated comparable adhesive strength to enamel as the other universal adhesive systems containing Bis-GMA.

Keywords: dendrimer; experimental monomer; experimental dental adhesive; G-IEMA

Citation: Alves, M.; Branco, F.; Delgado, A.H.S.; Brito, J.; Gonçalves, L.; Azul, A.M.; Cruz, J.V.e.; Polido, M. Bond Strength Properties of a Dental Adhesive with a Novel Dendrimer—G-IEMA. *Med. Sci. Forum* **2023**, *22*, 27. https://doi.org/10.3390/msf2023022027

Academic Editors: Nuno Taveira and Ana I. Fernandes

Published: 14 August 2023

1. Introduction

The major paradigm shift in adhesion to tooth structure occurred in 1955 with the introduction of phosphoric acid etching to enamel by Buonocore. Since then, the evolution of adhesive techniques has allowed dentists to adopt a minimally invasive philosophy in clinical practice [1,2]. Regarding material choice, currently, there is a wide offering of adhesive systems available for bonding to tooth tissues, using different adhesive strategies [3]. To simplify the use of these materials, universal adhesive (UA) systems were introduced, allowing clinicians to choose the best adhesive strategy according to different clinical scenarios, whether it is an etch-and-rinse, self-etch or selective enamel etching approach [4–6].

Replacing Bis-GMA while still improving the physicochemical and mechanical properties of adhesives has been researched in recent studies [7,8]. Specifically, some authors have examined the introduction of dendrimers as base constituents UAs. A second-generation dendrimer derived from isocyanatoethyl methacrylate, G-IEMA, has been investigated as a candidate for potential replacement of Bis-GMA-based systems [9,10]. Traditional linear crosslinking monomers can be replaced successfully by dendrimer G-IEMA without influencing the resulting properties. Not only did this monomer significantly improve the experimental UA's degree of conversion, but it was also responsible for reducing the co-polymer shrinkage and controlling water sorption [11]. Further to this, the same authors also observed that G-IEMA formulations could increase the bond strength to dentin, and later on, showed that they have promising interfacial properties [11,12].

However, to research and prove the beneficial applicability of G-IEMA, further studies are needed. It is therefore important to investigate the role of G-IEMA-based systems on the bond strength to enamel using two different adhesive strategies, while also evaluating their impact on the contact angle of enamel surfaces, which is as of yet unknown.

2. Materials and Methods

Two experimental adhesive systems, one with Bis-GMA (EM1) and another with G-IEMA, as a substitute for Bis-GMA (EM2), were developed in our lab. Two commercial adhesives, Futurabond® M+ (VOCO) (FUT) and ScotchbondTM Universal (3M ESPE) (SBU) were chosen as controls. Twenty healthy human permanent molars, obtained with informed consent (approved by the Ethics Committee of Egas Moniz School of Health & Science), were cut into halves and randomly divided into eight groups (n = 5) according to the application mode (self-etch or etch-and-rinse): FUT_ER, FUT_SE, SBU_ER, SBU_SE, EM1_ER, EM1_SE, EM2_ER, and EM2_SE. Afterwards, each specimen was polished with a 600 SiC grit paper for smear layer simulation and the adhesives were applied according manufacturer's directions. The etch-and-rinse method employed Octacid orthophosphoric acid (37%) (Clarben). Resin build-ups were conducted using the Schmidt Composite Nanohybrid (MADESPA), with increments shaped as rectangular prisms. The resin build-ups were applied in 2 mm increments, achieving a total height of 6 mm. The materials were light-cured using the ElliparTM DeepCure-S (3M ESPE) system, following the manufacturer's instructions, with a 40 s cure time specifically applied to the experimental universal adhesive systems. This system employs blue LEDs for the light-curing process. Its peak irradiance was 1200 mW/cm^2 which was confirmed with a radiometer. After processing, the specimens were kept in distilled water for 24 h at 37 °C. Beams (1 $\pm$ 0.2 mm^2) were obtained through additional sectioning and tested using a universal testing machine (μTBS) with a cross head speed of 0.5 mm/min until failure. The data analysis was performed using SPSS (version 28.0, IBM Corp., Armonk, NY, USA) with linear mixed models (LMMs) incorporating fixed effects, while maintaining a significance level of 5%.

3. Results

The effects of adhesive (p = 0.033) and method, or protocol, (p < 0.001) on the microtensile bond strength were significant and independent of one another. There was no interaction between the adhesive used and the technique adopted (p = 0.985) (Table 1). Independent of application procedure, SBU displayed a considerably greater μTBS than the experimental EM2 (p = 0.031). No variations were found between any other adhesive pairings.

Table 1. Linear mixed model analysis—Type III Tests of Fixed Effects considering the variables: (1) adhesive and (2) adhesive strategy or protocol.

	Type III Tests of Fixed Effects [a]			
Source	**Numerator df**	**Denominator df**	**F**	**Sig.**
Intercept	1	426	531.262	0.000
Adhesive	3	426.000	2.947	0.033
Strategy	1	426	79.606	0.000
Adhesive * strategy [b]	3	426.000	0.050	0.985

[a] Dependent variable: microtensile bond strength (MPa). [b] In the "Adhesive * strategy" term above, the asterisk (*) denotes the interaction between main factors Adhesive and strategy.

4. Discussion

Ongoing concerns about the use of biocompatible dental materials have called into question the incorporation of Bis-GMA in resin-based restorative materials, due to its Bisphenol A (BPA) content, which can elute and have systemic health implications [10]. The demand for new materials without Bis-GMA has emerged as a preventive measure to reduce exposure to Bisphenol A [13]. The same adhesive systems who formulated and subsequently evaluated the physicochemical properties and adhesion to dentin experimental universal adhesive systems without Bis-GMA, patented in Portugal (holder: Egas Moniz School of Health & Science). This study followed the same line of research, assessing the adhesion to enamel, which has not been studied until now [11,12].

According to the current scientific evidence, for enamel, the best adhesive strategy continues to be the use of orthophosphoric acid prior to the application of the adhesive system [13,14]. Several in vitro studies demonstrated that universal adhesive systems have higher values of adhesive resistance to enamel when used according to the etch-and-rinse (versus self-etch) protocol, as was observed in this study [5,15]. These results are justified by the reduced demineralization capacity of a universal adhesive system compared to orthophosphoric acid, resulting in an incomplete creation of microporosities, which inevitably reduces the micro-retention of the adhesive [2]. The chemical composition of the adhesive systems can also contribute to the differences observed in this study because, although the experimental adhesive systems were formulated according to the chemical composition of the commercial adhesives, the specific percentage of each component was not discriminated.

In this study, most of the adhesive systems used had a mild pH (Futurabond® M + and experimental adhesives), with Scotchbond™ Universal registering a more basic pH, falling within the ultra-mild category. Although a good adhesive behavior to dentin is associated with mild self-etching (pH $\cong$ 2), these solutions are unable to effectively condition the enamel, leading to increased microleakage; a mild pH is essential prior to etching dentin to obtain a micromechanical retention effective [16,17]. The pH of a universal adhesive system is a relatively important property because, while an acidic medium is required for the dissolution of the smear layer and smear plugs (opening the dentinal tubules), an excessively acidic adhesive system can remove excess calcium, decreasing its ability to adhere to 10-MDP, which becomes particularly important in adhering to dentin [16,17]. It is essential to select adhesive systems that contain 10-MDP, taking into account their molecular structure, their hydrophobic behavior and the characteristics of the adhesive interface that favor adhesion [18]. The microtensile bond strength to dentin using the same adhesive systems and protocols as the present study, there were no significant differences between the adhesive systems studied, suggesting that the experimental universal adhesive system without Bis-GMA could be used effectively in dentin [12].

5. Conclusions

The four universal adhesive systems examined (Futurabond® M+, ScotchbondTM Universal, an experimental universal adhesive system with Bis-GMA, and an experimental universal adhesive system without Bis-GMA) showed no statistically significant differences in adhesive strength to enamel when using either the etch-and-rinse or self-etch adhesive strategies. The experimental universal adhesive system without Bis-GMA exhibited comparable adhesive strength to enamel as the other universal adhesive systems containing Bis-GMA. The promising behavior of the experimental Bis-GMA-free universal adhesive system indicates the need for further investigations. These studies should focus on exploring the potential of the G-IEMA dendrimer as a substitute for Bis-GMA in the composition of adhesive systems.

6. Patents

This work resulted in a national patent, registered under No. 115,064—Formulation for a universal dental adhesive system containing a second-generation dendritic cross-linking monomer (2019); Vasconcelos e Cruz, J., Gonçalves, L. L. and Polido, M., Moniz School of Health & Science.

Author Contributions: Conceptualization, M.A. and M.P.; methodology, M.P.; software, J.B.; validation, A.M.A., M.P. and A.H.S.D.; formal analysis, J.B.; investigation, M.A., F.B. and J.V.e.C.; resources, L.G.; data curation, J.B.; writing—original draft preparation, A.H.S.D.; writing—review and editing, A.H.S.D.; visualization, J.V.e.C.; supervision, M.P.; project administration, L.G.; funding acquisition, M.P. All authors have read and agreed to the published version of the manuscript.

Funding: This research received no external funding.

Institutional Review Board Statement: Not applicable.

Informed Consent Statement: Not applicable.

Data Availability Statement: Data may be available upon request from the corresponding author.

Acknowledgments: The authors would like to express their gratitude to VOCO GmbH, Cuxhaven, Germany for providing the Futurabond M+ adhesive used in this study.

Conflicts of Interest: The authors declare no conflict of interest.

References

1. Buonocore, M.G. A simple method of increasing the adhesion of acrylic filling materials to enamel surfaces. *J. Dent. Res.* **1955**, *34*, 849–853. [PubMed]
2. Nagarkar, S.; Theis-Mahon, N.; Perdigão, J. Universal dental adhesives: Current status, laboratory testing, and clinical performance. *J. Biomed. Mater. Res.-Part B Appl. Biomater.* **2019**, *107*, 2121–2131. [CrossRef] [PubMed]
3. Van Meerbeek, B.; Yoshihara, K.; Van Landuyt, K.; Yoshida, Y.; Peumans, M. From Buonocore's Pioneering Acid-Etch Technique to Self-Adhering Restoratives. A Status Perspective of Rapidly Advancing Dental Adhesive Technology. *J. Adhes. Dent.* **2020**, *22*, 7–34. [PubMed]
4. Alex, G. Universal adhesives: The next evolution in adhesive dentistry? *Compend. Contin. Educ. Dent.* **2015**, *36*, 15–26. [PubMed]
5. Da Rosa, W.L.D.O.; Piva, E.; da Silva, A.F. Bond strength of universal adhesives: A systematic review and meta-analysis. *J. Dent.* **2015**, *43*, 765–776. [CrossRef] [PubMed]
6. Perdigao, J.; Kose, C.; Mena-Serrano, A.P.; De Paula, E.A.; Tay, L.Y.; Reis, A.L.A.D. A New Universal Simplified Adhesive: 18-Month Clinical Evaluation A New Universal Simplified Adhesive: 18-Month Clinical Evaluation. *Oper. Dent.* **2014**, *39*, 113–127. [CrossRef] [PubMed]
7. Nguyen, T.; Kim, B.; Kim, Y.J.; Lee, S.H.; Shin, S.; Cho, J.K. Synthesis of the bio-based alternative to Bis-GMA and its application to photo-polymerizable adhesives. *Int. J. Adhes. Adhes.* **2018**, *80*, 60–65.
8. Song, L.; Ye, Q.; Ge, X.; Misra, A.; Tamerler, C.; Spencer, P. Probing the neutralization behavior of zwitterionic monomer-containing dental adhesive. *Dent. Mater.* **2017**, *33*, 564–574. [CrossRef] [PubMed]
9. Papakonstantinou, A.E.; Eliades, T.; Cellesi, F.; Watts, D.C.; Silikas, N. Evaluation of UDMA's potential as a substitute for Bis-GMA in orthodontic adhesives. *Dent. Mater.* **2013**, *29*, 898–905. [CrossRef] [PubMed]
10. Yu, B.; Liu, F.; He, J.; He, Y.; Lin, Z. Preparation of Bis-GMA-Free Dental Restorative Composites with Dendritic Macromer (G-IEMA). *Adv. Polym. Technol.* **2015**, *34*, 1–6. [CrossRef]
11. Vasconcelos e Cruz, J.; Brito, J.; Polido, M.; Gonçalves, L.L. A new experimental adhesive system containing G-IEMA–physicochemical properties. *J. Adhes. Sci. Technol.* **2019**, *33*, 418–432. [CrossRef]
12. Vasconcelos e Cruz, J.; Polido, M.; Brito, J.; Gonçalves, L.L. Dentin Bonding and SEM Analysis of a New Experimental Universal Adhesive System Containing a Dendrimer. *Polymers* **2020**, *12*, 461. [CrossRef] [PubMed]
13. Fleisch, A.F.; Sheffield, P.E.; Chinn, C.; Edelstein, B.L.; Landrigan, P.J. Bisphenol A and related compounds in dental materials. *Pediatrics* **2010**, *126*, 760–768. [CrossRef] [PubMed]
14. Hanabusa, M.; Mine, A.; Kuboki, T.; Momoi, Y.; Van Ende, A.; Van Meerbeek, B.; De Munck, J. Bonding effectiveness of a new 'multi-mode'adhesive to enamel and dentine. *J. Dent.* **2012**, *40*, 475–484. [CrossRef] [PubMed]
15. Cuevas-Suárez, C.E.; Da Rosa, W.L.D.O.; Lund, R.G.; Da Silva, A.F.; Piva, E. Bonding performance of universal adhesives: An updated systematic review and meta-analysis. *J. Adhes. Dent.* **2019**, *21*, 7–26. [PubMed]
16. Perdigão, J.; Swift, E.J. Universal Adhesives. *J. Esthet. Restor. Dent.* **2015**, *27*, 331–334. [CrossRef] [PubMed]
17. Van Meerbeek, B.; Yoshihara, K.; Yoshida, Y.; Mine, A.; De Munck, J.; Van Landuyt, K.L. State of the art of self-etch adhesives. *Dent. Mater.* **2011**, *27*, 17–28. [CrossRef] [PubMed]
18. Carrilho, E.; Cardoso, M.; Ferreira, M.M.; Marto, C.M.; Paula, A.; Coelho, A.S. 10-MDP based dental adhesives: Adhesive interface characterization and adhesive stability-A systematic review. *Materials* **2019**, *12*, 790. [CrossRef] [PubMed]

medical sciences forum

MDPI

Proceeding Paper

The Effect of Treatment with Forsus Fatigue-Resistant Device on the Position of the Third Molars †

Joana Garrau [1], Pedro Mariano Pereira [1,2], José Brito [2] and Iman Bugaighis [1,3,*]

1 Department of Orthodontics, Egas Moniz School of Health and Science, Caparica, 2829-511 Almada, Portugal; joanagarrau@gmail.com (J.G.); pmarianop@sapo.pt (P.M.P.)
2 Egas Moniz Center of Interdisciplinary Research (CiiEM), Egas Moniz School of Health and Science, Caparica, 2829-511 Almada, Portugal; britojaa@hotmail.com
3 The Libyan Authority for Scientific Research, Tripoli P.O. Box 80045, Libya
* Correspondence: isbugaighis@yahoo.com
† Presented at the 6th International Congress of CiiEM—Immediate and Future Challenges to Foster One Health, Almada, Portugal, 5–7 July 2023.

Abstract: This study evaluated the influence of using the Forsus™ Fatigue-Resistant Device (FFRD) on the third molar (M3) position. Pre- and post-treatment panoramic radiographs of 28 individuals with class II malocclusion (ANB $\geq$ 4) treated with FFRD were compared to a matched control group (27 individuals) using the Tavano method. A mixed model repeated-measures ANOVA revealed the similar position of the M3s in both groups ($p > 0.05$) except for the significantly more proximal vertical position of the lower left M3 to the Menton plane in the FFRD group ($p = 0.010$). Therefore, the treatment with the FFRD device did not affect the position of M3s.

Keywords: Forsus; functional appliance; mandibular propulsion; third molar

Citation: Garrau, J.; Mariano Pereira, P.; Brito, J.; Bugaighis, I. The Effect of Treatment with Forsus Fatigue-Resistant Device on the Position of the Third Molars. *Med. Sci. Forum* **2023**, *22*, 32. https://doi.org/10.3390/msf2023022032

Academic Editors: Nuno Taveira and Ana I. Fernandes

Published: 15 August 2023

1. Introduction

Class II is one of the most frequent types of malocclusion, affecting about one-third of patients who seek orthodontic treatment [1]. The most common feature in class II malocclusion is a retruded mandible in relation to the craniofacial structure [2]. There are various types of treatment apparatus that allow the sagittal correction of the mandibular deficiency by holding the mandible in a more forward and downward position, therefore enabling the mandible to alter its postural position. These appliances enable the orofacial musculature to stretch, and the resulting reciprocal force is transmitted to the skeletal and dento-alveolar structures, resulting in a favourable alteration of the skeletal growth pattern and dento-alveolar tooth movement [3]. Ritto and Ferreira [4] categorized functional appliances into flexible, rigid or hybrid, according to the implemented force system to provide mandibular protraction. Out of those apparatuses, hybrid appliances combine flexible and rigid components with a spring system, aiming to move teeth by applying continuous force 24 h a day, replacing the conventional class II elastics which require patient compliance [5]. One of the most used hybrid functional appliances in treating class II malocclusion is the semirigid fixed functional appliance, Forsus™ Fatigue-Resistant Device (FFRD). This appliance has gained increased acceptance recently as a replacement for other class II treatment modalities [5].

Third molar impaction is a common finding accounting for 98% of all impacted teeth [6] and reaching an occurrence of 73% in young adults in Europe [7]. The aetiology of impaction is multifactorial including genetic or pathological factors, and a lack of required space to accommodate their size [8]. Orthodontic treatment for growing individuals frequently affects the third molars' eruption path. This effect is especially noticeable during non-extraction dentoalveolar treatment of cases with class II malocclusion [9]. Understanding this concern is essential to avoiding unpredictable side effects such as the third molars' impaction or altering its eruption path [10]. Therefore, this study aimed to evaluate the influence of the treatment of class II using the FFRD on the eruption path of the maxillary and mandibular third molars.

2. Material and Methods

Ethical approval of the present retrospective study was obtained from the Ethic Committee of the Egas Moniz School of Health and Science, and the participants/parents granted their informed consent. A convenience sample of 55 orthodontic patients participated in this study. The present sample was categorized into two groups: the first group included patients presented with dental and skeletal class II malocclusion (ANB $\geq 4°$ and a retruded mandible (SNB $\leq 76°$), treated with FFRD and fixed orthodontic appliance (n = 28; 14 males, 14 females, 13.6 years old, SD $\pm$ 2.4). The second group was the matched controls (n = 27: 12 males and 15 females, 13.2 years old, SD $\pm$ 1.5), who had a class I malocclusion, treated with a conventional fixed orthodontic appliance without using any propulsive mechanics.

Pre- and post-treatment digital panoramic radiographs were taken for each patient. All radiographs were taken using the same machine (Gendex Orthoralix 9200 DDE, Gendex Dental Systems, Des Plains, IL, USA). Radiographic analysis was performed using Auto CAD 2021 for Window. Landmarks and planes used in Tavano method [11] were determined on each panoramic radiograph (Figure 1). An intra-examiner reproducibility study was undertaken on eight panoramic radiographs by repeating the tracing after two weeks by the same operator. The results confirmed an excellent agreement between the two trials for all the variables ($\geq$95%).

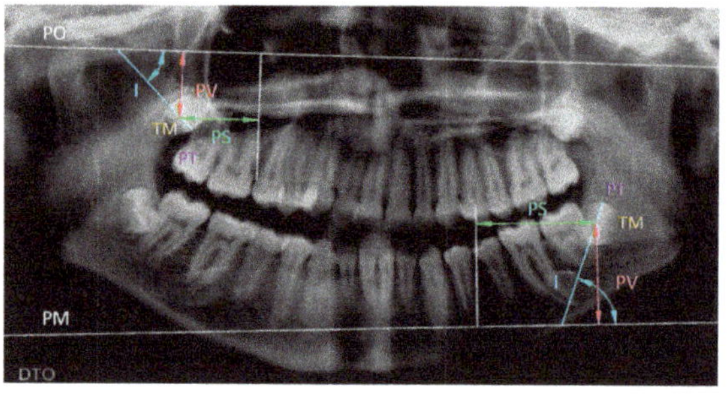

Figure 1. A panoramic radiograph illustrating the different reference planes used in determining the position of the maxillary and mandibular third molars: PO = Orbital plane, PM = Menton plane, PT = Transversal plane, TM = Midpoint of the longest mesio-distal distance of the third molars, I = Inclination, PS = Sagittal position, PV = Vertical position.

The inclination, the sagittal, and the vertical positions of the third molars were evaluated for each radiograph. Descriptive statistics and a mixed model repeated-measures ANOVA were used to determine the measurement differences between the two time points in each group at a significance level of 5%.

3. Results

The only statistically significant difference between the two groups was observed in the vertical position of the lower left third molar, which was more proximal to the Menton plane by -2.24 mm in the FFRD group compared to the controls at p = 0.010 (Table 1). However, all the other extracted measurements were similar in both groups (p > 0.05).

Table 1. Displays the *p* value of the different utilized measurements of the inclination, sagittal and vertical third molar position (18 = upper right third molar, 28 = upper left third molar, 38 = lower left third molar, and 48 = lower right third molar) between the group treated with Forsus^{TM} Fatigue-Resistant Device and controls. The asterisk symbol (0.010 *) indicates a statistically significant *p*-value.

	Third Molar	*p* Value
Inclination	18	0.603
	28	0.958
	38	0.452
	48	0.423
Sagittal position	18	0.292
	28	0.077
	38	0.666
	48	0.345
Vertical position	18	0.453
	28	0.916
	38	0.010 *
	48	0.290

4. Discussion

The mandibular third molar is the most prevalent impacted tooth, followed by the maxillary third molar. The third molars' size and shape dimorphism, variability in position, root formation, duration of calcification, lack of required space and altered eruption path make their eruption one of the most unpredictable events in the evolution of human dentition [8].

It has been demonstrated that FFRD produces relatively more dentoalveolar effects, a combination of mesialization of the lower molars and distalization of the upper molars, which substantially contribute to class II molar relationship correction. The literature reported an increase in the mandibular retromolar area due to the effect of FFRD [10,12]. On the other hand, there is a controversy concerning the influence of FFRD on the position of the maxillary third molar. On one side, Heinrichs et al. [13], confirmed the existence of a significant distalization of the maxillary third molars due to the use of FFRD, while on the other side, Jones et al. [14], observed a significant mesialization effect. In our FFRD group, there was a reduction in the angle of inclination of both the upper and lower third molars. This reduced inclination value was not significantly different compared to controls. This result might reflect the non-significant effect of FFRD on the inclination or the third molar, or it might be a false-negative finding due to using a convenience sample with no power calculation. According to our literature search, this study was the first to assess the effect of non-extraction treatment of class II malocclusion with FFRD on the position of the maxillary and mandibular third molars combined. Using the present findings to conduct another study based on a power calculation is recommended.

Within the limitation of this study, we could conclude that orthodontic treatment of class II malocclusion using FFRD device does not seem to influence the eruption path of third molars; accordingly, the probability of the eruption of third molars is multifactorial and does not rely only on orthodontic treatment with FFDR.

Author Contributions: Principal investigator, undertaking the research study to fulfil the requirements for the degree of specialist in orthodontics, methodology, original draft preparation, J.G.; supervision, conceptualization, writing up, P.M.P.; statistical analysis, J.B.; writing up, editing, I.B. All authors have read and agreed to the published version of the manuscript.

Funding: This research received no external funding.

Institutional Review Board Statement: The study was conducted in accordance with the Declaration of Helsinki and approved by the Institutional Review Board of Egas Moniz School of Health and Science, (protocol code 975, 27 May 2021).

Informed Consent Statement: Informed consent was obtained from all subjects involved in the study.

Data Availability Statement: Research data are available upon request.

Conflicts of Interest: The authors declare no conflict of interest.

References

1. Franchi, L.; Alvetro, L.; Giuntini, V.; Masucci, C.; Defraia, E.; Baccetti, T. Effectiveness of comprehensive fixed appliance treament used with the forsus fatigue resistant device in class II patients. *Angle Orthod.* **2011**, *81*, 678–683. [CrossRef] [PubMed]
2. McNamara, J.A. Components of Class II malocclusion in children 8–10 years of age. *Angle Orthod.* **1981**, *51*, 177–202. [PubMed]
3. Tendulkar, P.M.; Pradhan, T. Effects of fixed twin-block and forsus fatigue resistant device on mandibular third molar angulation–A comparative study. *Ind. J. Health Sci. Biomed. Res. KLEU* **2021**, *14*, 340–347. [CrossRef]
4. Ritto, A.K.; Ferreira, A.P. Fixed functional appliances- a classification. *Funct. Orthod.* **2000**, *17*, 12–30. [PubMed]
5. Vogt, W. The Forsus Fatigue Resistant Device. *J. Clin. Orthod.* **2006**, *40*, 368–377.
6. Padhye, M.N.; Dabir, A.V.; Girotra, C.S.; Pandhi, V.H. Pattern of mandibular third molar impaction in the Indian population: A retrospective clinico-radiographic survey. *Oral Surg. Oral Med. Oral Pathol. Oral Radiol. Endod* **2013**, *116*, e161–e166. [CrossRef] [PubMed]
7. Elsey, M.J.; Rock, W.P. Influence of orthodontic treatment on development of third molars. *Br. J. Oral Maxillofac. Surg.* **2000**, *38*, 350–353. [CrossRef]
8. Hatem, M.; Bugaighis, I.; Taher, E.M. Pattern of third molar impaction in Libyan population: A retrospective radiographic study. *Saudi J. Dent. Res.* **2016**, *7*, 7–12. [CrossRef]
9. Carter, K.; Worthington, S. Predictors of third molar impaction: A systematic review and meta-analysis. *J. Dent. Res.* **2015**, *95*, 1. [CrossRef] [PubMed]
10. Belma, I.; Zühre, Z.; Akarslan, K. Effects of Angle class II correction with the forsus fatigue resistant device on mandibular third molars-A retrospective study. *J. Orofac. Orthop.* **2021**, *82*, 403–412.
11. Tavano, O.; Ursi, W.; Almeida, R.; Henriques, J. Determinação de linhas de referências para medições angulares em radiografias ortopantomográficas. *Odontol. Atual* **1989**, *16*, 22–55.
12. Sakuno, A. Tomographic evaluation of dentoskeletal changes due to the treatment of class II malocclusion with Forsus appliance. *J. Oral Biol. Craniofac. Res.* **2019**, *9*, 277–279. [CrossRef] [PubMed]
13. Heinrichs, D.A.; Shammaa, I.; Martin, C.; Razmus, T.; Gunel, E.; Ngan, P. Treatment effects of fixed intermaxillary device to correction of class II malocclusion in growing patients. *Prog. Orthod.* **2014**, *15*, 145–151. [CrossRef] [PubMed]
14. Jones, G.; Buschang, P.H.; Kim, K.B.; Oliver, D.R. Class II non-extraction patients treated with the forsus fatigue resistant device versus intermaxillary elastics. *Angle Orthod.* **2008**, *78*, 332–338. [CrossRef] [PubMed]

medical sciences forum

MDPI

Proceeding Paper

External Cervical Resorption—The Commonly Misdiagnosed, Destructive Resorption—A Pilot Study †

Marta Alves Duarte * and João Albernaz Neves

Endodontics Department, Egas Moniz School of Health and Science, Caparica, 2829-511 Almada, Portugal; jalbernazneves@gmail.com

* Correspondence: martamatosalvs@gmail.com

† Presented at the 6th International Congress of CiiEM—Immediate and Future Challenges to Foster One Health, Almada, Portugal, 5–7 July 2023.

Abstract: External cervical resorption is a relatively uncommon, often misdiagnosed, destructive form of external resorption. It consists of an uncontrolled growth of resorptive tissue and consequent loss of mineralized tooth tissue, in an apical direction. The clastic cells are responsible for the process, which can lead to severe deterioration and, eventually, tooth loss. To date, however, an etiological factor has not yet been found. Dentists and students still have difficulties in diagnosing and treating it. In May 2023, a structural questionnaire about external cervical resorption was distributed via Google Forms to 104 people between dentists and fourth- and fifth-year students.

Keywords: external cervical resorption; external resorption; diagnose; treatment

Citation: Alves Duarte, M.; Albernaz Neves, J. External Cervical Resorption—The Commonly Misdiagnosed, Destructive Resorption—A Pilot Study. *Med. Sci. Forum* **2023**, *22*, 26. https://doi.org/10.3390/msf2023022026

Academic Editors: José Brito, Nuno Taveira and Ana I. Fernandes

Published: 14 August 2023

1. Introduction

External cervical tooth resorption is a relatively rare [1–3], often misdiagnosed [4,5], destructive form of external tooth resorption [6]. It is characterized by uncontrolled growth of resorptive tissue and consequent loss of mineralized tooth tissue in an apical direction [7–9]. For this reason, it can affect all areas of enamel, dentin and cementum [6]. The clastic cells are responsible for this process, which can lead to severe deterioration and eventual tooth loss [2,4,8,10]. However, there is no consensus on its origin and, to date, no etiological factor has been identified for this pathological resorption [6,7,10,11]. Even so, trauma [6], periodontal disease or orthodontic treatment may be predisposing factors [1,12]. In spite of this, there does not seem to be a significant difference in either gender or age [13]. Also, this pathological resorption is asymptomatic [3,13].

Usually, there are no obvious clinical signs [14]. However, there are some symptoms which are crucial to diagnosing this resorption. Firstly, as the name suggests, it occurs in the cervical part of the tooth [8], below the epithelial junction [1,12] but supra-osseous [11]. Furthermore, if the resorption is relatively small and the pulp is not involved, the sensibility tests are positive. In other words, the tooth is vital [6]. In general, it is shown as a restricted area, both clinically and radiographically. On radiographs, it may appear as a radiolucent [1], well-defined lesion [14], as shown in Figure 1. Generally, a radiopaque line separates the root canal from the cervical invasive resorption lesion [3]. In addition, in some cases it can be seen as a pinkish coronal discoloration [3,8]. However, for the diagnosis and assessment of a resorptive lesion, the use of cone beam computed tomography (CBCT) is recommended [4,11]. This helps to diagnose and assess the entry point [15]. For this reason, it may facilitate the selection of an appropriate treatment approach, especially in advanced cases [15].

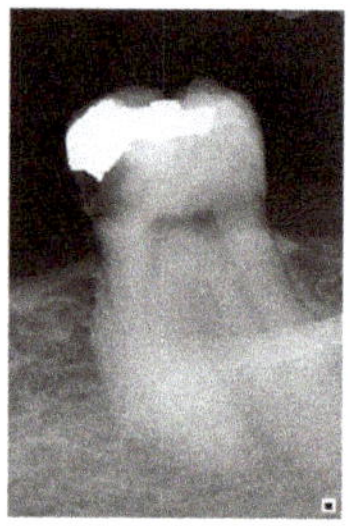

Figure 1. Cervical invasive resorption lesion, through radiograph.

If left untreated, it is expected to increase in size [16]. Therefore, treatment is mandatory, but the real question is: how? Nowadays, there is no well-established protocol for the treatment of cervical invasive reabsorption [4]. However, it should depend on the size, extent and dimension of the invasive area and its relationship to the root canal [8]. It can be treated externally, internally, with a combined approach externally and internally, externally with deliberate replantation or, when there is no other alternative, by extraction [4]. If possible, the infected tissue should be removed and sealed without compromising posterior rehabilitation. If the pulp is not involved and there is good access to the cavity, topical application of a 90% aqueous solution of trichloroacetic acid and manual or mechanical scaling [6,17] may be the answer. If the pulp is involved and the tooth is necrotic, endodontic treatment [6] is required as well as the previous treatment alternatives.

2. Materials and Methods

In this pilot study, a structured questionnaire was administered to one hundred and four people via Google Forms, in May 2023. The inclusion criteria were dentists and fourth- or fifth-year dentistry students. Of the one hundred and four responses, one hundred and one were usable. The questionnaire was used to assess the knowledge, perception and clinical experience towards external cervical resorption.

3. Results and Discussion

As shown in Figure 2a, among the volunteers, 99.0% claimed to be aware of tooth resorptions, although only 45.2%, as demonstrated in Figure 2b, were aware of external cervical resorptions. Of this last percentage, 21.8% were fourth-year students, 30.4% were fifth-year students and 47.8% were dentists. Thus, the results continue to progress as clinical experience grows. Only those who were aware of this pathology were included for further analysis.

(a) (b)

Figure 2. (**a**) Percentage of respondents who know what tooth resorption is. (**b**) Percentage of subjects aware of cervical invasive resorption.

As a result, the evidence shows that only 13% answered all the questions correctly, 30.4% gave one wrong answer and 56.6% gave at least two wrong answers. In addition, over 50% did not know how to radiographically identify a cervical invasive resorption.

In fact, only 45.7% recognized this resorption as a well-demarcated radiolucent area on the radiograph, while 53.3% mistakenly identified it as undefined. Furthermore, 56.5% identified the lesion as supra-osseous, 37.0% as infra-osseous and 6.5% as osseous. It can also be seen that 89.1% classified it as tooth resorption, but 8.7% as bone resorption and 2.2% as a tissue lesion. A total of 69.6% considered CBCT to be the best diagnostic test, while 30.4% considered periapical radiographs. A total of 89.1% described this pathological resorption as asymptomatic, although 10.9% described it as painful. In addition, 54.4% associated it with a pinkish discoloration, 15.2% with a yellowish discoloration, another 15.2% with a brownish discoloration, 13.0% with a reddish discoloration and 2.2% with a grey discoloration.

After analysis, 45.7% felt comfortable diagnosing cervical invasive resorption, although 54.3% did not (Figure 3a). As well demonstrated in Figure 3, among the interviewees who felt comfortable diagnosing this lesion (45.7%), only 5 people (23.8%) got all the answers correct. In fact, 42.9% gave one wrong answer and 33.3% gave at least two wrong answers (Figure 3b). In addition, only 21.7% felt comfortable treating this lesion, as opposed to 78.3% who said they were not comfortable (Figure 3c).

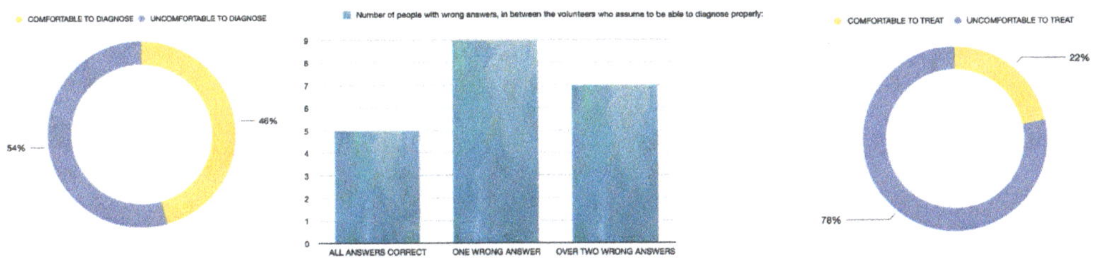

(a) (b) (c)

Figure 3. (**a**) Percentage of people, among those who know what an invasive cervical resorption is, who feel comfortable with the diagnosis. (**b**) This figure shows the number of people who gave all the correct answers, one wrong answer or at least two wrong answers, among the volunteers who consider themselves able to diagnose cervical invasive resorption. (**c**) This figure shows the percentage of volunteers who think they can diagnose invasive cervical resorption and who feel comfortable treating this external resorption.

4. Conclusions

Undoubtedly, external cervical resorption is a devastating form of external resorption and there are still difficulties among dentists and students in its diagnosis and treatment. For this reason, it remains crucial to educate this population about this type of pathology, so that it can be properly diagnosed and treated to avoid tooth loss and major complications.

Author Contributions: M.A.D. and J.A.N.; Conceptualization, M.A.D.; methodology, M.A.D.; validation, M.A.D. and J.A.N.; formal analysis, M.A.D.; investigation, M.A.D.; resources, M.A.D.; data curation, M.A.D. and J.A.N.; writing—original draft preparation, M.A.D.; writing—review and editing, M.A.D. and J.A.N.; visualization, M.A.D.; supervision, M.A.D. and J.A.N.; project administration, M.A.D.; funding acquisition, M.A.D. All authors have read and agreed to the published version of the manuscript.

Funding: This research received no external funding.

Institutional Review Board Statement: Not applicable.

Informed Consent Statement: Not applicable.

Data Availability Statement: Data Available at https://zenodo.org/record/8020464 (accessed on 9 June 2023).

Acknowledgments: This work was supported by Vanessa Machado, Madalena Caçador, Catarina Estácio, Matilde Guerra, Sandro Pinto, Mariana Machete, Rita Ferreira and Matilde Jardim. My thanks to Joana Costa for her incredible attention and availability.

Conflicts of Interest: The authors declare no conflict of interest.

References

1. De Brito, G.M.; Campos, P.S.F.; Mariz, A.C.R.; Simões, D.; Machado, A.W. Invasive Cervical Resorption of Central Incisor during Orthodontic Treatment. *Dent. Press J. Orthod.* **2020**, *25*, 49–58. [CrossRef] [PubMed]
2. Aiuto, R.; Fumei, G.; Lipani, E.; Garcovich, D.; Dioguardi, M.; Re, D. Conservative Therapy of External Invasive Cervical Resorption with Adhesive Systems: A 6-Year Follow-Up Case Report and Literature Review. *Case Rep. Dent.* **2022**, *2022*, 9620629. [CrossRef] [PubMed]
3. Heithersay, G.S. Clinical, Radiologic, and Histopathologic Features of Invasive Cervical Resorption. *Quintessence Int.* **1999**, *30*, 27–37. [PubMed]
4. Espona, J.; Roig, E.; Durán-Sindreu, F.; Abella, F.; Machado, M.; Roig, M. Invasive Cervical Resorption: Clinical Management in the Anterior Zone. *J. Endod.* **2018**, *44*, 1749–1754. [CrossRef] [PubMed]
5. Alqedairi, A. Non-Invasive Management of Invasive Cervical Resorption Associated with Periodontal Pocket: A Case Report. *World J. Clin. Cases* **2019**, *7*, 863–871. [CrossRef] [PubMed]
6. Heithersay, G.S. Invasive Cervical Resorption Following Trauma. *Aust. Endod. J.* **1999**, *25*, 79–85. [CrossRef] [PubMed]
7. Rabinovich, I.M.; Snegirev, M.V.; Golubeva, S.A.; Markheev, C.I. External cervical tooth root resorption. *Stomatologiia* **2022**, *101*, 73–78. [CrossRef] [PubMed]
8. Sarmento, E.B.; Tavares, S.J.; Thuller, K.A.; Falcao, N.P.; de Paula, K.M.; Antunes, L.A.; Gomes, C.C. Minimally Invasive Intervention in External Cervical Resorption: A Case Report with Six-Year Follow-Up. *Int. J. Burns Trauma* **2020**, *10*, 324–330. [PubMed]
9. Tsaousoglou, P.; Markou, E.; Efthimiades, N.; Vouros, I. Characteristics and Treatment of Invasive Cervical Resorption in Vital Teeth. A Narrative Review and a Report of Two Cases. *Br. Dent. J.* **2017**, *222*, 423–428. [CrossRef]
10. O'Mahony, A.; McNamara, C.; Ireland, A.; Sandy, J.; Puryer, J. Invasive Cervical Resorption and the Oro-Facial Cleft Patient: A Review and Case Series. *Br. Dent. J.* **2017**, *222*, 677–681. [CrossRef] [PubMed]
11. Rotondi, O.; Waldon, P.; Kim, S.G. The Disease Process, Diagnosis and Treatment of Invasive Cervical Resorption: A Review. *Dent. J.* **2020**, *8*, 64. [CrossRef] [PubMed]
12. Lo Giudice, G.; Matarese, G.; Lizio, A.; Lo Giudice, R.; Tumedei, M.; Zizzari, V.; Tetè, S. Invasive Cervical Resorption: A Case Series with 3-Year Follow-Up. *Int. J. Periodontics Restor. Dent.* **2016**, *36*, 103–109. [CrossRef]
13. Jeng, P.-Y.; Lin, L.-D.; Chang, S.-H.; Lee, Y.-L.; Wang, C.-Y.; Jeng, J.-H.; Tsai, Y.-L. Invasive Cervical Resorption-Distribution, Potential Predisposing Factors, and Clinical Characteristics. *J. Endod.* **2020**, *46*, 475–482. [CrossRef] [PubMed]
14. Vasconcelos, K.D.F.; de-Azevedo-Vaz, S.L.; Freitas, D.Q.; Haiter-Neto, F. CBCT Post-Processing Tools to Manage the Progression of Invasive Cervical Resorption: A Case Report. *Braz. Dent. J.* **2016**, *27*, 476–480. [CrossRef] [PubMed]
15. Shemesh, A.; Levin, A.; Hadad, A.; Itzhak, J.B.; Solomonov, M. CBCT Analyses of Advanced Cervical Resorption Aid in Selection of Treatment Modalities: A Retrospective Analysis. *Clin. Oral Investig.* **2019**, *23*, 1635–1640. [CrossRef] [PubMed]
16. Nosrat, A.; Dianat, O.; Verma, P.; Levin, M.D.; Price, J.B.; Aminoshariae, A.; Rizzante, F.A.P. External Cervical Resorption: A Volumetric Analysis on Evolution of Defects over Time. *J. Endod.* **2023**, *49*, 36–44. [CrossRef] [PubMed]
17. Heithersay, G.S. Treatment of Invasive Cervical Resorption: An Analysis of Results Using Topical Application of Trichloracetic Acid, Curettage, and Restoration. *Quintessence Int.* **1999**, *30*, 96–110.

medical sciences forum

MDPI

Proceeding Paper

Evaluation of the Morphology of Palatal Rugae in Portuguese Subjects †

Madalena Magalhães [1,*], Iman Bugaighis [1,2], Gunel Kizi [1,2], Inês Póvoas [1], Valter Alves [1] and Ana Delgado [1,2]

1 Department of Orthodontics, Egas Moniz School of Health & Science, Quinta da Granja, 2829-511 Caparica, Portugal; isbugaighis@yahoo.com (I.B.); gunelkizi@outlook.com (G.K.); ines-povoas@hotmail.com (I.P.); valtervpa@gmail.com (V.A.); anasintradelgado@gmail.com (A.D.)
2 Egas Moniz Center for Interdisciplinary Research (CiiEM), Egas Moniz School of Health & Science, Quinta da Granja, 2829-511 Caparica, Portugal
* Correspondence: madalenadsmagalhaes@gmail.com; Tel.: +351-910376206

† Presented at the 6th International Congress of CiiEM—Immediate and Future Challenges to Foster One Health, Almada, Portugal, 5–7 July 2023.

Abstract: Palatal Rugae (PR) are disproportionate bilateral protuberances of connective tissue found on each side of the medpalatine raphe posterior to the incisive papilla. The aim of this study was to investigate the pattern and sexual dimorphism of PR. The prevalence and pattern of PR of 120 Portuguese individuals attending the Orthodontic Clinic of Egas Moniz were categorized. The sample comprised 44.2% males and 55.8% females with similar PR frequencies among sex and age. The total number PR was 624. Straight and wavy PR morphologies were the most frequently observed. There was no sexual dimorphism related to the PR pattern.

Keywords: orthodontic population; palatal rugae; Portuguese individuals; straight palatal rugae

1. Introduction

Palatal Rugae are disproportionate bilateral protuberances of connective tissue found on both sides of the median palatine raphe posterior to the incisive papilla [1]. Comparable to fingerprints, PRs appear from birth, are distinctive to each subject, and are noticed in various configurations and forms. The number of PR ranges from three to five on each side of the midpalatine raphe [1,2].

PR has a number of physiological functions, such as assisting in suckling in babies and enhancing deglutition, food mastication, and speech [1,2].

This cross-sectional observational study aimed to investigate the anatomical morphological pattern and sexual dimorphism of palatal rugae (PR) in Portuguese subjects attending the orthodontic clinic at the Egas Moniz School of Health and Science.

2. Materials and Methods

This sample investigated orthodontic dental casts of 120 Portuguese individuals (mean age 19.9, SD = 6.7 years) attending the Orthodontic Clinic of Egas Moniz School of Health and Science. Each maxillary dental cast was assessed for the prevalence and pattern of PR morphology corresponding to the classification of Thomas and Kotze [3]: straight, wavy, curved, circular, divergent unification, convergent unification and cross-link. Each PR was determined by one operator using a graphite pencil (0.5 mm) under suitable light and magnification by one operator (Figure 1). Any PR less than 3 mm were excluded from the study.

Citation: Magalhães, M.; Bugaighis, I.; Kizi, G.; Póvoas, I.; Alves, V.; Delgado, A. Evaluation of the Morphology of Palatal Rugae in Portuguese Subjects. *Med. Sci. Forum* **2023**, *22*, 3. https://doi.org/10.3390/msf2023022003

Academic Editors: José Brito, Nuno Taveira and Ana I. Fernandes

Published: 8 August 2023

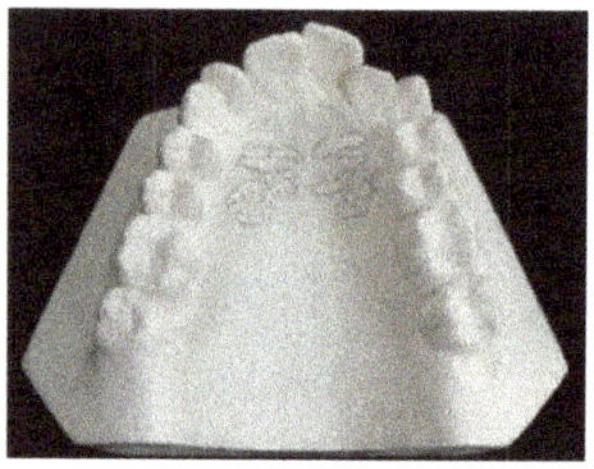

Figure 1. Maxillary study model showing delineated palatal.

An intra-observer reliability test was performed on 15 casts in two-week intervals. The intra-class correlation coefficient revealed an excellent level of reliability (≥0.97).

The data were analysed using Mann–Whitney U and Pearson's Chi-square test at $p < 0.05$.

3. Results

The sample comprised 44.2% males and 55.8% females (Figure 2). Chi-square tests determined a similar correlation between gender and the frequency of PRs ($p \geq 0.372$). The total number of PR included in the analysis was 624 rugae. The most frequent PR shape was straight (44.6%), followed by wavy 257 (40.7%) PR (Figure 3). Likewise, the prevalence of converging (11%), diverging (2.7%) and circular PR (1%) were the least frequent, while cross-linked PR did not exist in our group.

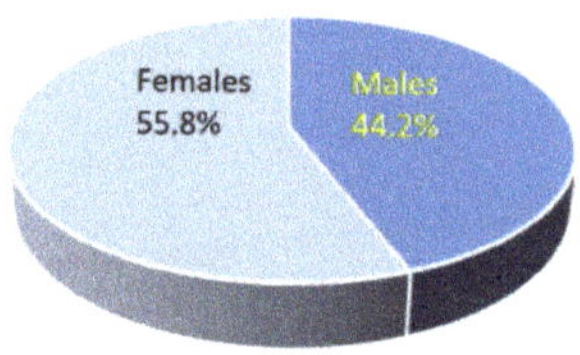

Figure 2. Percentage of males and females in the group.

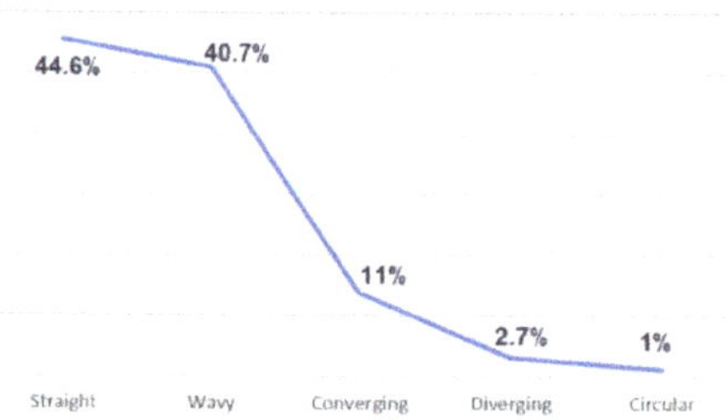

Figure 3. Percentage of different shapes of palatal rugae.

4. Discussion

This retrospective study evaluated the appearance, prevalence and sexual dimorphism of PR in a limited group of Portuguese individuals. Therefore, for the moment, we regard this investigation as a pilot study.

In agreement with our study, several other investigations on Portuguese individuals living in Porto [4] and other populations reported similar prevalence and distribution of PR in both sexes [1,2,5]. In comparison, sexual dimorphism was observed in other populations [6,7].

Straight (44.6%) and wavy (40.7%) PR were the most frequently observed shapes in the Portuguese subjects. Similarly, Santos and Caldas [4] reported a higher prevalence of straight first PR in their Portuguese sample living in Porto. However, their study analysed only each individual´s 1st and 3rd PRs. The occurrence of straight PRs in our sample was more than in Caucasians (15.2%) [1] and Libyans (14.28%) [2]. These variations might be attributed to the different ethnicities of the investigated groups.

The occurrence of diverging and converging PR combined (13.7%) was slightly less than in Caucasians [1] and relatively more than in Libyans [2].

This preliminary study explored the morphological discrepancy and sexual dimorphism of PR morphology in Portuguese individuals. A national research initiative involving larger groups is proposed to acquire a clearer picture of the morphological differences of PR in Portuguese individuals.

5. Conclusions

Straight and wavy PRs were the most common in Portuguese individuals. Furthermore, both sexes had a similar PR frequency and morphology; therefore, using PR to determine sex in the Portuguese subjects attending our clinic is not proposed.

Author Contributions: Conceptualization, M.M. and I.B.; methodology, M.M. and I.B.; software, M.M., I.B. and G.K.; validation, M.M., I.B. and V.A.; formal analysis, M.M. and I.B.; investigation, M.M., I.B. and V.A.; resources, I.B.; data curation, M.M. and I.B.; writing, original draft preparation, M.M., I.B., G.K. and I.P.; writing, review and editing, M.M., I.B., G.K. and I.P.; visualization, I.B.; supervision, I.B., V.A. and A.D.; project administration, I.B., V.A. and A.D. All authors have read and agreed to the published version of the manuscript.

Funding: This research received no external funding.

Institutional Review Board Statement: The study was conducted in accordance with the Declaration of Helsinki and approved by Egas Moniz School of Health and Science.

Informed Consent Statement: Informed consent was obtained from all subjects involved in the study.

Data Availability Statement: Not applicable.

Conflicts of Interest: The authors declare no conflict of interest.

References

1. Kapali, S.; Townsend, G.; Richards, L.; Parish, T. Palatal rugae patterns in Australian aborigines and Caucasians. *Aust. Dent. J.* **1997**, *42*, 129–133. [CrossRef] [PubMed]
2. Abdulmajid, Z.; Bugaighis, I. Evaluation of the morphology of palatal rugae in Libyan school children. *J. Dent. Res.* **2015**, *2*, 24.
3. Thomas, C.J.; Kotze, T.J. The palatal ruga pattern: A new classification. *J. Dent. Assoc. S. Afr.* **1983**, *38*, 153–157. [PubMed]
4. Santos, C.; Caldas, I.M. Palatal rugae pattern in a Portuguese population: A preliminary analysis. *J. Forensic Sci.* **2012**, *57*, 786–788. [CrossRef] [PubMed]
5. Nayak, P.; Acharya, A.B.; Padmini, A.T.; Kaveri, H. Differences in the palatal rugae shape in two populations of India. *Arch Oral Biol.* **2007**, *52*, 977–982. [CrossRef] [PubMed]
6. Fahmi, F.M.; Al-Shamrani, S.A.; Talic, Y.F. Rugae pattern in Saudi population sample of males and females. *SDJ* **2001**, *13*, 92–95.
7. Dohke, M.; Osato, S. Morphological study of the palatal rugae in Japanese. 1. Bilateral differences in the regressive evolution of the palatal rugae. *JPN J. Oral. Biol.* **1994**, *36*, 126–140. [CrossRef]

medical sciences forum

MDPI

Proceeding Paper

The Impact of Incisor Molar Hypomineralisation in a Paediatric Population †

Ana Margarida Matos [1,*], Ana Raquel Barata [1,2], Gunel Kizi [1,2] and Maria Irene Ventura [1,2]

1 Department of Paediatric, Egas Moniz School of Health & Science, Quinta da Granja, 2829-511 Caparica, Portugal; raquelgarciabarata@gmail.com (A.R.B.); gunelkizi@outlook.com (G.K.); imvcr.ireneventura@gmail.com (M.I.V.)
2 Egas Moniz Center for Interdisciplinary Research (CiiEM), Egas Moniz School of Health & Science, Quinta da Granja, 2829-511 Caparica, Portugal
* Correspondence: ana.margarida99@hotmail.com; Tel.: +351-915520032
† Presented at the 6th International Congress of CiiEM—Immediate and Future Challenges to Foster One Health, Almada, Portugal, 5–7 July 2023.

Abstract: Incisor Molar Hypomineralisation (IMH) is a qualitative alteration affecting one or more first permanent molars (FMPs) with the possible involvement of the permanent incisors, and is associated with several dental complications. The aim of this study was to assess the impact of IMH on the quality of life (QoL) of children/adolescents. A total of 56 children attending the Egas MonizDental Clinic (EMDC) were enrolled, 27 females and 29 males, and the most common age was 11 years. Most of them had a mild or moderate degree of IMH. The impact of IMH on the children's quality of life was not significant.

Keywords: Molar-incisor hypomineralisation; perception; impact; quality of life

Citation: Matos, A.M.; Barata, A.R.; Kizi, G.; Ventura, M.I. The Impact of Incisor Molar Hypomineralisation in a Paediatric Population. *Med. Sci. Forum* **2023**, *22*, 50. https://doi.org/10.3390/msf2023022050

Academic Editors: José Brito, Nuno Taveira and Ana I. Fernandes

Published: 13 September 2023

1. Introduction

In the late 1970s, opacities in the FMPs and, in more severe cases, enamel fragmentation were observed and analysed for the first time [1,2]. The term "Incisor Molar Hypomineralisation (IMH)" [3] was first introduced in Bergen at the European Academy of Paediatric Dentistry (EAPD) in 2000 [2,4,5]. The Academy defined IMH as a specific clinical condition, a qualitative defect arising during enamel development, with a systemic and multifactorial origin and affecting one or more FMPs, with possible involvement of the permanent incisors. IMH is associated with several dental complications, such as the rapid development of carious lesions, a loss of structure, poor restorations due to difficult adhesion of the restorative material and hypersensitivity [6,7]. Additionally, affected teeth are more sensitive to temperature and mechanical stimuli. In more severe cases, these stimuli even cause pain that can negatively affect the simplest activities of daily living, like brushing or eating [6]. Finally, there are negative social and aesthetic consequences when incisors are affected [6,7]. Today, it is a highly prevalent defect worldwide, with a global estimate of 17.5 million new cases each year [3], and no difference between men and women [8]. Dentists are faced with a challenging approach for patients with IMH because they are required to overcome the technical challenges associated with the intrinsic characteristics of hypomineralized enamel, and IMH is recognized as a potential public health problem worldwide.

2. Materials and Methods

The sample consisted of children with IMH attending the EMDC between January and May 2022. Participants were selected based on the following inclusion criteria: individuals attending the EMDC with at least one FPM affected by IMH; individuals with ages between 11 and 18 years; patients with no known medical pathologies; and informed consent, free, explained, and signed by the parent or legal guardian. The identification of IMH was based on the criteria established by the EAPD in 2003: white, yellow or brownish spots, mainly in the occlusal and vestibular areas of the tooth surface in at least one of the four PMPs; disintegrated enamel; atypical restorations; and tooth sensitivity and extractions. The questionnaire for the study was designed by the authors and divided into two parts, the first addressed to the child/adolescent and the second to the parent/guardian. The data were then stored in a password-protected remote access database and anonymized, in accordance with the authorization granted to the EMDC database by the National Data Protection Commission. In the second phase of the study, the data were subjected to descriptive and inferential statistical analysis in the second phase of the study; a significance level of 5% was set in the latter case. The study was approved by Egas Moniz Ethics Committee.

3. Results

A total of 56 children were enrolled, 27 females and 29 males (Table 1).

Table 1. Distribution of frequencies and percentages of individuals in the sample by gender.

Gender	Frequency (n)	Percentage (%)
Female	27	48.2
Male	29	51.8
Total	65	100

Regarding the complaints in the functional domain, which includes eating, drinking, talking, playing and sleeping, the percentages of those affected are all below 50% (eating: 16%, drinking: 44.6%, talking: 7.1%, playing: 1.8% and sleeping: 14.3%). When analysing social/emotional well-being, 28.5% of the children admitted to being ashamed of smiling, 12.5% admitted to having been teased and 16.1% had already sought treatment for the spots.

4. Discussion

Relating the self-perception to the age group, we found that 14–18 year olds had the highest awareness of the problem (46.4%), which is supported by the literature [9,10], and also the highest number of complaints, which, according to Freitas Fernandes, can be explained by the fact that this type of defect tends to worsen over time [11]. The percentage of children/adolescents who became aware of the problem themselves was 37.5% of children/adolescents became aware of it themselves, with females being the most aware and interested in the problem (44.4%, Table 2), due to their greater concern for their appearance [7,12].

Table 2. Distribution of frequencies and percentages by gender regarding the identification of stains.

Gender		Parent	MD	Child/Adolescent	Total
Female	Frequency (n)	4	11	12	27
	Percentage (%)	14.8	40.7	44.4	100
Male	Frequency (n)	4	16	9	29
	Percentage (%)	13.7	55.2	31	100

5. Conclusions

Child/adolescent perception of the disease varies by gender and age group. The impact of IMH on children's QoL was not significant.

Author Contributions: Conceptualization, A.M.M. and A.R.B.; methodology, A.M.M. and A.R.B.; software, A.M.M.; validation, A.M.M., A.R.B. and G.K.; formal analysis, A.M.M., A.R.B. and G.K.; investigation, A.M.M., A.R.B. and G.K.; resources, A.R.B. and G.K.; data curation, A.M.M.; writing—original draft preparation, A.M.M., A.R.B. and G.K.; writing—review and editing, A.M.M., A.R.B. and G.K.; visualization, A.R.B. and G.K.; supervision, A.R.B., G.K. and M.I.V.; project administration, A.R.B., G.K. and M.I.V. All authors have read and agreed to the published version of the manuscript.

Funding: This research received no external funding.

Institutional Review Board Statement: In response to the request for an opinion submitted to the Ethics Committee of Egas Moniz, entitled "The impact of incisor-molar hypomineralisation in a paeiatric population", was unanimously approved. The study was conducted in accordance with the Declaration of Helsinki, and approved by the Ethics Committee of the Egas Moniz Higher Education Cooperative (protocol code 1015, 24 February 2023).

Informed Consent Statement: Informed consent was obtained from all subjects involved in the study.

Data Availability Statement: Not applicable.

Conflicts of Interest: The authors declare no conflict of interest.

References

1. Jälevik, B.; Sabel, N.; Robertson, A. Can Molar Incisor Hypomineralization Cause Dental Fear and Anxiety or Influence the Oral Health-Related Quality of Life in Children and Adolescents?—A Systematic Review. *Eur. Arch. Paediatr. Dent.* **2022**, *23*, 65–78. [CrossRef] [PubMed]
2. Jälevik, B. Prevalence and Diagnosis of Molar-Incisor-Hypomineralisation (MIH): A Systematic Review. *Eur. Arch. Paediatr. Dent.* **2010**, *11*, 59–64. [CrossRef] [PubMed]
3. Schwendicke, F.; Elhennawy, K.; Reda, S.; Bekes, K.; Manton, D.J.; Krois, J. Global Burden of Molar Incisor Hypomineralization. *J. Dent.* **2018**, *68*, 10–18. [CrossRef] [PubMed]
4. Alanzi, A.; Faridoun, A.; Kavvadia, K.; Ghanim, A. Dentists' Perception, Knowledge, and Clinical Management of Molar-Incisor-Hypomineralisation in Kuwait: A Cross-Sectional Study. *BMC Oral Health* **2018**, *18*, 34. [CrossRef] [PubMed]
5. Bekes, K.; Amend, S.; Priller, J.; Zamek, C.; Stamm, T.; Krämer, N. Changes in Oral Health-Related Quality of Life after Treatment of Hypersensitive Molar Incisor Hypomineralization–Affected Molars with a Sealing. *Clin. Oral Investig.* **2021**, *25*, 6449–6454. [CrossRef] [PubMed]
6. Joshi, T.; Rahman, A.; Rienhoff, S.; Rienhoff, J.; Stamm, T.; Bekes, K. Impact of Molar Incisor Hypomineralization on Oral Health–Related Quality of Life in 8–10-Year-Old Children. *Clin. Oral Investig.* **2022**, *26*, 1753–1759. [CrossRef] [PubMed]
7. Dantas-Neta, N.B.; Moura, L.d.F.A.d.D.; Cruz, P.F.; Moura, M.S.; Paiva, S.M.; Martins, C.C.; de Lima, M.d.D.M. Impact of Molar-Incisor Hypomineralization on Oral Health-Related Quality of Life in Schoolchildren. *Braz. Oral Res.* **2016**, *30*. [CrossRef] [PubMed]
8. Zhao, D.; Dong, B.; Yu, D.; Ren, Q.; Sun, Y. The Prevalence of Molar Incisor Hypomineralization: Evidence from 70 Studies. *Int. J. Paediatr. Dent.* **2018**, *28*, 170–179. [CrossRef] [PubMed]
9. Da Silva, F.M.F.; Vasconcelos Cruz, C.; Leal, L.; De Castro Costa, M. Aesthetic Perception and Psychological Impact of Molar-Incisor Hypomineralisation aong Patients and Parents. *Dent. 3000* **2019**, *7*, 13–20. [CrossRef]
10. Leal, S.C.; Oliveira, T.R.M.; Ribeiro, A.P.D. Do Parents and Children Perceive Molar-Incisor Hypomineralization as an Oral Health Problem? *Int. J. Paediatr. Dent.* **2017**, *27*, 372–379. [CrossRef] [PubMed]
11. Freitas Fernandes, L.H.; Laureano, I.C.C.; Farias, L.; Andrade, N.M.; Soares Forte, F.D.; Barros Alencar, C.R.; Cavalcanti, A.L. Incisor Molar Hypomineralization and Quality of Life: A Population-Based Study with Brazilian Schoolchildren. *Int. J. Dent.* **2021**, *2021*, 6655771. [CrossRef] [PubMed]
12. Velandia, L.M.; Álvarez, L.V.; Mejía, L.P.; Rodríguez, M.J. Oral Health-Related Quality of Life in Colombian Children with Molar-Incisor Hypomineralization. *Acta Odontol. Latinoam.* **2018**, *31*, 7.

medical sciences forum

MDPI

Proceeding Paper

The Link between Intimate Partner Violence, Outness and Internalized Homophobia in an LGBT+ Sample †

Filipa R. Freire [1,*], Cristina Soeiro [2] and Telma C. Almeida [2]

1 Egas Moniz School of Health and Science, Campus Universitário, Caparica, 2829-511 Almada, Portugal
2 Egas Moniz Center for Interdisciplinary Research (CiiEM), Egas Moniz School of Health and Science, LabPSI—Laboratory of Psychology Egas Moniz, Caparica, 2829-511 Almada, Portugal; c.soeiro@netcabo.pt (C.S.); talmeida@egasmoniz.edu.pt (T.C.A.)
* Correspondence: frfreire12@gmail.com
† Presented at the 6th International Congress of CiiEM—Immediate and Future Challenges to Foster One Health, Almada, Portugal, 5–7 July 2023.

Abstract: Intimate Partner Violence can affect the LGBT+ population with the same prevalence as their heterosexual peers. This study explored how outness and internalized homophobia in the LGBT+ population correlate with intimate partner violence. The Internalized Homophobia Scale, the Conflict Tactics Scale 2 and the Outness Inventory were applied to 48 gay, lesbian, and bisexual men, women and non-binary participants. Higher levels of outness were positively correlated with internalized homophobia and sexual coercion. Higher levels of internalized homophobia were negatively correlated with victimization in negotiations with a partner.

Keywords: LGBT+; homosexuality; intimate partner violence

Citation: Freire, F.R.; Soeiro, C.; Almeida, T.C. The Link between Intimate Partner Violence, Outness and Internalized Homophobia in an LGBT+ Sample. *Med. Sci. Forum* **2023**, *22*, 24. https://doi.org/10.3390/msf2023022024

Academic Editors: José Brito, Nuno Taveira and Ana I. Fernandes

Published: 14 August 2023

1. Introduction

Intimate partner violence (IPV) is a prevalent issue affecting many romantic couples worldwide, regardless of gender or sexuality [1]. Most reported cases of IPV are among heterosexual couples, typically with a woman as the victim and a man as an offender [2,3]. Nevertheless, IPV can exist in the gay, lesbian, and bisexual populations in similar prevalence and violence typology compared to their heterosexual counterparts [4]. When trying to find an explanation for causes or links to IPV in same-gender couples, there are two unique variables in the research: internalized homophobia, and outness.

Outness is the comfort level the individual feels in publicly assuming their same-gender attraction, their bisexuality or homosexuality, and being "out" to friends, family, strangers, and peers [5]. Internalized homophobia is a gay, lesbian, or bisexual person's negative and internalized feelings regarding their sexuality or just gay, lesbian, or bisexuality in their peers, and feelings of shame to be attracted to the same gender or being disgusted by gayness, lesbianism, and bisexuality [6]. Results indicate that internalized homophobia is linked to IPV in LGBT+ couples, affects IPV experiences, and may be a predictor of IPV in these relationships [7]. Studies show that higher levels of outness are linked to lower stress levels and better self-esteem, and can cause positive emotions, reducing IPV [8].

Research also shows that higher internalized homophobia, which includes feelings of low self-esteem and sexuality acceptance, leads to lower levels of outness and decreased probability for an individual to "come out," as well as the reverse, in which individuals who are more "out" are likely to experience low levels of internalized homophobia [9].

This study aims to verify the relationship between outness, internalized homophobia, and IPV among LGBT+ couples. Furthering the research on this topic and the correlation between these two variables (outness and internalized homophobia) will provide a better understanding of the more unique and exclusive characteristics of intimate partner violence in LGBT+ couples and, therefore, also provide more knowledge for professionals to build better resources that are designed to help LGBT+ individuals who suffer from this specific type of violence.

In this study, we expect to find positive correlations between internalized homophobia and intimate partner violence and negative correlations between IPV and outness. We also expect to find positive correlations between internalized homophobia and outness.

2. Materials and Methods

This study comprised 48 LGBT+ adult individuals (39.6% men, 48.8% women, and 16.7% non-binary). A total of 43.8% identified as bisexual or attracted to all genders, while 56.25% identified as homosexual. An amount of 87.5% of participants were aged 18 to 30, and only 12.5% were over 31. Data were collected by sharing online links via Google Forms on LGBT+ groups and communities within several social media websites (such as Instagram, Facebook, and Twitter) using the following: a sociodemographic questionnaire; the Outness Inventory [10], which includes 3 factors (outness to family members, outness when presenting themselves to the world around them and outness to their religious community); the Internalized Homophobia Scale [11], which includes 2 factors (internal factor regarding person's sexuality, and external factor regarding how other people's sexuality is perceived); and the Conflict Tactics Scale 2 [12], which includes 5 factors (negotiation, psychological aggression, physical abuse without injury, physical abuse with injury and sexual coercion). Each factor is separately evaluated for levels of perpetration, as well as levels of victimization.

Informed consent for this study was requested on the online form's first page before submitting these data. No identifiable data were collected. Data were analyzed using IBM SPSS (Statistical Package for Social Sciences) version 27.

3. Results and Discussion

Participants showed above-middle score levels of outness around the middle scores of the scale (around 35), with a total mean score of 35.104 (SD = 13.178). For the Internalized Homophobia Scale, participants showed high levels [10], with a total mean score of 122.083 (SD = 13.733). The mean for the scores of "Perpetration" was 69.333 (SD = 49.133), and for the scores of "Victimization", it was 64 (SD = 49.310) in the Conflict Tactics Scale 2 (CTS2).

Regarding the Conflict Tactics Scale 2, we found significant positive correlations between "Negotiation (aggressor)" and "Negotiation (victim)" (r = 0.801; p = 0.01), between "Negotiation (aggression)" and "Physical aggression with injury (victim)" (r = 0.288; p = 0.047), and between "Negotiation (aggression)" and "Victimization Total" score (r = 0.772; p = 0.01), between "Negotiation (victim)" and "Aggression Total" score (r = 0.768 p = 0.01), between "Psychological aggression (aggressor)" and "Psychological aggression (victim)" (r = 743; p = 0.01) and between "Psychological aggression (aggressor)" and "Victimization total" score (r = 0.460; p = 0.001). We found positive correlations between "Psychological aggression (victim)" and "Aggression Total" score (r = 0.385; p = 0.007), between "Physical aggression without injury (aggressor)" and "Physical aggression without injury (victim)" (r = 0.593; p = 0.01), between "Physical aggression without injury (aggressor)" and "Physical aggression with injury (victim)" (r = 0.492; p = 0.01) and between "Physical aggression without injury (aggressor)" and "Victimization Total score (r = 0.335; p = 0.020). We found significant positive correlations between "Physical aggression without injury (victim)" and "Physical aggression with injury (aggressor)" (r = 0.524; p = 0.01), between "Sexual coercion (aggressor)" and "Sexual coercion (victim)" (r = 0.481; p = 0.001), between "Physical aggression with injury (victim)" and total aggression scores (r = 0.317; p = 0.028), between "Physical aggression with injury (aggressor)" and "Victimization Total" score (r = 0.307; p = 0.034), and between the "Aggression Total" score and "Victimization Total" score (r = 0.841; p = 0.01).

We found significant positive correlations between "Out to world" and "Sexual coercion (aggressor)" (r = 0.454; p = 0.001) and between "Out to world" and the "External" factor of the Internalized Homophobia Scale (r = 0.344; p = 0.017). We also verified significant positive correlations between the Outness Inventory total scores and the "Sexual

Coercion (aggressor)" (r = 0.390; p = 0.006). Results show significant negative correlations between the "External" factor of the Internalized Homophobia Scale and "Negotiation (victim)" (r = −0.341; p = 0.018). In this sense, higher levels of outness or openness about one's sexuality to the rest of the world correlate with high levels of "External" internalized homophobia (such as negative feelings regarding sexuality and same-gender attraction of lesbian, gay, or bisexual peers). Results also indicate that outness to the world, in general, correlates with perpetrating more acts of sexual coercion upon a partner. Lastly, higher "External" internalized homophobia was negatively correlated with victimization during negotiations with a partner; so, higher levels of this type of internalized homophobia correlate with lower levels of victimization during this type of conflict, and vice versa.

Regarding the correlation between outness and internalized homophobia, a positive correlation was found between outness and the external factor of internalized homophobia. This contradicts the literature since it was expected to find negative correlations between the two variables [9].

Results showed that higher levels of outness lead to higher levels of IPV, which contradicted the literature as it was expected to find that higher levels of outness lead to more positive feelings and well-being, which could potentially lead to decreased levels of violence. In comparison, lower levels of outness, lower levels of comfort with one's sexuality, and more negative feelings could potentially lead to increased levels of violence [8].

Results also found no significant correlations between internalized homophobia and most of the CTS2′s factors, which measure the types and prevalence of intimate partner violence. While expectations were to find that higher levels of internalized homophobia and negative feelings regarding one's sexuality (such as anger and shame) correlate with higher levels of violence [7], no such results were found. This could be because a bigger sample of IPV victims and perpetrators would be needed for more accurate statistical results, or it could be due to the complexity of LGBT+ issues, as there are still many characteristics and experiences of this population that need to be explored in the literature for a better understanding of social issues such as IPV within the LGBT+ community [13].

We recommend that future studies use a large and diverse LGBT+ sample that includes more LGBT+ individuals with IPV experiences. This could provide a more accurate statistical analysis, expanding on the current study and potentially explaining some of our results that were inconsistent with the literature. To add to this, and although not the focus of this study, a few participants left ending comments on how they had experienced IPV not with partners of the same gender but instead with partners of the opposite gender in the past. In this sense, it could be relevant for future studies to explore the LGBT+ experience of IPV with partners of the opposite gender, which was not explored in this study. The quality of information about these types of intimate partner violence can also be enhanced using qualitative approaches, such as individual interviews.

Author Contributions: Conceptualization, F.R.F., C.S. and T.C.A.; methodology, F.R.F., C.S. and T.C.A.; software, F.R.F.; validation, F.R.F., C.S. and T.C.A.; formal analysis, F.R.F.; investigation, F.R.F.; resources, F.R.F., C.S. and T.C.A.; data curation, F.R.F.; writing—original draft preparation, F.R.F.; writing—review and editing, F.R.F., C.S. and T.C.A.; visualization, F.R.F.; supervision, C.S. and T.C.A.; project administration, F.R.F. and C.S.; funding acquisition, F.R.F., C.S. and T.C.A. All authors have read and agreed to the published version of the manuscript.

Funding: This research received no external funding.

Institutional Review Board Statement: The study was conducted following the Declaration of Helsinki, and approved by the Institutional Review Board of Egas Moniz School of Health and Science.

Informed Consent Statement: Informed consent was obtained from all subjects involved in the study.

Data Availability Statement: Data availability is unavailable due to privacy and ethical restrictions.

Acknowledgments: The authors would like to express the most profound gratitude to all the participants who voluntarily provided their information to this study.

Conflicts of Interest: The authors declare no conflict of interest.

References

1. Catalano, S.M. *Intimate Partner Violence, 1993–2010*; US Department of Justice, Office of Justice Programs, Bureau of Justice Statistics: Washington, DC, USA, 2012.
2. Cezario, A.C.F.; Fonseca, D.S.; do Carmo Lopes, N.; Lourenço, L.M. Violência entre parceiros íntimos: Uma comparação dos índices em relacionamentos hetero e homossexuais. *Temas Em Psicol.* **2015**, *23*, 565–575.
3. Richards, A. Domestic violence and the male victim: A study of the prevalence of domestic violence in homosexual male partnerships, and the laws and social policies that exist to control and prevent it. *Plymouth Law Crim. Justice Rev.* **2020**, *12*, 118–140. [CrossRef]
4. Alejo, K. Long-term physical and mental health effects of domestic violence. *Themis Res. J. Justice Stud. Forensic Sci.* **2014**, *2*, 5.
5. Pierre, M.S.; Senn, C.Y. External barriers to help-seeking encountered by Canadian gay and lesbian victims of intimate partner abuse: An application of the barriers model. *Violence Vict.* **2010**, *25*, 536–552.
6. Rostosky, S.S.; Riggle, E.D.; Gray, B.E.; Hatton, R.L. Minority stress experiences in committed same-sex couple relationships. *Prof. Psychol. Res. Pract.* **2007**, *38*, 392.
7. Balsam, K.F. Nowhere to hide: Lesbian battering, homophobia, and minority stress. *Women Ther.* **2001**, *23*, 25–37.
8. Morris, J.F.; Waldo, C.R.; Rothblum, E.D. A model of predictors and outcomes of outness among lesbian and bisexual women. *Am. J. Orthopsychiatry* **2001**, *71*, 61–71. [PubMed]
9. Weber-Gilmore, G.; Rose, S.; Rubinstein, R. The impact of internalized homophobia on outness for lesbian, gay, and bisexual individuals. *Prof. Couns. Res. Pract.* **2011**, *1*, 163–175.
10. Mohr, J.; Fassinger, R. *Outness Inventory [Database Record]*; Retrieved from PsycTESTS; APA: Washington, DC, USA, 2000. [CrossRef]
11. Ross, M.W.; Rosser, B.R. Measurement and correlates of internalized homophobia: A factor analytic study. *J. Clin. Psychol.* **1996**, *52*, 15–21. [CrossRef]
12. Straus, M.A.; Hamby, S.L.; Boney-McCoy, S.U.E.; Sugarman, D.B. The revised conflict tactics scales (CTS2) development and preliminary psychometric data. *J. Fam. Issues* **1996**, *17*, 283–316. [CrossRef]
13. Edwards, K.M.; Sylaska, K.M. The perpetration of intimate partner violence among LGBTQ college youth: The role of minority stress. *J. Youth Adolesc.* **2013**, *42*, 1721–1731. [CrossRef] [PubMed]

medical sciences forum

MDPI

Proceeding Paper

Digital Compounding in Pharmacies: A Pilot Stability Study †

Ana I. Fernandes [1,*], Gabriela G. Pereira [2] and João F. Pinto [2]

1 Egas Moniz Center for Interdisciplinary Research (CiiEM), Egas Moniz School of Health & Science, 2829-511 Caparica, Portugal
2 iMed.ULisboa—Research Institute for Medicines, Faculdade de Farmácia, Universidade de Lisboa, Av. Prof. Gama Pinto, 1649-003 Lisboa, Portugal; garrastazugp@gmail.com (G.G.P.); jfpinto@ff.ul.pt (J.F.P.)
* Correspondence: aifernandes@egasmoniz.edu.pt; Tel.: +351-212946823
† Presented at the 6th International Congress of CiiEM—Immediate and Future Challenges to Foster One Health, Almada, Portugal, 5–7 July 2023.

Abstract: The manufacture of medicines on demand for a particular patient, at the point of care, may be achieved via 3D printing, improving therapeutic outcomes, medication adherence and patient safety. Tablets are often printed using a combination of hot-melt extrusion and fused deposition modelling. In this work, paracetamol-loaded hydroxypropylcellulose filaments were produced; their extrudability and printability were aided by a plasticizer and a lubricant. For printability, 11% humidity was the ideal storage condition for filaments up to 6 months. The tablets produced complied with the uniformity of mass and content requirements, and showed delayed drug release; these characteristics were maintained after in-use stability testing for 30 days.

Keywords: 3D printing; digital compounding; fused deposition modelling (FDM); humidity; paracetamol; pharmacies; storage stability; in-use stability; tablet

1. Introduction

The patient-centric design and manufacture of medicines may be achieved via 3D printing (3DP), an emergent technology that allows the digital customization of the dose and/or the design of the dosage form to tailor drug release [1]. Likewise, the on-demand manufacture of dosage units containing several drugs, according to a distinct dose and prescription regimen, simplifies life and increases compliance and the safety of polymedicated individuals.

Due to its simplicity, cost-effectiveness and versatility, fused deposition modelling (FDM) is one of the most popular 3DP techniques, comprising the successive layering of a molten drug-loaded thermoplastic polymer. FDM has been extensively used in the preparation of simple or complex dosage forms [2], with a variety of doses, shapes and release kinetics. However, besides being inadequate for heat-sensitive drugs [1], printers have to be fed with filaments, which are commercially unavailable. As such, these have to be produced beforehand using hot-melt extrusion (HME), increasing the duration and complexity of the manufacture. Depending on the formulation and processing parameters used, the integration of HME and FDM may not be straightforward [1], and may require evaluation and tuning. For compounding, it is also critical to expedite the process.

Since FDM printers are economical, compact and easy to operate, the decentralization of production from large pharmaceutical industries to community pharmacies and local hospitals is possible [3], being viable even in remote locations and when supply chains are disrupted. In addition to the potential of digital compounding to meet individual medical/clinical needs, the technology will also add value to the role of the pharmacist.

The production of medicines on demand at the dispensing site would benefit from stocking filaments (pre-produced in-house or industrially from a range of polymeric matrices and different drugs/drug loads for dose flexibility) that would be fed into the printer

Citation: Fernandes, A.I.; Pereira, G.G.; Pinto, J.F. Digital Compounding in Pharmacies: A Pilot Stability Study. *Med. Sci. Forum* **2023**, 22, 9. https://doi.org/10.3390/msf2023022009

Academic Editors: José Brito and Nuno

Published: 9 August 2023

when and as required, according to the needs of a particular patient. The stability of filaments is a major concern, particularly of those produced using cellulosic matrices, which are hygroscopic and lose printability due to the plasticizing effect of water [2].

Therefore, the aim of this work was to establish formulation and processing parameters, as well as storage conditions for filaments, to maximize shelf-life while maintaining printability for the production of tablets, tailored to a specific patient, at the point of care. The stability of filaments was evaluated over time under different moisture conditions; the in-use stability of the 3D-printed tablets was also evaluated up to 30 days in defined relative humidity (RH) environmental settings.

2. Materials and Methods

Paracetamol (PCT; AcoFarma, Madrid, Spain) was used as a model drug, and hydroxypropylcellulose (HPC; Klucel LF®, Ashland, Wilmington, DE, USA), Soluplus® (SLP; BASF, Ludwigshafen, Germany) and magnesium stearate (MgS; Roic Farma, Barcelona, Spain) were used as excipients. Lithium chloride (11% RH), magnesium nitrate (53% RH), sodium chloride (75% RH), potassium nitrate (93% RH), silica gel (below 40% RH) and HPLC-grade methanol were supplied by Sigma Aldrich (Darmstadt, Germany).

Six different drug-loaded filaments (F1–F6; Table 1 for composition) were produced by HME using a single-screw extruder (Noztek Pro, Noztek, Shoreham, UK; nozzle die diameter of 1.5 mm, screw speed of 20 rpm, extruding temperature of 100 ± 2 °C and 65 ± 2 °C).

Table 1. Composition (% *w*/*w*) of the formulations used to produce filaments (F) and tablets (T).

Filament/Tablet	HPC	Soluplus	Magnesium Stearate	Paracetamol
F1/T1	54.00	15.00	1.00	30.00
F2/T2	54.00	14.00	2.00	30.00
F3/T3	52.50	12.50	5.00	30.00
F4/T4	39.50	9.50	1.00	50.00
F5/T5	39.50	8.50	2.00	50.00
F6/– [1]	37.50	7.50	5.00	50.00

[1] Filament F6 could not be printed.

Tablets were printed from the corresponding PCT-loaded filaments using a commercial FDM 3D printer (DeltaWASP 2040 Turbo2, Massa Lombarda, Italy). Disk-shaped tablets with a mean diameter of 6.0 mm and a thickness of 3.00 mm were digitally designed (3D Sprint Software v2.11, 3D Systems, Rock Hill, SC, USA) and the .stl file generated was imported into Cura (Ultimaker B.V., v15.04.2, Utrecht, The Netherlands) for slicing. Printing parameters were set to the following: 180 °C and 220 °C extrusion temperature (respectively, for 30 and 50% PCT); 90 mm/s extrusion speed; 150 mm/s traveling speed; 2 shells, 100% infill; and 0.20 mm layer thickness. Batches of 10 or 20 tablets from each filament were prepared.

The sensitivity of the filaments to moisture was evaluated via storage at 25 °C under controlled RH environments provided by desiccators containing saturated salt solutions. Filaments were collected at day 0, 7, 14, 21, 30 and 180. The in-use stability of tablets was evaluated over a period of 30 days in sealed flasks at 11% RH/25 °C and ambient RH/25 °C in the presence and absence of silica gel.

The mean mass of tablets was determined by individually weighing 10 units, using a calibrated analytical balance (Sartorius). The mean diameter and thickness were measured (n = 10) in two different parallels using a digital caliper (Toolland, 150mm/6″, Shanghai, China). Likewise, filaments were measured in ten different positions for diameter estimation.

PCT was quantified in individual tablets, or pieces of filaments using HPLC [4]. The drug content was calculated as the amount of drug per tablet, or mass unit in filaments.

In vitro drug release studies of the 3DP tablets were performed using a paddle apparatus (AT7 Sotax AG, Aesch, Switzerland; 50 rpm, 900 mL of phosphate buffer, pH 6.8 as the dissolution media, at 37 ± 0.5 °C). Samples (5.0 mL) were withdrawn at time intervals and quantified using HPLC (HP Agilent 1100, Waldbronn, Germany; λ max 247 nm). From the dissolution profiles, the time required for 50% drug release (t_{50}), the dissolution rate (DR) and the similarity (f_2) factor were calculated.

3. Results and Discussion

Compatibility between HME and FDM is critical for the success of 3DP, and a good extrusion behavior does not guarantee accurate and reliable printing. Thus, initial experiments considered the adjustment of the formulation and processing parameters for both techniques. HPC and PCT were used as a model cellulose-based matrix and drug, respectively. To ease the extrusion process and printability, SLP as a solubilizer/plasticizer, and MgS as a hydrophobic lubricant were also introduced in the formulations (Table 1).

3.1. Production of Drug-Loaded Filaments and Effect of Moisture on Printability

Filaments with two drug loads were produced (Table 1) in order to offer dose flexibility. All the formulations were easily hot-melt extruded under the conditions tested, and PCT-loaded filaments F1–F6 were obtained. These were off-white, smooth and flexible, with a mean diameter of ≈1.75 mm, in accordance with the printer manufacturer tolerance. The compliance of dimensions is vital for adequate printer feeding. The drug content of the filaments was assayed, at different locations, before the printing process used to evaluate the effectiveness of the components' blending and to discard drug loss during extrusion was performed. All the filaments showed drug loadings close to the theoretical value (between 99.90–102.08%).

Printability was ascertained via the ability to print tablets with simple and more complex geometries. The temperature was as low as possible to allow constant flow out of the printer's nozzle and below the degradation temperature of PCT (240 °C). To maintain extrudability and printability, the drug fraction could not exceed 50%; equally, filaments containing 50% PCT (F4 and F5) required a higher printing temperature (220 °C).

Immediately after production, all the filaments were consistently squeezed by the printer's gears and ruptured, precluding printing to finish. Hence, the filaments were left to settle for 7 days under different controlled RH values to evaluate the influence of moisture on printability. Under high RH (55, 75 and 95%), the filaments became sticky and lost shape due to water uptake, which plasticized the matrix and modified its mechanical behavior. Appropriate apparent elasticity–brittleness was obtained after storage at 11% RH for filaments F1–F5; filament F6, which had a slightly smaller diameter, could not be fed into the printer.

Filaments F1–F5 retained their dimensions, drug content, and adequate rheology for printing up to 6 months under the same storage (11% RH) and printing conditions.

3.2. Characterization and In-Use Stability of Tablets

Filaments F1–F5 could be printed into tablets (T1–T5), whose average mass, dimensions, and drug content are shown in Table 2. Besides being uniform in their dimensions and shape (in agreement with the digital template), the tablets complied with the Eur. Pharm. [4] regarding the intra-batch uniformity of mass, regardless of the batch size produced. The drug content of the tablets was within the range 100 ± 5% of the theoretical value, suggesting no drug loss.

Considering that tablets are produced on demand and in small batches as a result of personalized prescriptions, stability testing was performed for 30 days in a desiccator (11% RH) and under conditions that mimicked tablet usage by the patient. After this time, the tablets maintained identical masses, dimensions, and amounts of PCT, regardless of the storage conditions, suggesting the good and desirable stability of both the matrix and drug.

Table 2. Properties of tablets 3D-printed from filaments stored at 11% RH for 7 days, immediately after production.

Sample	Mass (mg) [1]	Diameter (mm) [1]	Thickness (mm) [1]	Drug Content (mg/tablet) [2]
T1	268 ± 0.005	5.91 ± 0.07	2.95 ± 0.022	83.20 ± 0.002
T2	252 ± 0.012	5.95 ± 0.05	3.00 ± 0.014	75.61 ± 0.004
T3	277 ± 0.011	5.98 ± 0.02	3.00 ± 0.029	84.48 ± 0.003
T4	251 ± 0.005	5.97 ± 0.04	3.00 ± 0.014	124.09 ± 0.002
T5	247 ± 0.012	5.96 ± 0.04	3.00 ± 0.017	126.40 ± 0.006

[1] n = 10 for mass and dimensions; [2] n = 3 for drug content.

The results of the in vitro dissolution testing of tablets, immediately after printing and after 30 days storage in silica gel, are presented in Figure 1A,B. The use of large—molecular-weight polymers, such as HPC, resulted in extended, but complete, drug release patterns. Due to the swelling of HPC, the matrix expanded in the dissolution media, followed by the steady erosion of the polymer, leading to a somewhat complex release behavior.

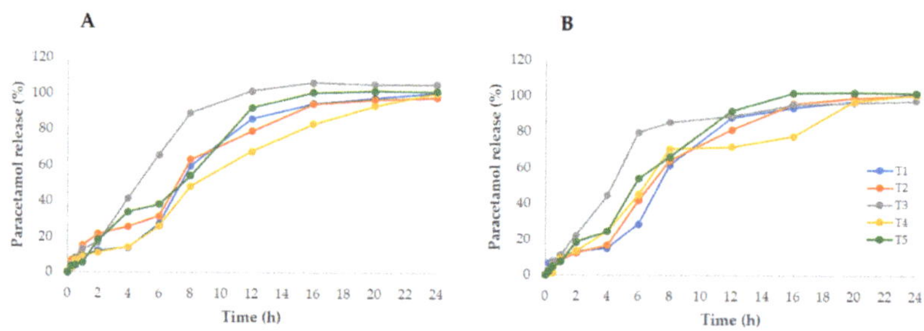

Figure 1. PCT release from 3D-printed tablets (T1–T5) immediately after printing (**A**) and after 30 days of storage in a bottle containing silica gel (**B**).

The dissolution profiles enabled the kinetic parameters (t_{50} and DR) and similarity factors (f_2) to be calculated, which aided a comparison between the different formulations. As expected, the decrease in HPC (as the PCT content increased) resulted in greater DR (T1–T3 > T4–T5), suggesting that, indeed, the matrix controls drug release; as for t_{50}, the opposite trend could be observed. Formulation T3, with the highest MgS content, stood out for the lowest t_{50} (≈4.73 h), despite its hydrophobic nature. At day 30, a comparison with the initial corresponding release profile resulted in $f_2 > 50$, indicating similar curves.

These results, although preliminary, contribute to the yet limited knowledge about the stability of both filaments and tablets manufactured via FDM–3D printing. The formulation developed may also be used as a polymeric carrier for drugs other than PCT, provided the necessary adjustments are made.

Author Contributions: Conceptualization, G.G.P., A.I.F. and J.F.P.; methodology, investigation and formal analysis, G.G.P.; writing—review and editing, A.I.F.; resources, project administration and funding acquisition, A.I.F. and J.F.P. All authors have read and agreed to the published version of the manuscript.

Funding: This work was supported by the Fundação para a Ciência e a Tecnologia, grant number PTDC/CTM CTM/30949/2017 (Lisboa 010145 Feder 030949).

Institutional Review Board Statement: Not applicable.

Informed Consent Statement: Not applicable.

Data Availability Statement: Data is not publicly available.

Acknowledgments: Ashland is acknowledged for providing the Klucel LF® samples.

Conflicts of Interest: The authors declare no conflict of interest.

References

1. Dumpa, N.; Butreddy, A.; Wang, H.; Komanduri, N.; Bandari, S.; Repka, M.A. 3D printing in personalized drug delivery: An overview of hot-melt extrusion-based fused deposition modeling. *Int. J. Pharm.* **2021**, *600*, 120501. [CrossRef] [PubMed]
2. Melocchi, A.; Uboldi, M.; Maroni, A.; Foppoli, A.; Palugan, L.; Zema, L.; Gazzaniga, A. 3D printing by fused deposition modeling of single- and multi-compartment hollow systems for oral delivery—A review. *Int. J. Pharm.* **2020**, *579*, 119155. [CrossRef] [PubMed]
3. Beer, N.; Kaae, S.; Genina, N.; Sporrong, S.K.; Alves, T.L.; Hoebert, J.; De Bruin, M.L.; Hegger, I. Magistral Compounding with 3D Printing: A Promising Way to Achieve Personalized Medicine. *Ther. Innov. Regul. Sci.* **2023**, *57*, 26–36. [CrossRef] [PubMed]
4. EDQM (Ed.) *Council of Europe European Pharmacopoeia*, 10th ed.; EDQM: Strasbourg, France, 2019; ISBN 9789287175250.

medical sciences forum

MDPI

Proceeding Paper

Intra-Rater and Inter-Rater Reliability of the Kinvent Hand-Held Dynamometer in Young Adults †

Manuel Barbosa de Almeida, Carlos Oliveira, Guilherme Ornelas, Tiago Soares, Jean Souto, Ana Rita Póvoa, Luciano Maia Alves Ferreira and Ana Laura Ricci-Vitor *

Egas Moniz Center for Interdisciplinary Research (CiiEM), Egas Moniz School of Health & Science, Campus Universitário, Quinta da Granja, 2829-511 Almada, Portugal; mbalmeida@egasmoniz.edu.pt (M.B.A.); 115788@alunos.egasmoniz.edu.pt (C.O.); 115922@alunos.egasmoniz.edu.pt (G.O.); 115969@alunos.egasmoniz.edu.pt (T.S.); jjarrier@egasmoniz.edu.pt (J.S.); rpovoa@egasmoniz.edu.pt (A.R.P.); lucianomaia@egasmoniz.edu.pt (L.M.A.F.)

* Correspondence: lricci@egasmoniz.edu.pt

† Presented at the 6th International Congress of CiiEM—Immediate and Future Challenges to Foster One Health, Almada, Portugal, 5–7 July 2023.

Abstract: Health professionals like physiotherapists require additional choices in portable dynamometers to conduct evaluations pre- and post-intervention in order to accurately measure the efficacy of treatments and patient progression and to adjust rehabilitation goals. New dynamometers have arrived on the market, but there is no evidence for performance of the Kinvent. This study aimed to investigate intra- and inter-rater reliability (ICC one-or-two-way-random-model/Bland–Altman) of the Kinvent hand-held dynamometer in the muscle groups of the lower and upper limbs (COSMIN guidelines). The Kinvent showed a good to excellent intra- and inter-rater reliability for almost all the upper and lower limbs movements assessed. Ankle dorsiflexion was moderate in all assessments.

Keywords: muscle strength; muscle strength dynamometer; upper extremity; lower extremity; dynamometry; intra-rater reliability; inter-rater reliability; muscle performance

Citation: de Almeida, M.B.; Oliveira, C.; Ornelas, G.; Soares, T.; Souto, J.; Póvoa, A.R.; Ferreira, L.M.A.; Ricci-Vitor, A.L. Intra-Rater and Inter-Rater Reliability of the Kinvent Hand-Held Dynamometer in Young Adults. *Med. Sci. Forum* **2023**, *22*, 12. https://doi.org/10.3390/msf2023022012

Academic Editors: José Brito, Nuno Taveira and Ana I. Fernandes

Published: 9 August 2023

1. Introduction

Muscle force assessment is often essential in the physical examination of patients [1–3]. In the clinical setting, manual muscle testing stands as the main approach for assessing strength levels. When evaluating muscle strength, it is crucial to consider multiple factors, such as test standardization, proper positioning, the careful observation of the patient's technique during the test, and ensuring an environment to prevent any pain/discomfort that might hinder the participant from achieving a maximal contraction [3]. However, the influence of physiotherapist skill and experience on the results has sparked controversy regarding this test's reliability as an assessment technique and has led to the development of instruments capable of eliminating this problem, such as hand-held dynamometers (HHDs) and isokinetic equipment [1–3]. Isokinetic dynamometry is considered the gold standard in strength assessment with well-documented validity and reliability [4,5]. However, it is equipment that is expensive, time consuming, and difficult to access, making its clinical use questionable. In the light of this, HHDs represent an answer to the aforementioned disadvantages, providing an affordable, portable, and time-efficient solution for measuring isometric strength [1–3]. In recent studies, HHDs have demonstrated validity and reliability when compared to isokinetic dynamometers [1,2,6]. Although studies have reported valid and reliable results when compared to the gold standard, new pieces of equipment have been arrived on the market for the use of professionals such as physiotherapists. In this proceeding review, one study using Kinvent [7] is included in the examined literature, but it does not include lower limbs assessments and the increasing use of this device among physiotherapists requires the determination of its psychometric values. Since other brands of HHD have yielded good results, our hypothesis is that Kinvent will also exhibit good intra- and inter-rater reliability for young adults. Therefore, the study objective was to assess the intra- and inter-rater reliability of the Kinvent HHD in the muscle groups of the lower and upper limbs.

2. Materials and Methods

2.1. Study Design

This study had a cross-sectional design and was carried out as a two-day assessment. The first day included (1) the warmup and (2) three acceptable measurements (out of five attempts, excluding the highest and the lowest) of the maximum voluntary isometric contraction for each muscle group, assessed by both Rater 1 (R1) and Rater 2 (R2). After 48 h, the described procedure was repeated for Rater 1. Intra-rater reliability was assessed across the two days of assessment for Rater 1. Inter-rater reliability was determined by comparing the mean of the three measurements of different movements for each participant made in one day between the two raters. The order, pairing, and assessment of the raters' sequences between upper or lower limb muscles group were randomized each week on the website randomization.com. The primary outcome was the maximum voluntary isometric contraction, which was concealed from the raters. This procedure followed the COSMIN guidelines.

2.2. Participants and Setting

Students from Egas Moniz School of Health & Science, Caparica, Almada, Portugal, representing a convenience sample, were invited to participate in this study during their internships at the Egas Moniz University Clinic in Almada. Entry criteria included participants who reported themselves as healthy on the Physical Activity Readiness Questionnaire for Everyone (PAR-Q+), were of either sex, and were between 18 to 30 years. Participants with a history of major injuries in the three months preceding the testing, permanent impairments, or medical conditions that would hinder participation in exercise were excluded from the study.

2.3. Procedure

The measurements were assessed using the Kinvent (KFORCE Pro, v. 5.4.9) HHD by three physiotherapy professors/raters from the Egas Moniz School of Health Sciences. Participants underwent training and received instructions on how to use the Kinvent prior to data collection. This training was provided by an experienced practitioner (with more than 20 years as a physiotherapist and who had worked with the Kinvent for 2 years), who was familiar with this piece of equipment.

During the assessments, the participants were instructed to exert maximum effort during an isometric contraction to assess isometric strength. To do this, the examiner stabilizes the dynamometer in a fixed position while the participant applies maximal effort against it. Instructions to gradually increase the contraction for one second in order to avoid position and stabilization errors and to "push as hard as you can, as hard as you can" for five seconds were given in a loud voice before and during the test, respectively. Participants were directed to perform three consecutive actions with maximal effort with their dominant limb, with each action lasting five seconds, followed by an interval of 30 s of rest between repetitions to avoid fatigue.

For the lower limbs muscle groups, participants warmed up on a static cycle ergometer, while for the upper limbs, they used a pulley for five minutes before proceeding with strength measurements. The evaluated movements of the lower limbs were hip flexion and abduction, knee flexion and extension, and ankle dorsiflexion on participants' dominant leg. For the upper limbs, the movements were shoulder abduction and flexion, elbow extension and flexion, and wrist extension and flexion. The testing sequence progressed from proximal to distal joints.

2.4. Statistics

Statistical Package for Social Sciences (SPSS®) 26 software (IBM Corp, Armonk, NY, USA) was used for reliability by computing intra-class correlation coefficients (ICCs). ICCs were calculated using a single measurement—specifically, the mean value obtained during testing—to assess intra-rater reliability. An ICC (one-way random model) was employed for evaluating intra-rater reliability, while an ICC (two-way random model) was utilized for assessing inter-tester reliability. The ICC values were classified as moderate (0.50–0.75), good (0.76–0.90), and excellent (>0.90). Bland–Altman plots (BA) were used to complete inter-rater analysis and define absolute reliability.

3. Results

A total of twelve individuals, seven of whom were women (58.34%), of 21.83 ± 3.16 years and 22.76 ± 2.03 kg/m^2, participated in the study. The intra- and inter-rater reliabilities are described in Table 1. The BA confirmed all measures to be strongly or very strongly correlated, except for hip flexion. The upper and lower limits of agreement presented, in general, non-concordant measures.

Table 1. Intra-rater reliability and inter-rater reliability for upper and lower limbs.

Movement (kgf)	Rater 1—Day1 (Mean ± SD)	Rater 1—Day2 (Mean ± SD)	Intra-Rater Reliability		Rater 2—Day2 (Mean ± SD)	Inter-Rater Reliability	
			ICC (95% CI)	p		ICC (95% CI)	p
Upper limbs							
Shoulder Flex.	19.13 ± 4.58	16.93 ± 5.90	0.707(0.273–0.905)	0.003	16.00 ± 6.13	0.922 (0.273–0.905)	<0.001
Shoulder Abd.	18.80 ± 4.86	16.98 ± 6.42	0.850 (0.575–0.945)	<0.001	16.14 ± 5.65	0.917 (0.739–0.975)	<0.001
Elbow Flex.	26.69 ± 8.98	25.38 ± 10.03	0.951 (0.854–0.985)	<0.001	25.30 ± 9.68	0.985 (0.948–0.996)	<0.001
Elbow Ext.	17.77 ± 6.60	16.49 ± 6.69	0.912 (0.734–0.973)	<0.001	16.39 ± 6.69	0.961 (0.872–0.989)	<0.001
Wrist Flex.	17.88 ± 6.28	17.51 ± 7.17	0.854 (0.586–0.955)	<0.001	17.59 ± 6.46	0.894 (0.675–0.968)	<0.001
Wrist Ext.	15.73 ± 5.24	14.84 ± 5.55	0.874 (0.636–0.962)	<0.001	15.65 ± 6.70	0.929 (0.773–0.979)	<0.001
Lower limbs							
Hip Flex.	32.27 ± 9.68	31.94 ± 7.44	0.796 (0.453–0.936)	<0.001	32.08 ± 8.97	0.900 (0.692–0.970)	<0.001
Hip Abd.	35.15 ± 10.26	34.00 ± 8.75	0.840 (0.552–0.951)	<0.001	33.70 ± 9.36	0.916 (0.734–0.975)	<0.001
Knee Flex.	18.90 ± 5.19	18.50 ± 4.95	0.939 (0.811–0.982)	<0.001	18.19 ± 4.87	0.863 (0.594–0.959)	<0.001
Knee Ext.	57.74 ± 17.68	59.06 ± 15.67	0.909 (0.726–0.972)	<0.001	54.02 ± 14.78	0.876 (0.604–0.965)	<0.001
Ankle Dorsiflex.	26.20 ± 7.08	23.86 ± 5.76	0.707 (0.392–0.926)	<0.001	22.85 ± 3.97	0.722 (0.281–0.911)	0.03

Abbreviations: kgf = kilogram-force; SD = standard deviation; ICC = intraclass correlation coefficient; CI = confidence interval; Flex. = flexion; Abd. = abduction; Ext. = extension. There were no missing data.

4. Discussion

The intra-rater reliability showed a similar correlation classification between the upper and lower limbs. However, for the inter-rater reliability, the upper limb correlation classifications were higher than those of the lower limbs. In addition, ankle dorsiflexion was moderate in all assessments. A systematic review [2] demonstrated that this HHD is a valid and reliable instrument for measuring muscle strength, except in regard to the largest joints, such as the knee. This could also explain the wider CI in some measurements. Also, a recent study [8] found a good-to-excellent correlation in isometric lower limb strength and power in a healthy population, with particular emphasis on proximal muscle groups. Our results are in accordance with these findings.

Some limitations need to be considered such as the small sample used (effect size 0.68, α error 0.05, power 80%) and the lack of an available isokinetic dynamometer to compare the results with the gold standard. Another limitation was related to the interval between assessments, which was 48 h. Despite our results, we cannot neglect its influence on correlations, and it will be increased in future experiments.

Despite the limitations, this study brings some strengths, as health professionals like physiotherapists need more HHD options to perform assessments before and after interventions to quantify treatment effectiveness and patient progression and to adjust rehabilitation goals [1–3]. Our research also suggests that Kinvent is a good instrument for use in clinical practice, since it yielded good to excellent intra- and inter-rater reliability for almost all movements assessed. The acquisition of strength measurements using this manual dynamometer was a quick and simple procedure that is able to provide useful information for clinical practice [3]. Another important note concerns the care taken to reduce bias in our methodology through a pilot study; patient and examiner training; random allocation; and statistical analysis methods.

5. Conclusions

The Kinvent showed a good to excellent intra and inter-rater reliability for almost all upper and lower limb movements assessed. Ankle dorsiflexion presented moderate values in all assessments.

Author Contributions: Conceptualization and methodology: M.B.A., A.L.R.-V., J.S., L.M.A.F. and A.R.P.; formal analysis and resources: M.B.A., A.L.R.-V. and L.M.A.F.; investigation, M.B.A., A.L.R.-V. and A.R.P.; data curation, A.L.R.-V.; writing—original and draft preparation: M.B.A., A.L.R.-V., C.O., G.O. and T.S.; writing—review, editing and supervision, M.B.A. and A.L.R.-V.; project administration, M.B.A. All authors have read and agreed to the published version of the manuscript.

Funding: This research received no external funding.

Institutional Review Board Statement: Egas Moniz Institute Ethics Commission approved this study prior to its start with the number CE-1088.

Informed Consent Statement: All subjects involved in the study singed an informed consent statement in accordance with the Helsinki Declaration of the World Medical Association.

Data Availability Statement: Not applicable.

Acknowledgments: We would like to thank the Laboratório de Avaliação Física e Funcional em Fisioterapia (LAFFFi) by Sausport for all their support and FCT/MCTES for the financial support to CiiEM (UIDB/04585/2020) through national funds and to the students who participated in this study.

Conflicts of Interest: The authors declare no conflict of interest.

References

1. Chamorro, C.; Armijo-Olivo, S.; De la Fuente, C.; Fuentes, J.; Chirosa, L.J. Absolute reliability and concurrent validity of hand held dynamometry and isokinetic dynamometry in the hip, knee and ankle joint: Systematic review and meta-analysis. *Open Med.* **2017**, *12*, 359–375. [CrossRef] [PubMed]
2. Stark, T.; Walker, B.; Phillips, J.K.; Fejer, R.; Beck, R. Hand-held dynamometry correlation with the gold standard isokinetic dynamometry: A systematic review. *PMR* **2011**, *3*, 472–479. [CrossRef] [PubMed]
3. Clarke, M.; Ni Mhuircheartaigh, D.; Walsh, G.; Walsh, J.; Meldrum, D. Intra-tester and inter-tester reliability of the MicroFET 3 hand-held dynamometer. *Physiother. Pract. Res.* **2011**, *32*, 13–18. [CrossRef]
4. Contreras-Díaz, G.; Chirosa-Ríos, L.J.; Chirosa-Ríos, I.; Intelangelo, L.; Jerez-Mayorga, D.; Martinez-Garcia, D. Reliability of Isokinetic Hip Flexor and Extensor Strength Measurements in Healthy Subjects and Athletes: A Systematic Review and Meta-Analysis. *Int. J. Environ. Res. Public Health* **2021**, *18*, 11326. [CrossRef] [PubMed]
5. Sørensen, L.; Oestergaard, L.G.; van Tulder, M.; Petersen, A.K. Measurement Properties of Isokinetic Dynamometry for Assessment of Shoulder Muscle Strength: A Systematic Review. *Arch. Phys. Med. Rehabil.* **2021**, *102*, 510–520. [CrossRef] [PubMed]
6. Hirano, M.; Katoh, M.; Gomi, M.; Arai, S. Validity and reliability of isometric knee extension muscle strength measurements using a belt-stabilized hand-held dynamometer: A comparison with the measurement using an isokinetic dynamometer in a sitting posture. *J. Phys. Ther. Sci.* **2020**, *32*, 120–124. [CrossRef] [PubMed]
7. Olds, M.; McLaine, S.; Magni, N. Validity and Reliability of the Kinvent Handheld Dynamometer in the Athletic Shoulder Test. *J. Sport Rehabil.* **2023**, *1*, 1–9. [CrossRef] [PubMed]
8. Mentiplay, B.F.; Perraton, L.G.; Bower, K.J.; Adair, B.; Pua, Y.-H.; Williams, G.P.; McGaw, R.; Clark, R.A. Assessment of lower limb muscle strength and power using hand-held and fixed dynamometry: A reliability and validity study. *PLoS ONE* **2015**, *10*, e0140822. [CrossRef] [PubMed]

medical sciences forum

MDPI

Proceeding Paper

Comparison between Digital and Paper Handwriting—A Contribution to Graphoscopic Analysis †

Vlad Tatarescu [1,2], Daniela Eusébio [1,3], Mariana M. Louro [4] and Alexandra Bernardo [1,3,*]

1 Laboratório de Ciências Forenses e Psicológicas Egas Moniz, 2829-511 Almada, Portugal; vlad.tatarescu@gmail.com (V.T.); danielaeusebio99@icloud.com (D.E.)
2 Mestrado em Tecnologias Laboratoriais em Ciências Forenses, Egas Moniz School of Health and Science, 2829-511 Almada, Portugal
3 Egas Moniz Center for Interdisciplinary Research (CiiEM), Egas Moniz School of Health & Science, 2829-511 Almada, Portugal
4 Licenciatura em Ciências Forenses e Criminais, Instituto Universitário Egas Moniz, 2829-511 Almada, Portugal; marianalouro1@gmail.com
* Correspondence: abernardo@egasmoniz.edu.pt
† Presented at the 6th International Congress of CiiEM—Immediate and Future Challenges to Foster One Health, Almada, Portugal, 5–7 July 2023.

Abstract: The growing development of new technologies and the evolution of communication devices has led the world's population to gradually replace traditional writing on paper and pen with digital handwriting. This shift in writing instruments and media presents new challenges for calligraphy specialists. The aim of this study is to compare digital handwriting and traditional writing on paper and to explore the possible influence of different supports on the analysis of handwriting. This study's design consists of the analysis and comparison, by three calibrated forensic experts, of the handwriting of 10 participants, 4 of whom provided anonymous samples for authorship identification purposes. Preliminary results of the comparative analysis in both media revealed some differences in letter size, spacing, dots, ovals, and initial and terminal strokes, and similarities in graphic impulses, inclination, commas, punctuation, accentuation, calligraphic case, speed, and letter shapes. Concerning authorship identification, a 75% success rate was achieved.

Keywords: graphoscopy; forensic handwriting analysis; questioned documents

Citation: Tatarescu, V.; Eusébio, D.; Louro, M.M.; Bernardo, A. Comparison between Digital and Paper Handwriting—A Contribution to Graphoscopic Analysis. *Med. Sci. Forum* **2023**, *22*, 46. https://doi.org/10.3390/msf2023022046

Academic Editors: José Brito, Nuno Taveira and Ana I. Fernandes

Published: 29 August 2023

1. Introduction

The current technological era promotes the rapid transmission of information worldwide; however, it raises security issues that require the adaptation of traditional security elements (such as signatures and handwriting itself) to new digital writing instruments. In this sense, several studies have explored the eventual equivalence between an individual's signature and handwriting in both supports (paper and digital), with the aim of developing tools and methodologies that assist in graphoscopic analysis. In general, differences have been described, for example, in terms of speed (faster in the digital support), size (larger in the digital support) and pressure (lower in the digital support) [1,2]. In addition to these differences, significant differences between handwriting/signatures have also been reported when written on different software and hardware resources, which makes standardization difficult when it comes to digital samples, especially when the individual writes/signs in different positions [3–5]. Thus, the present work aims to contribute to the description of the common and particular characteristics of handwriting in each medium (paper and digital).

2. Materials and Methods

This pilot study adopts a design, presented in Figure 1, involving 10 forensic students as volunteers, between the ages of 18 and 20. The participants provided handwriting samples, including four anonymous samples for authorship identification (2 positive controls and 2 negative controls). All participants willingly agreed to participate in the study and provided their written informed consent. To ensure anonymity, participants were coded with randomly generated numbers.

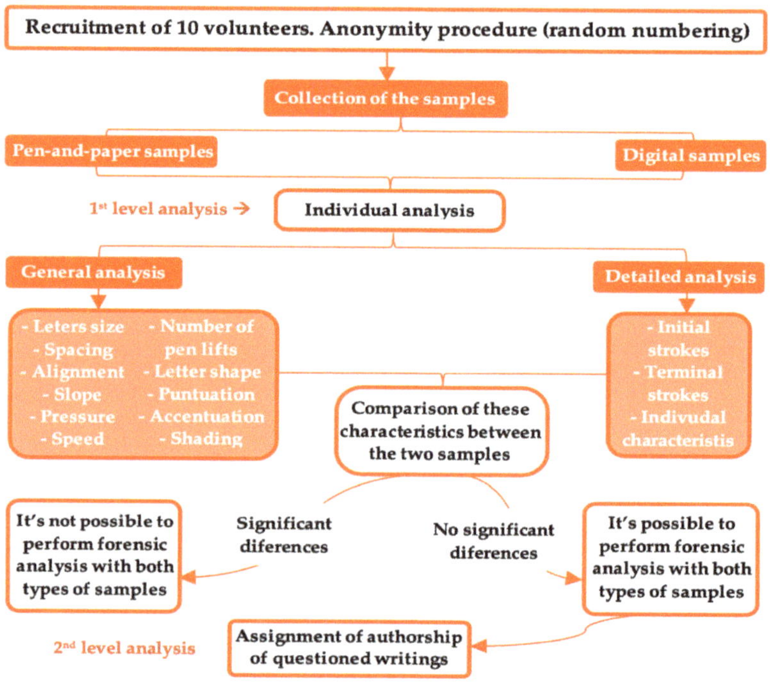

Figure 1. Pilot study design.

Participants were asked to write a text three times (once in uppercase and twice in lowercase) on a piece of paper, and to repeat the process using a digital medium, namely an 8th generation iPad® and an Apple Pencil 1st gen. Blue ink pens and individual sheets of white A4 paper were used, in accordance with the Judiciary Police guidelines (LPC-FEM-MD012, LPC-FEM-MD016 and LPC-FEM-MD017).

Forensic handwriting analysis was performed at two levels by three independent calibrated experts. The first level involved an analysis of the handwriting and comparison of the general and detailed characteristics between the two supports for 8 participants, and the second level involved the graphoscopic analysis on both supports with the aim of assigning the graphic identity of 4 anonymous samples.

3. Results

The comparison of the handwriting from eight participants revealed notable differences between the pen-and-paper and digital media. Several recurring differences were observed, including discrepancies in the opening of ovals, periods, spacing, and letter size. The detailed analysis highlighted disparities, particularly in the initial and terminal strokes. It should be noted that no significant changes in digital writing pressure were detected, as indicated by the figures presented in Figure 2.

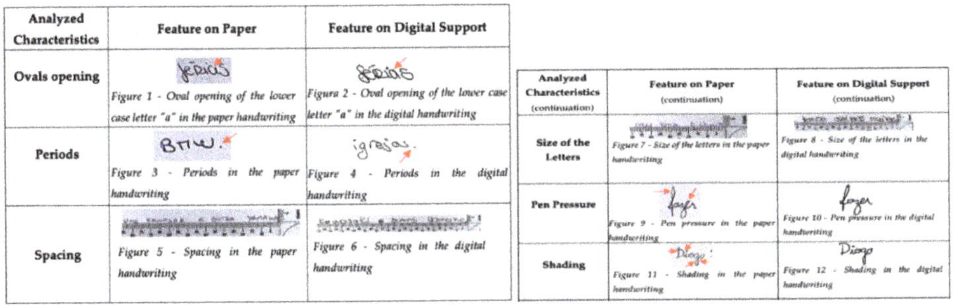

Analyzed Characteristics	Feature on Paper	Feature on Digital Support
Ovals opening	*Figure 1 - Oval opening of the lower case letter "a" in the paper handwriting*	*Figura 2 - Oval opening of the lower case letter "a" in the digital handwriting*
Periods	*Figure 3 - Periods in the paper handwriting*	*Figure 4 - Periods in the digital handwriting*
Spacing	*Figure 5 - Spacing in the paper handwriting*	*Figure 6 - Spacing in the digital handwriting*

Analyzed Characteristics (continuation)	Feature on Paper (continuation)	Feature on Digital Support (continuation)
Size of the Letters	*Figure 7 - Size of the letters in the paper handwriting*	*Figure 8 - Size of the letters in the digital handwriting*
Pen Pressure	*Figure 9 - Pen pressure in the paper handwriting*	*Figure 10 - Pen pressure in the digital handwriting*
Shading	*Figure 11 - Shading in the paper handwriting*	*Figure 12 - Shading in the digital handwriting*

Figure 2. Handwritings differences observed between paper and digital supports.

For instance, when examining the opening of ovals, a distinct contrast was observed between paper writing and digital support. In paper writing, the oval exhibited an open shape at the top right, while in digital writing, it appeared closed at the top right. Similarly, in the case of periods, the shape differed between paper writing (dot shape) and digital writing (straight line). Furthermore, the comparison indicated that the spacing and letter size in paper handwriting were generally smaller than those in the digital medium. This finding aligns with the observations made by other authors [3]. In terms of the initial and terminal strokes, there were a few discernible distinctions. The terminal stroke of certain letters and the initial stroke of dashes exhibited variations between paper handwriting (straight line) and digital writing (hook-shaped). On the other hand, it was possible to establish a similarity relationship between certain graphical and dynamic characteristics of the writings performed on paper and the digital support. Thus, similarities were observed in graphic impulses, inclination, commas, punctuation, accentuation, calligraphic case, baseline, speed, and letter shapes. An example of these similarities can be observed in Figure 3.

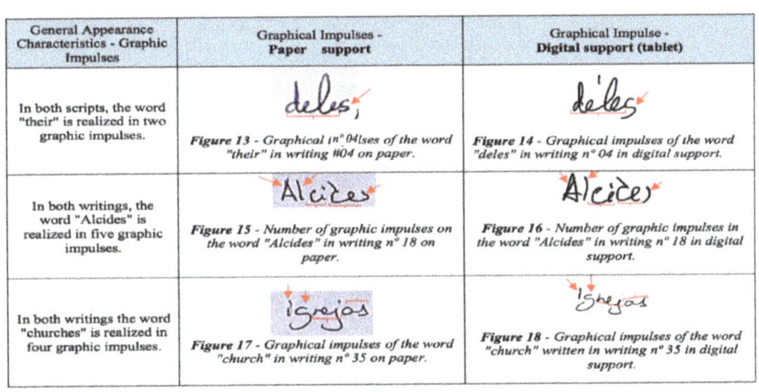

General Appearance Characteristics - Graphic Impulses	Graphical Impulses - Paper support	Graphical Impulse - Digital support (tablet)
In both scripts, the word "their" is realized in two graphic impulses.	*Figure 13 - Graphical in° 04lses of the word "their" in writing #04 on paper.*	*Figure 14 - Graphical impulses of the word "deles" in writing n° 04 in digital support.*
In both writings, the word "Alcides" is realized in five graphic impulses.	*Figure 15 - Number of graphic impulses on the word "Alcides" in writing n° 18 on paper.*	*Figure 16 - Number of graphic impulses in the word "Alcides" in writing n° 18 in digital support.*
In both writings the word "churches" is realized in four graphic impulses.	*Figure 17 - Graphical impulses of the word "church" in writing n° 35 on paper.*	*Figure 18 - Graphical impulses of the word "church" written in writing n° 35 in digital support.*

Figure 3. Comparison of characteristics of writing performed on paper and digital supports.

Regarding the second level of graphoscopic analysis, namely the assignment of authorship of questioned writings, it was possible to attribute an identity to one out of four anonymous handwriting samples and conclude that two anonymous samples did not match any of the eight samples previous analyzed. The obtained results suggest that it is possible to compare texts in both formats, as similarities were identified in 16 parameters (inclination, commas, periods, question marks, accentuation, calligraphic style, opening of ovals in lowercase letters "a", "g", "p", and "q", speed, letter size, letter shape, lowercase letter "d", lowercase letter "f", uppercase letter "E") between the anonymous writing and its corresponding paper and digital formats. Consequently, it is possible to attribute samples 18 and 67 to the same author. Figure 4 provides two examples of similarities between the two writings concerning appearance characteristics.

	Analysed characteristics	Anonymus sample nº 18	Anonymus sample nº 67
Paper support	**Oval opening Lower-case "a"** In both scripts, the letter oval is mostly open in the upper right area.	 *Figure 19* - *Opening of the lower case oval of letter "a" in writing nº18 on paper.*	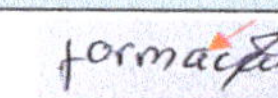 *Figure 20* - *Opening of the lower case oval of letter "a" in writing nº 67 on paper.*
Digital support	**Letter Size** In both scripts, the size of the letters is approximately 5 mm.	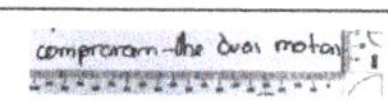 *Figure 21* - *Size of the letters of writing nº 18 in digital support.*	 *Figure 22* - *Size of the letters of writing nº 67 in digital support.*

Figure 4. Comparison between paper and digital handwriting samples.

4. Conclusions

This pilot study indicates the feasibility of comparing handwriting in both supports and evaluates handwriting authorship with a 75% success rate. However, it is crucial to emphasize that a validated methodology is essential in supporting the comparison between the supports and in the attribution of handwriting authorship.

Author Contributions: Conceptualization, A.B.; methodology, A.B.; software, M.M.L., D.E. and V.T.; validation, M.M.L., D.E. and V.T.; formal analysis, M.M.L., D.E. and V.T.; investigation, M.M.L., D.E. and V.T.; writing—original draft preparation, V.T.; writing—review and editing, A.B.; supervision, A.B.; project administration, A.B. All authors have read and agreed to the published version of the manuscript.

Funding: This research received no external funding.

Informed Consent Statement: Informed consent was obtained from all subjects involved in the study.

Data Availability Statement: The data presented in this study are available on request from the corresponding author. The data are not publicly available due to data protection.

Acknowledgments: Laboratório de Ciências Forenses e Criminais Egas Moniz (LCFPEM) and Egas Moniz School of Health & Science.

Conflicts of Interest: The authors declare no conflict of interest.

References

1. Gerth, S.; Dolk, T.; Klassert, A.; Fliesser, M.; Fischer, M.H.; Nottbusch, G.; Festman, J. Adapting to the surface: A comparison of handwriting measures when writing on a tablet computer and on paper. *Hum. Mov. Sci.* **2016**, *48*, 62–73. [CrossRef] [PubMed]
2. Heckeroth, J.; Kupferschmid, E.; Dziedzic, T.; Kalantzis, N.; Čakovská, B.G.; Fernandes, C.; Branco, M.J.; Spjuth, K.A.; Kerkhoff, A.; Vaccarone, P.; et al. Features of digitally captured signatures vs. pen and paper signatures: Similar or completely different? *Forensic Sci. Int.* **2021**, *318*, 110587. [CrossRef] [PubMed]
3. Zimmer, J.; Kalantzis, N.; Dziedzic, T.; Heckeroth, J.; Kupferschmid, E.; Fernandes, C.; Čakovská, B.G.; Branco, M.J.; Spjuth, K.A.; Vaccarone, P.; et al. The challenge of comparing digitally captured signatures registered with different software and hardware. *Forensic Sci. Int.* **2021**, *327*, 110945. [CrossRef] [PubMed]
4. Harralson, H.H. *Developments in Handwriting and Signature Identification in the Digital Age*; Anderson Publishing: New York, NY, USA, 2013.
5. Geistová Čakovská, B.; Kalantzis, N.; Dziedzic, T.; Fernandes, C.; Zimmer, J.; Branco, M.J.; Heckeroth, J.; Spjuth, K.A.; Kupferschmid, E.; Vaccarone, P.; et al. Recommendations for capturing signatures digitally to optimize their suitability for forensic handwriting examination. *J. Forensic Sci.* **2021**, *66*, 743–747. [CrossRef] [PubMed]

medical sciences forum

MDPI

Proceeding Paper

Dental Students' Digital Competence Evaluation: Preliminary Results of a Cohort Study †

Mariana Morgado [1], José João Mendes [1,2] and Luís Proença [1,2,*]

[1] Clinical Research Unit (CRU), Egas Moniz Center for Interdisciplinary Research (CiiEM), Egas Moniz School of Health & Science, 2829-511 Almada, Portugal; mmorgado@egasmoniz.edu.pt (M.M.); jmendes@egasmoniz.edu.pt (J.J.M.)

[2] Evidence-Based Hub, Egas Moniz Center for Interdisciplinary Research (CiiEM), Egas Moniz School of Health & Science, 2829-511 Almada, Portugal

* Correspondence: lproenca@egasmoniz.edu.pt

† Presented at the 6th International Congress of CiiEM—Immediate and Future Challenges to Foster One Health, Almada, Portugal, 5–7 July 2023.

Abstract: Digital competence, a crucial aspect of online education, was studied among dental students at the Egas Moniz School of Health & Science. This study, in line with the Horizon Europe Strategic Plan 2021–2024, aimed to understand students' digital skills and competences in the context of higher education. A questionnaire was administered to 503 participants, covering various topics such as socio-demographic characteristics, digital learning participation and the impact of the COVID-19 pandemic. The results highlight the importance of digital learning and its future role in clinical practice. While students perceived their digital competence level as high/intermediate, an assessment test indicated a lower level of competence. These findings highlight the need to address challenges and improve digital competence in education for social and economic advancement.

Keywords: digital competence; dental students; higher education; online learning

Citation: Morgado, M.; Mendes, J.J.; Proença, L. Dental Students' Digital Competence Evaluation: Preliminary Results of a Cohort Study. *Med. Sci. Forum* **2023**, *22*, 25. https://doi.org/10.3390/msf2023022025

Academic Editors: José Brito, Nuno Taveira and Ana I. Fernandes

Published: 14 August 2023

1. Introduction

Digital competence, which encompasses essential information and communication technology (ICT) skills, plays a vital role in the success of online educational activities [1]. As the world becomes increasingly interconnected and technology-driven, it is imperative for educational institutions to equip students with the necessary digital skills to navigate the digital landscape effectively. This is particularly true in the field of dentistry, where technological advances and digital transformation have revolutionized the ways in which oral health professionals practice and engage with patients [2].

The present study focuses on characterizing and profiling the digital competence of dental students at the Egas Moniz School of Health & Science (EMSHS), within the broader context of the Horizon Europe Strategic Plan 2021–2024 [3]. In line with Cluster 6 ('Digital, Industry and Space'), the research aims to contribute to the advancement of digital technologies, particularly in the fields of education and healthcare.

In recent years, the integration of technology in higher education has become increasingly widespread, and the COVID-19 pandemic has further accelerated the adoption of digital learning platforms and tools [4]. The transition to online education has highlighted the importance of digital competence among students, as they need to navigate virtual classrooms, engage with online resources and use technology for effective learning [5].

The specific aims of this study are to assess the digital skills and competencies of dental students, identify the challenges they face in the digital learning process, and explore their perceptions of the future role of digital health in clinical practice. By investigating these factors, the study aims to contribute to the development of strategies and interventions to enhance the digital competence of dental students, ultimately improving their online learning experience and preparing them for the digital transformations taking place in the field of dentistry.

This article presents the preliminary results of a cohort study involving 503 dental students at EMSHS. A comprehensive questionnaire was used, covering various dimensions.

In the following sections of this article, we will describe the methodology used, present the results of this study and discuss the implications for educational institutions and the wider field of dentistry. By examining the current state of digital competence among dental students, we can take steps to foster their digital competence and maximize the potential of technology to advance universal health coverage.

2. Methods

2.1. *Study Design*

This research used a cohort study design to assess the digital competence of dental students at EMSHS. This study aimed to understand students' digital competence, as well as to identify the challenges they faced in the learning process.

2.2. *Participants*

A total of 503 dental students participated in this study. Participants were selected using convenience sampling to ensure representation from different academic years and programs within the curricular plan.

2.3. *Data Collection*

Data were collected using an anonymous structured questionnaire consisting of 44 items, administered electronically to ensure efficient and standardized data collection. The questionnaire covered various aspects, including socio-demographic characteristics, residential context, family background, digital characteristics, involvement in digital learning, and the impact of the COVID-19 pandemic. Additionally, digital competence was evaluated using DigComp to assess their proficiency in utilizing digital technologies and skills.

2.4. *Measures*

The questionnaire included items designed to assess participants' digital competence, such as their ability to use technology in an integrated way for work and study, their critical evaluation of the technologies they use, and their active engagement in digital culture. Participants were also asked to self-assess their digital competence.

2.5. *Data Analysis*

Descriptive statistical analyses were conducted in order to summarize the demographic characteristics of the participants, as well as their digital learning preferences and self-perceived digital competence. The data collected on digital competence were compared with the participants' self-assessments to identify any gaps or discrepancies.

2.6. *Ethical Considerations*

Ethical approval from the Egas Moniz Ethics Committee was obtained prior to conducting the study. Participants were provided with information about the study's purpose and procedures, emphasizing their voluntary participation. Strict confidentiality and anonymity were maintained during data collection and analysis.

3. Results

Regarding the importance of digital learning, the findings indicate a high level of recognition among dental students. A significant majority (78.5%) of participants acknowledged the pivotal role digital health will play in their future clinical practice. This finding emphasizes their awareness of the increasing integration of technology in the field and the necessity to acquire digital skills for professional success.

The study found that over half (51.3%) of dental students use digital skills daily in their learning environment. They particularly prefer class transcripts and video-based support for digital learning, indicating that they appreciate the convenience and accessibility of digital resources, which enhance their learning experience.

Interestingly, despite most students having high or medium self-perceived digital competence (94.0%), the objective assessment test showed lower competence levels. Only 26.6% rated themselves highly skilled after the assessment. This highlights the limitations of self-assessment in accurately measuring digital competence and stresses the need for objective assessments and targeted training to address identified gaps effectively.

The results also shed light on the challenges faced by dental students during the shift to online education due to the COVID-19 pandemic. The rapid adoption of digital learning platforms has required quick adjustments in educational practices. Understanding the specific challenges in this digital learning environment is crucial for institutions to develop effective strategies and interventions. Further research focused on exploring the unique challenges that dental students encounter in digital learning could inform the design of tailored support mechanisms.

4. Discussion

This study explores the digital competence of dental students and its implications for online education and future practice. The findings shed light on various aspects of dental students' digital competence, including skills, preferences, and self-perceived competencies, thus having significant implications for educational institutions and in this field.

The findings can guide targeted interventions to address skill gaps, enhance online learning, and prepare students for technology-driven healthcare practice. This study aligns with broader goals of digital transformation in education and healthcare, contributing to societal and economic progress through effective use of technology [6].

The implications of this research extend beyond dental education. This study aligns with the broader goals of digital transformation in education and healthcare, promoting the effective use of digital technologies for social and economic progress. By identifying the digital competence levels of dental students, educational institutions can develop targeted interventions to improve digital competence. This, in turn, will enhance the overall online learning experience and prepare students for the evolving digital landscape in the field of dentistry.

Future research could use more diverse sampling methods and incorporate objective measures to provide a comprehensive understanding of digital competence among dental students.

Limitations

There are several limitations to this study. Firstly, the convenience sampling method may introduce selection bias towards a potential overrepresentation of certain groups or characteristics within the sample, limiting the generalizability of the findings to the wider population of dental students. Secondly, the self-reported nature of the data may be subject to response bias and overestimation of digital competence. Finally, the preliminary results do not allow for causal inference or the assessment of changes in digital competence over time.

5. Conclusions

This study offers valuable insights into the landscape of digital competence among dental students, underscoring the significance of digital skills and competences within the realm of online education and future dental practice.

The results reveal a noteworthy discrepancy between the participants' self-perceived digital competences and their actual performance as evaluated by the digital competence assessment test. This suggests a tendency for participants to overestimate their digital competence, highlighting the importance of objective assessments for an accurate understanding of individuals' digital competence, as well as promoting self-awareness and training.

The findings also underscore the imperative for educational institutions to develop strategies aimed at bridging gaps in digital competences and enhancing the overall learning experience. By fostering digital competence among dental students, institutions can ensure that future oral health professionals are well-equipped to effectively utilize technology for improved healthcare practices and contribute to the advancement of digital transformation within the field.

Author Contributions: Conceptualization, L.P., J.J.M. and M.M.; methodology, M.M. and L.P.; formal analysis, L.P.; investigation, M.M.; resources, J.J.M.; data curation, M.M.; writing—original draft preparation, L.P. and M.M.; writing—review and editing, J.J.M., L.P. and M.M.; supervision, L.P. and J.J.M. All authors have read and agreed to the published version of the manuscript.

Funding: This research received no external funding.

Institutional Review Board Statement: The study was conducted in accordance with the Declaration of Helsinki, and approved by the Egas Moniz Ethics Committee (protocol code 959 and 25 March 2021) for studies involving humans.

Informed Consent Statement: Informed consent was obtained from all subjects involved in the study.

Data Availability Statement: The data used to support the findings of this study are available from the corresponding author (L.P.) upon request.

Conflicts of Interest: The authors declare no conflict of interest.

References

1. Audrin, C.; Audrin, B. Key Factors in Digital competence in Learning and Education: A Systematic Literature Review Using Text Mining. *Educ. Inf. Technol.* **2022**, *27*, 7395–7419. [CrossRef]
2. Dickenson, A.; Tebbutt, J.; Abdulhussein, H. An Overview of Digital Readiness in Dentistry—Are We Ready? *Br. Dent. J.* **2022**, *233*, 87–88. [CrossRef] [PubMed]
3. European Commission. Directorate General for Research and Innovation. In *Horizon Europe: Strategic Plan 2021–2024*; Publications Office of the European Union: Luxembourg, 2021.
4. Akram, H.; Yingxiu, Y.; Al-Adwan, A.S.; Alkhalifah, A. Technology Integration in Higher Education During COVID-19: An Assessment of Online Teaching Competencies Through Technological Pedagogical Content Knowledge Model. *Front. Psychol.* **2021**, *12*, 736522. [CrossRef] [PubMed]
5. DeCoito, I.; Estaiteyeh, M. Transitioning to Online Teaching During the COVID-19 Pandemic: An Exploration of STEM Teachers' Views, Successes, and Challenges. *J. Sci. Educ. Technol.* **2022**, *31*, 340–356. [CrossRef] [PubMed]
6. Haleem, A.; Javaid, M.; Qadri, M.A.; Suman, R. Understanding the Role of Digital Technologies in Education: A Review. *Sustain. Oper. Comput.* **2022**, *3*, 275–285. [CrossRef]

medical sciences forum

MDPI

Proceeding Paper

Precision of Tooth Size Measurement in Digital Models Acquired by Intraoral Scanning and by Scanning of Plaster Models Versus Conventionally Cast Models †

Matilde Jardim [1], Pedro Mariano Pereira [1,2,*], Luís Proença [2,3] and Iman Bugaighis [1,*]

1 Department of Orthodontics, Egas Moniz School of Health and Science, Caparica, 2829-511 Almada, Portugal; matildeajardim@gmail.com
2 Egas Moniz Center for Interdisciplinary Research (CiiEM), Egas Moniz School of Health and Science, Caparica, 2829-511 Almada, Portugal; lproenca@egasmoniz.edu.pt
3 Quantitative Methods for Health Research Unit, Egas Moniz School of Health and Science, Caparica, 2829-511 Almada, Portugal
* Correspondence: pmarianop@sapo.pt (P.M.P.); isbugaighis@yahoo.com (I.B.)
† Presented at the 6th International Congress of CiiEM—Immediate and Future Challenges to Foster One Health, Almada, Portugal, 5–7 July 2023.

Abstract: This study compared the precision of extracting mesio-distal tooth size of each participant, from plaster models, digital models, and by scanning plaster models using the iTero Element 5D Plus scanner. Ten participants were included in the study. Descriptive analysis was carried out and the reproducibility of measurements for the three methods was assessed with the intraclass correlation coefficient (ICC). The overall reproducibility of mesio-distal tooth size measurements on digital models was comparable to direct measurements on plaster models and the digital replicate of the study models. A trend towards lower reproducibility was found for measurements of posterior maxillary teeth in digital models.

Keywords: intraoral scanners; digital models; model scanning; 3D models; model analysis; orthodontics; precision; accuracy

Citation: Jardim, M.; Mariano Pereira, P.; Proença, L.; Bugaighis, I. Precision of Tooth Size Measurement in Digital Models Acquired by Intraoral Scanning and by Scanning of Plaster Models Versus Conventionally Cast Models. *Med. Sci. Forum* **2023**, *22*, 20. https://doi.org/10.3390/msf2023022020

Academic Editors: José Brito, Nuno Taveira and Ana I. Fernandes

Published: 11 August 2023

1. Introduction

Traditionally, diagnostic measurements were obtained from plaster models with a digital calliper. The literature recognizes manual measurements using callipers as accurate and reliable [1–3]. In recent years, conventional plaster models have been replaced by digital models that are becoming more common due to the benefits related with storage, retrieval, reproduction, and communication. On that account, the clinician can have easy access to diagnostic and routine measurements and perform various analyses [4].

Different methods can obtain digital models. Several studies have evaluated the reliability of 3D digital models, obtained with extra-oral scanners on plaster models such as 3Shape D250 3-dimensional (3D) scanner (3Shape, Copenhagen, Denmark) [3,5], Ortho Insight 3D laser scanner (Motion View LLC, Chatanooga, TN, USA) [6], and R500 3Shape scanner (3Shape, Copenhagen, Denmark) [7]. Scans with the 3Shape D250 showed high precision values with mean deviations of 10 μm and, root mean square errors of 20 μm [5].

Three dimensional (3D) digital models can also be acquired with intraoral scanners. Pelliteri et al. [7] studied the accuracy and efficiency of 3D digital models using three different intraoral scanners, Carestream CS3600 (Onex Corporation, Rochester, NY, USA), CEREC Omnicam(Dentsply Sirona, Charlotte, NC, USA) and Trios 3Shape (3Shape, Copenhagen, Denmark). Variations between the three intraoral scanners and the polyvinyl siloxane (PVS) impression scan ranged from 100 to 200 μm.

Intraoral scanners, such as iTero, can also be used to scan plaster models. Reported values of mean deviations of 25 µm and root mean square errors of 51 µm [5] confirm that the iTero intraoral scan can generate 3D digital models of plaster models with precision.

The availability of several software applications enables clinicians to import digital models and move their generated images around three axes of rotation as well as measuring tooth size and arch width and length [4] and calculate Bolton ratios [6], such as OrthoCAD (Cadent, Carlstadt, NJ, USA), E-models (GeoDigm Corporation, Falcon Heights, MN, USA), Ortho AnalyzerTM (3Shape, Copenhagen, Denmark), SureSmile (OraMetrix, Richardson, TX, USA), Maestro3D (AGE Solutions, Pisa, Italy), NemoCast (Nemotec, Madrid, Spain), and DigiModel (OrthoProof, Nieuwegein, Netherlands) [4,6,8,9].

Most studies found that even though the reproducibility of digital models is acceptable for clinical diagnosis and investigation measurement, precision varies across studies according to the clinician's experience and ability and the visualization tools offered by different software. Therefore, the precision and accuracy of different measurements and techniques should be evaluated in the practice of evidence-based clinical orthodontics [2].

The aim of this study was to determine and compare the precision of mesio-distal tooth size measurements in plaster models using a digital calliper with measurements obtained with NemoCast software in digital models acquired through intraoral scanning and by scanning plaster models using the iTero Element 5D Plus scanner.

2. Materials and Methods

The present pilot study was undertaken at the Egas Moniz School of Health and Science. Ethical approval was secured from the Ethics Committee, and consent forms were obtained from the participants. The study sample consisted of 10 patients with full permanent dentition randomly selected from the clinical archives of the postgraduate orthodontics program. Each case had available dental study models and intraoral digital scans. Digital models were acquired both through intraoral scanning and by scanning the plaster models using the iTero Element 5D Plus scanner (Align Technology, San Jose, CA, USA) (Figure 1).

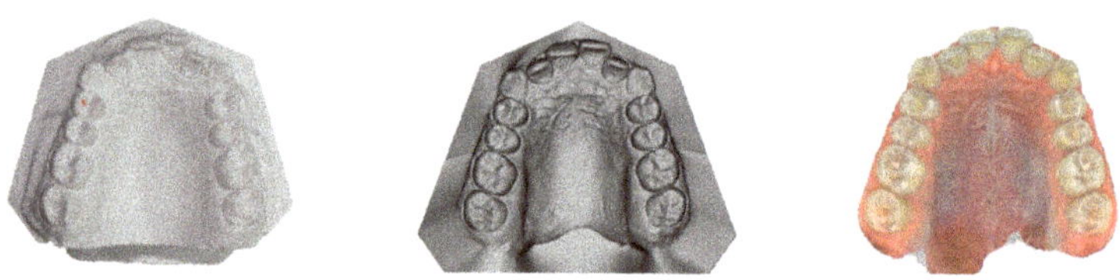

Figure 1. Dental models used in the study. Left to right: Conventional plaster study model, 3D scanning of plaster model, 3D intraoral scan.

The iTero Element 5D Plus acquisition method is based on parallel confocal microscopic features colour scanning and does not require opacification. The scanning procedure started from the occlusal surface, progressing to the buccal and lingual surfaces. A sample size calculation determined that a 10 to 13 dental study models are required to obtain a statistical power of 95% [4].

One examiner (M.J.) was trained in using the three different methods, measuring with a digital calliper, scanning the plaster models and measuring of the 3D virtual images with NemoCast software v2021. The measurements from conventional dental casts were obtained with a 145 mm digital calliper (Hammacher, Solingen, Germany) with 0.01 mm accuracy. Digital model images were analyzed by using Nemocast software (Figure 2). Measurements were recorded in a Microsoft Excel spreadsheet (Microsoft, Redmond, DC, USA) and analyzed with SPSS Statistics version 28.0 (SPSS, Chicago, IL, USA). Descriptive analysis of the mean error of mesio-distal tooth size duplicate measurements was carried out and the reproducibility of measurements assessed with the intraclass correlation coefficient (ICC).

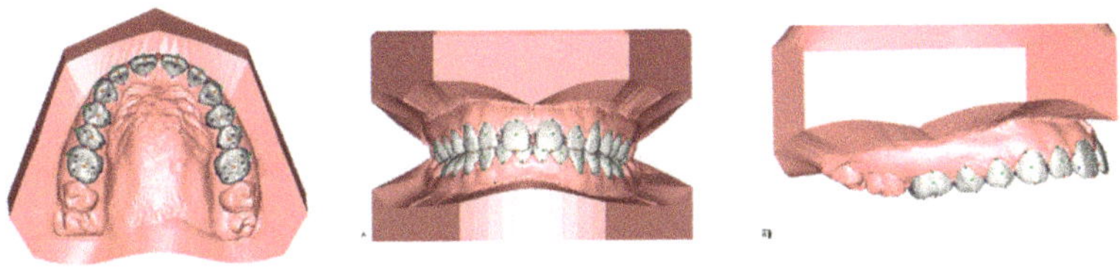

Figure 2. Mesiodistal tooth measurements from the Nemocast software. Mesial point landmarks are marked as a red circle and distal point landmarks as a blue circle.

3. Results

The intra-examiner mean random errors of mesio-distal tooth size ranged from 0.07 to 0.20 mm for measurements obtained with a digital calliper on plaster models, from 0.07 to 0.29 mm for measurements obtained with Nemocast in digital models acquired through iTero intraoral scanning, and from 0.07 to 0.28 mm for measurements obtained with Nemocast in digital models acquired using iTero scanning of plaster models.

Intraclass correlation coefficients (ICC) for mesio-distal tooth size measurements (Table 1) ranged from 0.630 to 0.990 for the digital calliper (mean ICC = 0.921), from 0.524 to 0.965 for Nemocast/iTero intraoral scanning (mean ICC = 0.885), and from 0.281 to 0.971 for Nemocast/iTero plaster scanning (mean ICC = 0.855).

Table 1. Reproducibility of tooth size measurements evaluated by intraclass correlation coefficient.

	Digital Calliper/Plaster Model	Nemocast/Intraoral Scan	Nemocast/Plaster Scan	Mean
Upper Arch				
16	0.908	0.693	0.685	0.762
15	0.972	0.935	0.909	0.939
14	0.951	0.915	0.804	0.890
13	0.630	0.943	0.840	0.804
12	0.931	0.930	0.962	0.941
11	0.990	0.944	0.935	0.956
21	0.969	0.965	0.971	0.968
22	0.974	0.864	0.957	0.932
23	0.927	0.900	0.490	0.772
24	0.961	0.941	0.942	0.948
25	0.933	0.853	0.648	0.811
26	0.847	0.665	0.281	0.598
Lower Arch				
36	0.989	0.947	0.914	0.950
35	0.979	0.912	0.835	0.909
34	0.890	0.915	0.961	0.922
33	0.904	0.867	0.902	0.891
32	0.895	0.965	0.961	0.940
31	0.905	0.524	0.861	0.763
41	0.919	0.937	0.968	0.941
42	0.878	0.932	0.965	0.925
43	0.915	0.952	0.900	0.922
44	0.932	0.921	0.915	0.923
45	0.967	0.927	0.955	0.950
46	0.931	0.888	0.949	0.923

4. Discussion

The reproducibility of mesiodistal tooth measurements can be affected by numerous variables, including inclination, rotation, interproximal contacts, and anatomical differences. In this study, repeated measurements had acceptable correlations and mean errors were determined to be within clinically acceptable limits for all methods.

Even though the iTero intraoral scan is not optimized for plaster model scanning, results were comparable to iTero intraoral scan in vivo. Difficulty in capturing the morphology of a study model seem to be related to the presence of undercut areas. This problem can be solved by positioning the cast model from several different angles, rather than relocating the scanner while capturing scanned images of the occlusal, buccal, and lingual surface, allowing the production of a 360° model with greater accuracy [10].

In this study, the largest errors tended to be found in maxillary arch scans. This finding is in agreement with a previous study in which full arch scans showed a greater mean error in maxillary scans compared to mandibular scans [11]. The lower reproducibility that was observed in this study in the measurement of some posterior teeth has been previously attributed to errors in scanning [11]. Images acquired by several scanners showed distortion patterns, which can compromise the accuracy of linear measurements, with a tendency for greater deviation in the molar area [7]. The 3D and 2D error analyses also displayed a trend of greater distortion of digital scans compared to conventional measurements in the molar region [7].

5. Conclusions

(1) The overall reproducibility of mesiodistal tooth size measurements on digital models was comparable to direct measurements with a calliper on plaster models. (2) Similar reproducibility was found for measurements performed on digital models acquired by scanning plaster models and by intraoral scanning. (3) A trend towards lower reproducibility was found for measurements of posterior maxillary teeth in digital models.

Author Contributions: M.J. is the principal investigator, collecting study data, taking direct measurements, using the software, validating the measurements, collecting references, and writing the proceeding draft. P.M.P.: conceptualization, supervision, co-editing and co-writing up, and revising the final draft of the Proceeding. L.P.: undertaking the statistical analysis. I.B.: second supervisor, co-editing and co-writing up the proceeding. All authors have read and agreed to the published version of the manuscript.

Funding: This research received no external funding.

Institutional Review Board Statement: This study was approved by the Ethic Committee of the Egas Moniz School of Health and Science.

Informed Consent Statement: Informed consent was obtained from all subjects involved in the study.

Data Availability Statement: Data are available upon request.

Conflicts of Interest: The authors declare no conflict of interest.

References

1. Quimby, M.L.; Vig, K.W.; Rashid, R.G.; Firestone, A.R. The accuracy and reliability of measurements made on computer-based digital models. *Angle Orthod.* **2004**, *74*, 298–303. [PubMed]
2. Leifert, M.F.; Leifert, M.M.; Efstratiadis, S.S.; Cangialosi, T.J. Comparison of space analysis evaluations with digital models and plaster dental casts. *Am. J. Orthod. Dentofac. Orthop.* **2009**, *136*, e1–e4; discussion 16. [CrossRef] [PubMed]
3. Naidu, D.; Scott, J.; Ong, D.; Ho, C.T. Validity, reliability and reproducibility of three methods used to measure tooth widths for bolton analyses. *Aust. Orthod. J.* **2009**, *25*, 97–103. [PubMed]
4. Sousa, M.V.; Vasconcelos, E.C.; Janson, G.; Garib, D.; Pinzan, A. Accuracy and reproducibility of 3-dimensional digital model measurements. *Am. J. Orthod. Dentofac. Orthop.* **2012**, *142*, 269–273. [CrossRef] [PubMed]
5. Flügge, T.V.; Schlager, S.; Nelson, K.; Nahles, S.; Metzger, M.C. Precision of intraoral digital dental impressions with iTero and extraoral digitization with the iTero and a model scanner. *Am. J. Orthod. Dentofac. Orthop.* **2013**, *144*, 471–478. [CrossRef] [PubMed]
6. Kim, J.; Lagravére, M.O. Accuracy of Bolton analysis measured in laser scanned digital models compared with plaster models (gold standard) and cone-beam computer tomography images. *Korean J. Orthod.* **2016**, *46*, 13–19. [CrossRef] [PubMed]
7. Pellitteri, F.; Albertini, P.; Vogrig, A.; Spedicato, G.A.; Siciliani, G.; Lombardo, L. Comparative analysis of intraoral scanners accuracy using 3D software: An in vivo study. *Prog. Orthod.* **2022**, *23*, 21. [CrossRef] [PubMed]
8. Naidu, D.; Freer, T.J. Validity, reliability, and reproducibility of the iOC intraoral scanner: A comparison of tooth widths and Bolton ratios. *Am. J. Orthod. Dentofac. Orthop.* **2013**, *144*, 304–310. [CrossRef]

9. Camardella, L.T.; Ongkosuwito, E.M.; Penning, E.W.; Kuijpers-Jagtman, A.M.; Vilella, O.V.; Breuning, K.H. Accuracy and reliability of measurements performed using two different software programs on digital models generated using laser and computed tomography plaster model scanners. *Korean J. Orthod.* **2020**, *50*, 13–25. [CrossRef]
10. Hajeer, M.Y.; Millett, D.T.; Ayoub, A.F.; Siebert, J.P. Applications of 3D imaging in orthodontics: Part I. *J. Orthod.* **2004**, *31*, 62–70. [CrossRef] [PubMed]
11. Moon, Y.G.; Lee, K.M. Comparison of the accuracy of intraoral scans between complete-arch scan and quadrant scan. *Prog. Orthod.* **2020**, *21*, 36. [CrossRef] [PubMed]

medical sciences forum MDPI

Proceeding Paper

The Treatment of Anterior Open Bite Associated with a Thumb Sucking Habit: A Case Report †

Laura Amorim *, Dinis Pereira, Ana Margarida Ramos, Margarida Fernandes, Paulo Fernandes-Retto and Ana Sintra Delgado

Clinic Orthodontics Consultation, Egas Moniz Center for Interdisciplinary Research (CiiEM), Egas Moniz School of Health & Science, Caparica, 2829-511 Almada, Portugal; dpereira@egasmoniz.edu.pt (D.P.); ramos.margarida@hotmail.com (A.M.R.); margarida.l.a.fernandes@gmail.com (M.F.); dpr.dental@gmail.com (P.F.-R.); adelgado@egasmoniz.edu.pt (A.S.D.)

* Correspondence: laura_amorim12@hotmail.com

† Presented at the 6th International Congress of CiiEM—Immediate and Future Challenges to Foster One Health, Almada, Portugal, 5–7 July 2023.

Abstract: Anterior open bite malocclusion, defined as a lack of vertical overlap or contact between the maxillary and mandibular incisors, has a multifactorial etiology. A 17-year-old female patient came to the Egas Moniz University Clinic complaining of anterior open bite. The patient had posterior crossbite on the left side, second premolar agenesis, the presence of a primary molar, and a thumb-sucking habit. This case report describes a nonsurgical orthodontic treatment of an anterior open bite, applying a fixed appliance and bonded acrylic expander.

Keywords: open bite; posterior crossbite; thumb sucking; orthodontic treatment

Citation: Amorim, L.; Pereira, D.; Ramos, A.M.; Fernandes, M.; Fernandes-Retto, P.; Delgado, A.S. The Treatment of Anterior Open Bite Associated with a Thumb Sucking Habit: A Case Report. *Med. Sci. Forum* **2023**, *22*, 2. https://doi.org/10.3390/msf2023022002

Academic Editors: José Brito, Nuno Taveira and Ana I. Fernandes

Published: 8 August 2023

1. Introduction

Anterior open bite malocclusion is characterized by the lack of definition when there is no contact in the anterior dental region between the dental arches when the posterior teeth are in occlusion [1] Anterior open bite impacts oral function and can influence the quality of life of orthodontic patients [2].

Open bite is a malocclusion with several etiologic factors, such as a thumb-sucking habit, lingual interposition habit, mouth breathing, or due to genetic influence. A detailed diagnosis and anamnesis in patients with anterior open bite is very important to uncover the etiology of the anterior open bite [3].

The aim of this article is to report a case of an open bite with a thumb-sucking habit treated with a nonsurgical orthodontic treatment.

2. Materials and Methods

A 17-year-old female patient came to the Egas Moniz University Clinic complaining of anterior open bite. The patient had a bilateral molar and canine class I, posterior crossbite on the left side, lower left second premolar agenesis with the presence of a primary molar, and a straight facial profile. Furthermore, the patient had a severe transverse maxillary deficiency (Pont's Premolar Index: −7.3 mm; Pont's Molar Index: −13.1 mm), proclined upper and lower incisors, and a hyperdivergent class I skeletal pattern. The patient had a thumb-sucking habit during sleep. Regarding the functional examination, she reported bilateral pain on the Temporalis muscle and TMJ palpation.

The patient refused the first treatment plan proposed, combining the orthodontic treament with an orthognathic surgery, as well as the extraction of the primary molar. Conversely, the patient accepted a non-invasive approach to also correct the posterior discrepancy and its effects. The orthodontic treatment was initiated after obtaining signed informed consent from the legal guardian of the patient.

3. Results

The orthodontic treatment intruded the molars and corrected the posterior discrepancy, with an extraction of the third molars to treat the anterior open bite. Firstly, the orthodontic treatment involved a multibracket mandibular fixed appliance and a bonded acrylic maxillary expander with a tongue grid to intrude the molars and correct the thumb-sucking habit. Secondly, the upper multibracket fixed appliance was bonded. Upper removable and lower fixed retainers were placed at the end of the treatment. The tongue posture was corrected using miofunctional speech therapy. The treatment duration was 29 months.

4. Discussion

According to the literature, open bite malocclusion presents a prevalence of 6% in permanent dentition and 3% in mixed dentition [4,5].

In adult patients, severe anterior open bite cases are often treated with orthognatic surgery [6]. Notwithstanding this, orthodontic treatment with molar intrusion and the correction of posterior discrepancy appears to be a good alternative, with founded long-term stability [3,7,8].

In this case report, the anterior open bite was treated with orthodontic camouflage and without any orthognathic surgery. Treatment stability was present with a follow up of 24 months (Figure 1).

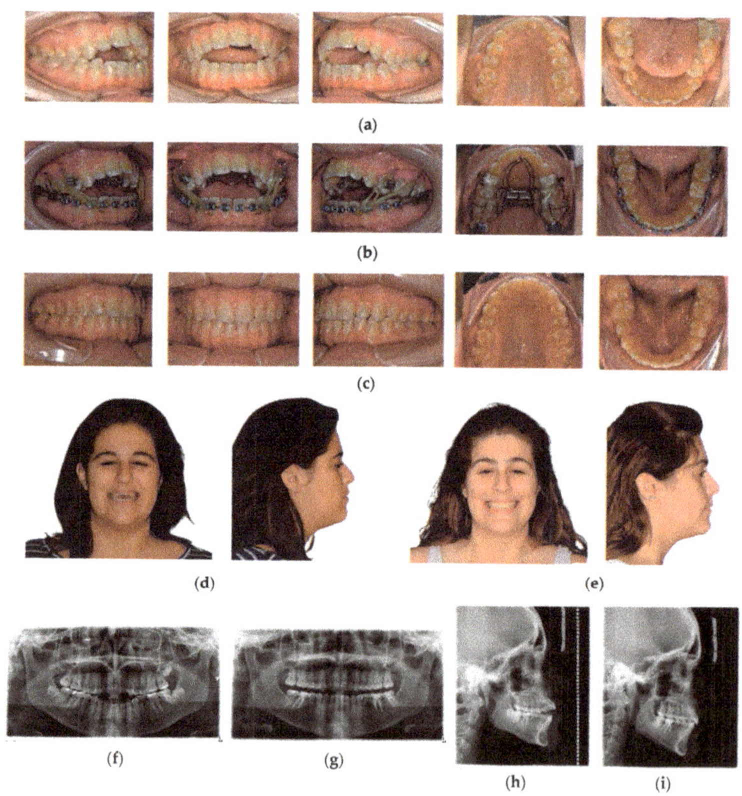

Figure 1. Before and after orthodontic treatment records: (**a**) Initial intra-oral photographs; (**b**) Intra-oral photographs during orthodontic treatment; (**c**) Final intra-oral photographs; (**d**) Initial facial photographs; (**e**) Final facial photographs; (**f**) Initial orthopantomography; (**g**) Final orthopantomography; (**h**) Initial cephalometric radiography; and (**i**) Final cephalometric radiography.

Author Contributions: Conceptualization, L.A., D.P., A.M.R., M.F., P.F.-R. and A.S.D.; methodology, L.A., D.P., M.F., P.F.-R. and A.S.D.; software, L.A.; validation, L.A., D.P., M.F., P.F.-R. and A.S.D.; formal analysis, L.A., D.P., A.M.R., M.F., P.F.-R. and A.S.D.; investigation, L.A., D.P., A.M.R., M.F., P.F.-R. and A.S.D.; resources, L.A., D.P., A.M.R., M.F., P.F.-R. and A.S.D.; data curation, L.A., D.P., A.M.R., M.F., P.F.-R. and A.S.D.; writing—original draft preparation, L.A., D.P. and A.M.R.; writing—review and editing, L.A., D.P., A.M.R., M.F., P.F.-R. and A.S.D.; visualization, L.A.; supervision, D.P., M.F., P.F.-R. and A.S.D.; project administration, L.A., D.P., A.M.R., M.F., P.F.-R. and A.S.D. All authors have read and agreed to the published version of the manuscript.

Funding: This research received no external funding.

Institutional Review Board Statement: Not applicable.

Informed Consent Statement: Informed consent was obtained from all subjects involved in the study.

Data Availability Statement: Data is unavailable due to privacy restrictions.

Conflicts of Interest: The authors declare no conflict of interest.

References

1. Sandler, P.J.; Madahar, A.K.; Murray, A. Anterior open bite: Aetiology and management. *Dent. Updat.* **2011**, *38*, 522–532. [CrossRef] [PubMed]
2. Curto, A.; Albaladejo, A.; Alvarado-Lorenzo, A. Oral-Health-Related Quality of Life (OHRQoL) and Anterior Open Bite inAdult Patients: A Case-Control Study. *Healthcare* **2022**, *10*, 129. [CrossRef] [PubMed]
3. Rijpstra, C.; Lisson, J.A. Etiology of anterior open bite: A review. *J. Orofac. Orthop.* **2016**, *77*, 281–286. [CrossRef] [PubMed]
4. Lombardo, G.; Vena, F.; Negri, P.; Pagano, S.; Barilotti, C.; Paglia, L.; Colombo, S.; Orso, M.; Cianetti, S. Worldwide prevalence of malocclusion in the different stages of dentition: A systematic review and meta-analysis. *Eur. J. Paediatr. Dent.* **2020**, *21*, 115–122. [PubMed]
5. Alhammadi, M.S.; Halboub, E.; Fayed, M.S.; Labib, A.; El-Saaidi, C. Global distribution of malocclusion traits: A systematic review. *Dental. Press J. Orthod.* **2018**, *23*, 40.e1–40.e10. [CrossRef] [PubMed]
6. Reichert, I.; Figel, P.; Winchester, L. Orthodontic treatment of anterior open bite: A review article—Is surgery always necessary? *Oral Maxillofac. Surg.* **2014**, *18*, 271–277. [CrossRef] [PubMed]
7. Vela-Hernández, A.; López-García, R.; García-Sanz, V.; Paredes-Gallardo, V.; Lasagabaster-Latorre, F. Nonsurgical treatment of skeletal anterior open bite in adult patients: Posterior build-ups. *Angle Orthod.* **2017**, *87*, 33–40. [CrossRef] [PubMed]
8. Theodoridou, M.Z.; Zarkadi, A.E.; Zymperdikas, V.F.; Papadopoulos, M.A. Long-term effectiveness of non-surgical open-bite treatment: A systematic review and meta-analysis. *Prog. Orthod.* **2023**, *24*, 18. [CrossRef] [PubMed]

medical sciences forum

MDPI

Proceeding Paper

Selection of a Hydroxypropylcellulose Grade for 3D-Printable Paroxetine Formulations by Fused Deposition Modelling †

Sara Figueiredo [1], João F. Pinto [1], Fátima G. Carvalho [2] and Ana I. Fernandes [3,*]

1 iMed.ULisboa—Research Institute for Medicines, Faculdade de Farmácia, Universidade de Lisboa, Av. Prof. Gama Pinto, 1649-003 Lisboa, Portugal; sara.figueiredo@ff.ulisboa.pt (S.F.); jfpinto@ff.ul.pt (J.F.P.)
2 Infosaúde—Laboratório de Estudos Farmacêuticos, R. Ferrarias Del Rei 6A, 2730-036 Barcarena, Portugal; fatimag.carvalho@anf.pt
3 Egas Moniz Center for Interdisciplinary Research (CiiEM), Egas Moniz School of Health & Science, 2829-511 Almada, Portugal
* Correspondence: aifernandes@egasmoniz.edu.pt; Tel.: +351-212946823
† Presented at the 6th International Congress of CiiEM—Immediate and Future Challenges to Foster One Health, Almada, Portugal, 5–7 July 2023.

Abstract: This work presents the preliminary development of paroxetine-based formulations containing hydroxypropylcellulose (HPC) polymers suitable for hot-melt extrusion coupled to fused deposition modeling (3D printing). Two grades of HPC (54% *w*/*w*), Klucel™ LF and Klucel™ GF, were tested in a polymeric formulation of paroxetine (30% *w*/*w*) and adjuvants (16% *w*/*w* of dicalcium dihydrate phosphate, magnesium stearate, and triethylcitrate; 10:1:5 ratio). Both formulations exhibited a release of almost 100% of paroxetine after 12 h, but the drug released from the Klucel™ LF formulation was quicker and closer to the formulations available in the market.

Keywords: hydroxypropylcellulose (HPC) grade; paroxetine (PRX); printability; 3D-printed tablet; fused deposition modelling (FDM); hot-melt extrusion (HME)

Citation: Figueiredo, S.; Pinto, J.F.; Carvalho, F.G.; Fernandes, A.I. Selection of a Hydroxypropylcellulose Grade for 3D-Printable Paroxetine Formulations by Fused Deposition Modelling. *Med. Sci. Forum* **2023**, *22*, 17. https://doi.org/10.3390/msf2023022017

Academic Editors: José Brito and Nuno Taveira

Published: 10 August 2023

1. Introduction

Three-dimensional printing (3DP) has recently been attracting the attention of the pharmaceutical community because it allows patient-centric design and production of dosage forms, according to the individual needs of a specific patient [1]. Fused Deposition Modelling (FDM), one of the most used 3DP techniques, relies on the previous production of a drug-containing thermoplastic polymeric filament. The FDM 3D-printer is fed with the filament, which is molten at high temperature, extruded, and continuously deposited on the printer plate, layer by layer, building the 3D-printed dosage form. Hot-melt extrusion (HME) is the most interesting method to manufacture the filaments, using existing pharmaceutical-grade polymers [2].

The success of FDM for medicine customization depends on several factors, such as the choice of the adequate polymeric matrix, according to the intended drug release profile. Recently, cellulose-derived polymers have been increasingly used for filament preparation by HME. Cellulose ethers encompass a category of polymers designed by the linking of cellulose to alkyl substituents, such as methyl (methylcellulose, MC), ethyl (ethylcellulose, EC) and propyl (hydroxypropylcellulose, HPC) groups.

Among the cellulosic polymers, HPC (Figure 1) has been extensively studied, in several works, for application in integrated HME-FDM 3DP [3]. Typically, HPC polymers exhibit plasticity and hydrophobicity, high solubility in water and organic solvents, and a low Tg (0–120 °C, which tends to decrease with increasing moisture due to the plasticizer effect of water).

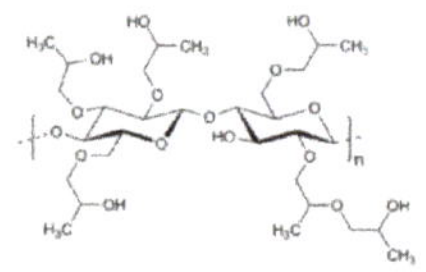

Figure 1. Chemical structure of HPC with a 2.5 degree of substitution. Reprinted with permission from Ref. [4]. 2017, Ashland Inc.

Since HPC polymers are marketed with different viscosities and molecular weight grades (Klucel™ ELF, EF, LF, JF, GF, MF, and HF; Table 1) they can be used in pharmacy to modulate the drug release profile [4]. In fact, the drug release rate depends on the polymer viscosity, which in turn is affected by the molecular weight and temperature.

Table 1. Properties and applications of typical pharmaceutical HPC polymers (adapted with permission from Ref. [4]. 2017, Ashland Inc.).

HPC Grade	Viscosity (mPa·s)	Molecular Weight (Da)	Usual Pharmaceutical Applications
Klucel™ HF Pharm	1500–3000	1,150,000	Controlled-release matrix
Klucel™ MF Pharm	4000–6500	850,000	Controlled-release matrix
Klucel™GF Pharm [1]	150–400	370,000	Controlled-release matrix
Klucel™ JF Pharm	150–400	140,000	Controlled-release matrix
Klucel™LF Pharm [1]	75–150	95,000	Immediate-release binder/Film-coating
Klucel™ EF Pharm	300–600	80,000	Immediate-release binder/Film-coating
Klucel™ ELF Pharm	150–225	40,000	Immediate-release binder/Film-coating

[1] Klucel™ GF and LF grades, marked in bold, were tested in this work. Klucel is a trademark of Ashland Inc.

As these matrix polymers have not been developed specifically for 3DP applications, it is crucial to evaluate their properties both alone and in the presence of the drug and adjuvants. Polymers can affect the characteristics of the final formulation, such as their aqueous solubility, erosion, and/or swelling properties. The mechanical and rheological properties of the filaments determine the quality and behavior (e.g., immediate or delayed release) of the final dosage form produced and are largely dependent on the matrix. Furthermore, drug release can be adjusted by the addition of excipients (e.g., disintegrants, surfactants, and/or pore builders) and by printing specimens with different infills or geometries [3].

Based on a comparative study of the drug dissolution profile, this work reports the selection of the most suitable grade of HPC polymer to modulate the release of paroxetine (PRX; used for the treatment of major depression, generalized anxiety, and related disorders) from 3D-printed tablets obtained by HME coupled to FDM 3DP.

2. Materials and Methods

PRX (Lusifar, Lisbon, Portugal) was used as a model drug; as matrix-forming polymers, hydroxypropylcellulose (HPC; Klucel™ LF and Klucel™ GF Pharm, Ashland Inc., Schaffhausen, Switzerland) were used. Magnesium stearate (MgS) (Roic Pharma, Terrasa, Barcelona, Spain), dicalcium dihydrate phosphate (CaP) (Budenheim, Rheinstrasse, Germany), and triethylcitrate (TEC) (Sigma Aldrich, Darmstadt, Germany) were used as excipients. Paroxetine film-coated tablets 20 mg (Tecnimede Group, Sintra, Portugal) were used as commercial references.

The extrusion of physical mixtures of the raw materials was performed in a single-screw extruder (Noztec Pro, Noztek, Shoreham, UK) at temperatures of 120 °C and 90 °C (barrel with two heating sections), at a constant screw speed (10 rpm). Two different polymeric formulations (HPC™ LF and HPC™ GF) were considered. Tablets were 3D-printed by FDM (3D printer Delta WASP 20 40 Turbo 2, Massa Lombarda, Italy) from PRX-loaded filaments, according to a digital template (3D Sprint Software v2.11, 3D Systems,

Rock Hill, SC, USA) and exported as a stereolithography (.stl) file into Cura (v15.04.2, Ultimaker B.V., Utrecht, The Netherlands). The tablets (10 mm diameter × 3 mm thick cylinders; 0.7 mm layer width × 1.4 mm wall thickness; 100% infill) were printed at a temperature of 200 °C and a 60 mm/s printing speed.

In vitro dissolution of the 3D-printed tablets was performed, and kinetic parameters, such as the time required for 50% drug release ($t_{50\%}$) and the dissolution rate (DR), were calculated [5].

3. Results and Discussion

Previous work demonstrated the feasibility of PRX-based formulations to be ex-truded by HME into filaments, which could be used to manufacture 3D-printed tablets by FDM. The polymeric formulation containing PRX (30% *w*/*w*), HPC (54% *w*/*w*), and excipients (16% *w*/*w* of CaP, MgS, and TEC) exhibited the most adequate behavior, among those studied, for coupling both technologies [6].

Nevertheless, it remained unclear which HPC polymer grade (Klucel™ LF and Klucel™ GF) was the most suitable to use. In terms of the manufacturing process (extrudability and printability), no significant differences were observed by the use of any of the HPC polymer grades considered. Thus, the selection criterion was directed towards the quality attributes of the 3D-printed dosage forms produced. In particular, it was defined that the best HPC candidate would be the one capable of producing a drug release profile closer to that of the commercial PRX tablets produced by tableting.

Both polymeric formulations containing Klucel™ LF and Klucel™ GF were used to produce filaments, and 3D-printed tablets were manufactured. A comparative study of the in vitro dissolution profile of both 3D-printed dosage forms was carried out (Figure 2), and the kinetic parameters were evaluated (Table 2).

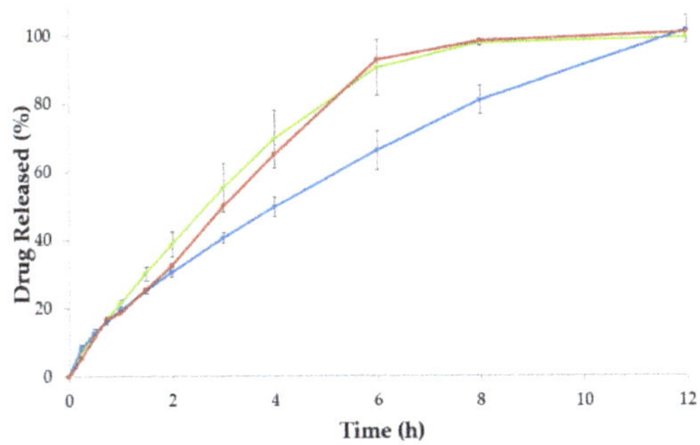

Figure 2. Dissolution profiles of the 3D-printed PRX tablets prepared with Klucel™ LF (green) and Klucel™ GF (blue); commercial tablets (red) were used as a reference (*n* = 3).

Table 2. Dissolution parameters of 3D-printed tablets produced by HME coupled to FDM.

Formulation	$t_{50\%}$ (min)	DR (mg·min^{-1})	f_2	Similarity
Klucel™ LF	2.681	0.184	71.46	Yes *
Klucel™ GF	4.027	0.174	48.20	No

* Criterion defined for f_2: 50–100; (*n* = 3).

The dissolution exhibited a profile typically associated with controlled release formulations, particularly useful in the treatment of psychiatric diseases and related to the use of the HPC polymer in both formulations, regardless of its grade. For the polymeric formulation composed of Klucel™ LF, the release of ≥85% of PRX was observed for ≈6 h of the test, reaching the steady state (release close to 100%) after 8 h. The release profile was superimposable to that of the commercial formulation obtained by a conventional tableting process; the similarity of the dissolution profiles between both formulations was supported by an f_2 test, which exhibited a value higher than 50 (71, i.e., good similarity). This finding suggested that coupling HME and FDM technologies can produce 3D-dosage forms with drug release kinetics similar to those commercially available. Although the dissolution profiles were similar, the kinetics of drug release are amenable to adjustments leading to complete overlapping of the profiles, namely by modulating the polymer:PRX ratio or the adjuvants present in the formulation.

On the contrary, the polymeric formulation containing Klucel^{TM} GF presented a slower PRX release rate, as inferred by higher $t_{50\%}$ values and lower DR results, when compared to the Klucel^{TM} LF-based polymeric formulation. In this scope, the increase in the HPC viscosity associated with the higher molecular weight of Klucel^{TM} GF, impaired the release of the drug. Likewise, the dissolution profile was not comparable with the commercial formulation, despite the f_2 factor being close to 50 (48, i.e., almost similar). In fact, more than 8 h were required for $\geq$85% of the PRX to be released from 3D-printed tablets composed of Klucel^{TM} GF polymeric formulation (even though it reached 100% at the end of dissolution test). The potential for further extending drug action with this formulation is apparent and is the subject of future investigation.

Overall, this work supports the selection of the Klucel^{TM} LF polymeric matrix as the best option, among those studied, to manufacture 3D-printed PRX tablets by integrated HME-FDM, as a therapeutic strategy in the treatment of psychiatric diseases. 3DP is proven to be capable of mimicking the drug release of commercial formulations with the added value of possible customization according to the patient needs.

Author Contributions: Conceptualization, S.F., A.I.F. and J.F.P.; methodology, investigation, and formal analysis, S.F.; writing—original draft preparation, S.F.; writing—review and editing, A.I.F.; resources, F.G.C., A.I.F. and J.F.P.; project administration, and funding acquisition, A.I.F. and J.F.P. All authors have read and agreed to the published version of the manuscript.

Funding: This work was supported by the Fundação para a Ciência e a Tecnologia, grant number PTDC/CTM CTM/30949/2017 (Lisboa 010145 Feder 030949).

Institutional Review Board Statement: Not applicable.

Informed Consent Statement: Not applicable.

Data Availability Statement: Data are not publicly available.

Acknowledgments: Klucel™ HPC samples courtesy of Ashland Inc.

Conflicts of Interest: The authors declare no conflict of interest.

References

1. Alomari, M.; Mohamed, F.H.; Basit, A.W.; Gaisford, S. Personalised dosing: Printing a dose of one's own medicine. *Int. J. Pharm.* **2015**, *494*, 568–577. [CrossRef] [PubMed]
2. Zhang, J.; Feng, X.; Patil, H.; Tiwari, R.V.; Repka, M.A. Coupling 3D printing with hot-melt extrusion to produce controlled-release tablets. *Int. J. Pharm.* **2017**, *519*, 186–197. [CrossRef] [PubMed]
3. Pereira, G.C.; Figueiredo, S.; Fernandes, A.I.; Pinto, J.F. Polymer selection for hot-melt extrusion coupled to fused deposition modelling in pharmaceutics. *Pharmaceutics* **2020**, *12*, 795. [CrossRef] [PubMed]
4. Klucel™ Hydroxypropylcellulose: Physical and Chemical Properties. Ashland Pharmaceutical Technology Report, PC-11229.3, 2017. Available online: https://www.ashland.com (accessed on 19 May 2023).
5. U.S. Department of Health and Human Services. *Guidance for Industry: Dissolution Testing of Immediate Release Solid Oral Dosage Forms*; Food and Drug Administration: Rockville, MD, USA, 1993.
6. Figueiredo, S.; Pinto, J.F.; Carvalho, F.G.; Fernandes, A.I. Tuning of Paroxetine 3D-Printable Formulations for Fused Deposition Modelling. *Med. Sci. Forum* **2021**, *5*, 17. [CrossRef]

medical sciences forum

MDPI

Proceeding Paper

Analysis of the Colour Change in Composite Resins When Exposed to Colouring Agents Possible to Be Found in the Oral Cavity †

Sara Salgueiro [1], Joana Vasconcelos e Cruz [1,2], José Brito [1,2], Ana Mano Azul [1,2] and Mário Polido [1,2,*]

1 Egas Moniz School of Health & Science, Campus Universitário, Quinta da Granja, Monte de Caparica, 2829-511 Almada, Portugal; sara.vsalgueiro@gmail.com (S.S.); joanavcruz@gmail.com (J.V.e.C.); jbrito@egasmoniz.edu.pt (J.B.); anazul2@gmail.com (A.M.A.)

2 Egas Moniz Center for Interdisciplinary Research (CiiEM), Campus Universitário, Quinta da Granja, Monte de Caparica, 2829-511 Almada, Portugal

* Correspondence: mpolido@egasmoniz.edu.pt

† Presented at the 6th International Congress of CiiEM—Immediate and Future Challenges to Foster One Health, Almada, Portugal, 5–7 July 2023.

Abstract: Currently, composite resins are widely used as aesthetic restorative materials. The success of any aesthetic restoration depends on the stability of the material's colour. Colour change in composite resin restorations is one of the most frequent reasons for their replacement. The aim of this study is to evaluate the change in colouration, using the CIE L*a*b* colour system, of nanohybrid and microhybrid composite resins when they are exposed to potential staining solutions over a period of 14 days.

Keywords: colour change; composite resin; aesthetics; pigmentation

Citation: Salgueiro, S.; e Cruz, J.V.; Brito, J.; Azul, A.M.; Polido, M. Analysis of the Colour Change in Composite Resins When Exposed to Colouring Agents Possible to Be Found in the Oral Cavity. *Med. Sci. Forum* **2023**, *22*, 18. https://doi.org/10.3390/msf2023022018

Academic Editors: Nuno Taveira and Ana I. Fernandes

Published: 10 August 2023

1. Introduction

Colour represents one of the most relevant aesthetic parameters in dentistry [1]. To evaluate and determine the colour, the instrumental method can be used; it is objective and can be quantified, such as by using spectrophotometers and colorimeters [2–4]. The CIE L*a*b* colour system measures the value and chroma through three parameters, "L", "a", and "b". "L" represents brightness, "a" the red–green region, and "b" the yellow–blue region [5]. The total colour difference (ΔE) is calculated using the formula $\Delta E = [(\Delta L)^2 + (\Delta a)^2 + (\Delta b)^2)]^{\frac{1}{2}}$ [5,6].

Composite resins play a key role in dental restoration, especially in anterior regions, due to their aesthetic capabilities. Inadequate colouration and chromatic instability in a composite resin restoration are among the most prevalent reasons for its replacement [6–8]. This study aims to evaluate the alteration of the colouration of composite resins when they are exposed to solutions that may cause pigmentation during a period of 14 days.

2. Materials and Methods

Three composite resins were used: Clearfil Majesty™ ES-2 Premium (Ma) (Kuraray Noritake Dental Inc.—Okayama, Japan), Elegance Composite Universal (El) (Medicaline—Castellón, Spain), and Point 4™ (Pt) (Kerr—Orange, CA, USA). For each resin, shade A2 (Vita Scale) was chosen. A control solution (Cl) (artificial saliva) and 4 potentially pigmenting solutions were used: coffee (C) (Delta Cafés—Alentejo, Portugal), red fruit juice (Fv) (Compal—Carnaxide, Portugal), chlorhexidine—0.12% (Cx) (Laboratorios KIN—Barcelona, Spain), and an energy drink (Rb) (Red Bull GmbH—Fuschl, Austria). In total, 225 resin disks (10 × 2 mm) were prepared (75 discs for each resin). For each composite resin, 5 experimental groups (n = 15) were randomly formed. In the first 24 h after its confection, each group was placed in Cl and stored at 37 °C. In the subsequent 14 days, the discs were placed in their respective solutions. The samples (apart from the control group, as it was permanently immersed in the Cl solution) were incorporated into the potentially pigmenting solutions for a period of 5 min per day, for 14 days. During the remaining 23 h 55 min, the samples were kept in Cl, at 37 °C. In total, 2 colour measurements were performed on the discs with the SpectroShade^{TM} Micro (MHT—Verona, Italia): after the discs were initially immersed in Cl for 24 h (T0) and after 14 days (T2) following their exposure to the solutions. The CIE L*a*b* colour system was then applied.

3. Results

A two-way ANOVA was applied to evaluate the effect of the type of resin and the type of beverage to which the resins were exposed on the variations in four colour parameters (E, L, a, and b) between the 1st and 14th days of exposure (T = 2). The assumption of normality of the distribution of dependent variables in each subgroup defined by the two independent factors was assessed through the Shapiro–Wilk test, whose results proved its validity in the overwhelming majority of cases. Regarding the assumption of the homogeneity of variance of the dependent variables between subgroups, Levene's test revealed that this assumption was violated for all variables. However, this violation had no impact on the results because all sub-groups considered in the experimental design had the same size (n = 15). Statistical analysis was performed with the SPSS 28.0 software (IBM, SPSS Inc., Armonk, NY, USA) at a $p < 0.05$.

The application of the two-way ANOVA model to the data of the four variables produced the results shown in Table 1. Based on these results, we concluded that, for each of the dependent variables considered, there was a significant interaction between the two independent factors (effect represented in the table by the term "Type of Resin * Type of Beverage"). This result means that the effects of exposure to different types of beverages varied according to the type of resin, so comparisons between groups defined by one of the independent factors were conducted for fixed levels of the other factor, as described below, separately for each dependent variable.

Table 1. Evaluation of the effect of the type of resin and beverage to which the resins were exposed, on the variation that occurred in the ΔE, ΔL, Δa, and Δb colour parameters between the 1st and 14th days of exposure.

Tests of Between-Subjects Effects					
Dependent Variable: ΔE2					
Source	Type III Sum of Squares	df	Mean Square	F	Sig.
Type of Resin	31.375	2	15.688	66.572	<0.001
Type of Beverage	869.816	4	217.454	922.791	<0.001
Type of Resin * Type of Beverage	18.863	8	2.358	10.006	<0.001
Dependent Variable: ΔL2					
Source	Type III Sum of Squares	df	Mean Square	F	Sig.
Type of Resin	47.701	2	23.851	52.352	<0.001
Type of Beverage	720.732	4	180.183	395.502	<0.001
Type of Resin * Type of Beverage	15.951	8	1.994	4.377	<0.001
Dependent Variable: Δa2					
Source	Type III Sum of Squares	df	Mean Square	F	Sig.
Type of Resin	3.988	2	1.994	67.053	<0.001
Type of Beverage	69.039	4	17.260	580.360	<0.001
Type of Resin * Type of Beverage	3.642	8	0.455	15.308	<0.001
Dependent Variable: Δb2					
Source	Type III Sum of Squares	df	Mean Square	F	Sig.
Type of Resin	90.391	2	45.196	347.404	<0.001
Type of Beverage	381.685	4	95.421	733.472	<0.001
Type of Resin * Type of Beverage	10.934	8	1.367	10.505	<0.001

Figure 1a shows the data of the dependent variable ΔE between the 1st and 14th days of exposure, considering the type of beverage as fixed. Figure 1b shows the data of the dependent variable ΔE between the 1st and 14th days of exposure, considering the type of resin as fixed.

Figure 1. Data of the dependent variable ΔE between the 1st and 14th days of exposure. (**a**) Considering the type of beverage as fixed; (**b**) considering the type of resin as fixed.

All the composite resins showed significant colour variation. There were significant differences between the composite resins when exposed to the same solution. Similarly, there were significant differences between the solutions in their ability to cause colour changes in the same resin. The effects of exposure to the solutions differed according to the type of resin.

4. Discussion

Composite resins composed of larger inorganic particles are more susceptible to colour changes from extrinsic pigmentation (due to absorbing more water) than resins of smaller inorganic particles. This is not in line with the results obtained, since the nanohybrid composite resin El showed a greater chromatic variation than the Pt microhybrid composite resin, probably due to the difference in the composition of its organic matrix [4,9,10]. The nanohybrid composite resin Ma, which did not contain TEGDMA, showed the highest colour stability, followed by the Pt microhybrid composite resin, and finally the nanohybrid composite resin El, which contained TEGDMA, with the latter showing the lowest colour stability. The results obtained are in line with those of a study that compared a microhybrid and a nanohybrid resin, which contained TEGDMA, and the microhybrid resin showed greater colour stability [11]. Another study related a nanohybrid resin (which did not contain TEGDMA) and two microhybrid resins, and the nanohybrid resin showed higher chromatic stability [6]. This is in line with the results obtained.

There is also a study in which it was found that the evaluated microhybrid composite resin obtained greater chromatic stability compared to the nanohybrid one (which contained TEGDMA in its constitution) [9]. In the present study, these results were also verified, considering that the nanohybrid composite resin El (containing TEGDMA) demonstrated lower colour stability compared to the Pt microhybrid composite resin. Some investigations have described that resins containing TEGDMA exhibit greater colour changes, due to absorbing a greater amount of water [12,13]. This fact is in line with the results obtained in the present study, considering that the composite resins (El and Pt) containing TEGDMA in their compositions showed greater chromatic variation.

The analysis of the solutions showed that there was a highly significant difference between them. Their pigmentation capacity differed when applied to the same composite resin. The C caused the greatest alteration in all composite resins, being the only one that was perceptible through the human eye, with ΔE > 5, which agrees with other studies [4,6,9,10,14]. Other solutions did not cause a perceptible change in colour, with ΔE < 2. Regarding the red fruit solution, the values were not in line with the results of another study [14], possibly due to the different exposure times. The results obtained for the chlorhexidine solution coincided with those of another study [9], although the percentage of chlorhexidine was different (0.2%).

The results of the solution containing the energy drink were also in agreement with another study [15], although the exposure time to the solution varied.

5. Conclusions

The El resin showed the greatest chromatic change for the Cl, C, Cx, and Rb solutions. The Ma composite resin showed the greatest chromatic change for the Fv solution. The C solution caused the greatest chromatic alteration.

Author Contributions: Conceptualization, S.S. and M.P.; methodology, M.P.; software, J.B.; validation, A.M.A. and M.P.; formal analysis, J.B.; investigation, S.S.; resources, M.P. and J.B.; data curation, J.B.; writing—original draft preparation, S.S.; writing—review and editing, S.S., M.P., A.M.A. and J.B.; visualization, J.V.e.C.; supervision, M.P. and J.V.e.C.; project administration, M.P.; funding acquisition, M.P. All authors have read and agreed to the published version of the manuscript.

Funding: This research received no external funding.

Institutional Review Board Statement: Not applicable.

Informed Consent Statement: Not applicable.

Data Availability Statement: Data may be available upon request from the corresponding author.

Conflicts of Interest: The authors declare no conflict of interest.

References

1. Paravina, R.D.; Ghinea, R.; Herrera, L.J.; Bona, A.D.; Igiel, C.; Linninger, M.; Sakai, M.; Takahashi, H.; Tashkandi, E.; Del Mar Perez, M. Color Difference Thresholds in Dentistry. *J. Esthet. Restor. Dent.* **2015**, *27*, S1–S9. [CrossRef] [PubMed]
2. Borges, M.G.; Silva, G.R.; Neves, F.T.; Soares, C.J.; Faria-E-silva, A.L.; Carvalho, R.F.; Menezes, M.S. Oxygen Inhibition of Surface Composites and Its Correlation with Degree of Conversion and Colorstability. *Braz. Dent. J.* **2021**, *31*, 91–97. [CrossRef] [PubMed]
3. Ceci, M.; Viola, M.; Rattalino, D.; Beltrami, R.; Colombo, M.; Poggio, C. Discoloration of Different Esthetic Restorative Materials: A Spectrophotometric Evaluation. *Eur. J. Dent.* **2017**, *11*, 149–156. [PubMed]
4. Valizadeh, S.; Asiaie, Z.; Kiomarsi, N.; Kharazifard, M.J. Color Stability of Self-Adhering Composite Resins in Different Solutions. *Dent. Med. Probl.* **2020**, *57*, 31–38. [PubMed]
5. Şişmanoğlu, S.; Sengez, G. Effects of Acidic Beverages on Color Stability of Bulk-Fill Composites with Different Viscosities. *Odovtos Int. J. Dent. Sci.* **2022**, *24*, 90–99.
6. Poggio, C.; Vialba, L.; Berardengo, A.; Federico, R.; Colombo, M.; Beltrami, R.; Scribante, A. Color Stability of New Esthetic Restorative Materials: A Spectrophotometric Analysis. *J. Funct. Biomater.* **2017**, *8*, 26. [PubMed]
7. Mada, D.C.; Gasparik, C.; Irimie, A.I.; Mada, M.D.; Dudea, D.; Campian, R.S. Evaluation of Chromatic Changes of a Nanocomposite Resin Using the New Whitness Index. *Clujul Med.* **2018**, *91*, 222. [PubMed]
8. Usha, C.; Rao, S.; George, G. A Comparative Evaluation of the Staining Capacity of Microhybrid and Nanohybrid Resin-Based Composite to Indian Spices and Food Colorants: An in Vitro Study. *Indian J. Dent. Res.* **2018**, *29*, 201–205. [PubMed]
9. Hasani, E.; Baghban, A.A.; Sheikh-Al-Eslamian, S.M.; Sadr, A. Effect of Bleaching on Color Change of Composite after Immersion in Chlorhexidine and Coffee. *J. Conserv. Dent.* **2019**, *22*, 529–532. [PubMed]
10. Schroeder, T.; da Silva, P.B.; Basso, G.R.; Franco, M.C.; Maske, T.T.; Cenci, M.S. Factors Affecting the Color Stability and Staining of Esthetic Restorations. *Odontology* **2019**, *107*, 507–512. [CrossRef] [PubMed]
11. Chittem, J.; Sajjan, G.S.; Kanumuri, M.V. Spectrophotometric Evaluation of Colour Stability of Nano Hybrid Composite Resin in Commonly Used Food Colourants in Asian Countries. *J. Clin. Diagn. Res.* **2017**, *11*, ZC61–ZC65. [PubMed]
12. Tanthanuch, S.; Kukiattrakoon, B.; Peerasukprasert, T.; Chanmanee, N.; Chaisomboonphun, P.; Rodklai, A. The Effect of Red and White Wine on Color Changes of Nanofilled and Nanohybrid Resin Composites. *Restor. Dent. Endod.* **2016**, *41*, 130. [CrossRef] [PubMed]
13. Yikilgan, İ.; Akgul, S.; Hazar, A.; Kedıcı Alp, C.; Baglar, S.; Bala, O. The Effects of Fresh Detox Juices on Color Stability and Roughness of Resin-Based Composites. *J. Prosthodont.* **2019**, *28*, e82–e88. [CrossRef] [PubMed]
14. Alkhadim, Y.K.; Hulbah, M.J.; Nassar, H.M. Color Shift, Color Stability, and Post-Polishing Surface Roughness of Esthetic Resin Composites. *Materials* **2020**, *13*, 1376. [CrossRef] [PubMed]
15. Al-Dharrab, A. Effect of Energy Drinks on the Color Stability of Nanofilled Composite Resin. *J. Contemp. Dent. Pract.* **2013**, *14*, 704–711. [CrossRef] [PubMed]

medical sciences forum

Proceeding Paper

Effect of Irrigation with Sodium Hypochlorite on the Bond Strength to Dentin Using Different Bonding Protocols †

Sara Salgueiro [1,*], Miguel Rento [1], Luís Proença [1,2], Joana Costa [1,2], Inês Caldeira Fernandes [1,2], Inês Carpinteiro [1,2], Alexandra Pinto [1,2] and Ana Mano Azul [1,2]

1 Egas Moniz School of Health & Science, Campus Universitário, Quinta da Granja, Monte de Caparica, 2829-511 Almada, Portugal; miguel.rento@gmail.com (M.R.); lproenca@egasmoniz.edu.pt (L.P.); jcosta@egasmoniz.edu.pt (J.C.); inescfernandes@gmail.com (I.C.F.); icarpinteiro@egasmoniz.edu.pt (I.C.); xana.pinto.r@gmail.com (A.P.); anazul@egasmoniz.edu.pt (A.M.A.)

2 Egas Moniz Center for Interdisciplinary Research (CiiEM), Campus Universitário, Quinta da Granja, Monte de Caparica, 2829-511 Almada, Portugal

* Correspondence: sara.vsalgueiro@gmail.com

† Presented at the 6th International Congress of CiiEM—Immediate and Future Challenges to Foster One Health, Almada, Portugal, 5–7 July 2023.

Abstract: The study evaluates the influence of irrigation with sodium hypochlorite on the microtensile bond strength to dentin with different bonding protocols on pre-endodontic restorations. After endodontic opening of restored human molars, teeth were randomly divided into four experimental groups: group 1 was not irrigated and the access was restored, while the other groups were irrigated with sodium hypochlorite. Group 2 was restored, group 3 had the endodontic access walls instrumented, and group 4 had CoJet sandblasting and silane application prior to final restoration. The bond strength after irrigation showed higher values when silicatization is performed.

Keywords: pre-endodontic restorations; sodium hypoclorite; bond strength; dentin

Citation: Salgueiro, S.; Rento, M.; Proença, L.; Costa, J.; Fernandes, I.C.; Carpinteiro, I.; Pinto, A.; Azul, A.M. Effect of Irrigation with Sodium Hypochlorite on the Bond Strength to Dentin Using Different Bonding Protocols. *Med. Sci. Forum* **2023**, *22*, 19. https://doi.org/10.3390/msf2023022019

Academic Editors: José Brito, Nuno Taveira and Ana I. Fernandes

Published: 10 August 2023

1. Introduction

Pre-endodontic restorations are widely used to properly plan the rehabilitation treatment of the tooth, before considering endodontic treatment [1–3]. To achieve this goal, it is essential to remove carious tooth lesions, old restorations and/or root canal posts, and to perform an adhesive dentistry prior to endodontic treatment. The primary goal is to achieve an optimal contour of the four walls, improve the structural strength and functionality of the affected teeth, and prevent fractures. It also helps to create better conditions for rubber dam isolation and effective irrigation [2–4]. Another major benefit that contributes to the effectiveness and success of endodontic treatment is the prevention of contamination by microorganisms from the carious lesion, saliva, and blood [2,5].

Root canal disinfection is critical to the success of endodontic treatment, and the use of irrigants such as sodium hypochlorite (NaOCl) is generally recommended [5]. In the case of pre-endodontic restorations, the question arises as to whether endodontic irrigation with NaOCl throughout the entire endodontic procedure affects the integrity of the hybrid layer and the adhesive bond strength of this restoration to dentin. No studies have determined, so far, whether it is necessary to remove pre-endodontic restorations after irrigation with NaOCl, and, if its removal is not necessary, it is important to determine the ideal bonding protocols to achieve an effective seal. The aim of this study was to evaluate the influence of irrigation with NaOCl on microtensile bond strength (μTBS) to dentin, using different bonding protocols on pre-endodontic restorations.

2. Materials and Methods

This study was approved by the Ethics Committee of the Egas Moniz School of Health & Science, Portugal (no. 943) and was performed in accordance with ISO/TS 11405/2015 [6]

and the Academy of Dental Materials [7]. Twenty intact human third molars were scaled and cleaned of all debris, stored in a 1% chloramine T (v/v) solution at 4 °C for one week and kept in artificial saliva at 4 °C until use. Two cross-sections were made of each tooth using a hard tissue microtome (Accutom-50, Struers A/S, Ballerup, Denmark) at low speed, with water refrigeration. Pulp remnants were removed, and the pulp chamber filled with cyanoacrylate glue (Zapit, Dental Ventures of America, Corona, CA, USA). A smear layer was then created by polishing with a 600-grit silicon carbide sandpaper on a polisher with constant refrigeration (Buehler Ltd., Lake Bluff, IL, USA), and the dentin was restored with orthophosphoric acid (Octacid), followed by Optibond FL™ adhesive system (Kerr, Orange, CA, USA), and a nanohybrid composite (Grandio SO—VOCO, GMBH, Cuxhaven, Germany), shade A3, according to the manufacturer's recommendations. An endodontic access cavity preparation (3 × 6 mm) was conducted using a conical truncated diamond drill bur (ISO 504) in a high-speed handpiece (Pana-Max2 (NSK, Tokyo, Japan) under refrigeration. After 24 h, the specimens were randomly assigned to four experimental groups (n = 5). Group 1 (G1) was not irrigated, and the access was restored, while the other groups were irrigated with NaOCl according to the Egas Moniz School of Health & Science protocol. Group 2 (G2) was restored after irrigation. In group 3 (G3), the endodontic access cavity preparation walls were instrumented with a high-speed diamond bur (ISO 504), and group 4 (G4) was treated with CoJet™ (3M, Saint Paul, MN, USA) (30-μm Al_2O_3 coated with SiO_2) followed by silane application, prior to final restoration. A range of 76–84 beams (1 ± 0.2 mm^2) were obtained from each five-teeth group, using a hard tissue microtome (Accutom-50, Struers A/S, Ballerup, Denmark), under constant water refrigeration, and immersed in artificial saliva at 37 °C for 24 h in an incubator (Memmert INE 400, Memmert, Germany) prior to testing. The cross-sectional area of the beams was measured using a digital calliper (MPI/E-101, Mitutoyo, Tokyo, Japan), and then they were fixed to Geraldeli-type jigs. The microtensile bond strength (μTBS) test was performed in a universal testing machine (Shimadzu Autograph Ag-IS, Tokyo, Japan) using a 1 kN load cell at 1 mm/min until fracture.

The means of microtensile bond strength were obtained for each group, considering the beam as the experimental unit for the inferential analysis. Since normality and homoscedasticity were verified, a one-way ANOVA test was applied, followed by Tuckey's HSD ($\alpha = 0.05$), using IBM SPSS Statistic v.27.0 software (IBM, SPSS Inc., Armonk, NY, USA).

3. Results

The means and standard deviations of microtensile bond strength are represented in Table 1. The microtensile values ranged between 40.6 (±32.1) MPa, in G1, and 53.6 (±23.8) MPa, in G4. ANOVA revealed statistically significant differences for μTBS mean values among groups ($p = 0.001$). Further statistical analysis using Tukey's HSD showed that the microtensile mean values for G3 and G4 were significantly higher ($p < 0.05$) than those of G1.

Table 1. Descriptive statistics of the microtensile bond strength (μTBS) according to experimental groups.

Experimental Groups	*n*	Beams	Mean ± Standard Deviation
G1	5	78	40.6 ± 32.1
G2	5	84	45.3 ± 23.0
G3	5	78	50.9 ± 20.5
G4	5	76	53.6 ± 23.8

4. Discussion

Pre-endodontic restorations are often performed prior to endodontic treatment and help to create suitable conditions for endodontic treatment, but there is still no consensus on whether to maintain or remove the pre-endodontic restoration for the final rehabilitation of the tooth [8,9]. At the end of the endodontic treatment, a better bond strength of the final restoration to the dentin walls increases the marginal sealing, the mechanical resistance to masticatory stress, and the durability of the restorations for the long-term clinical success of the restorative treatment [10,11].

Firstly, it is important to understand whether irrigation with NaOCl affects the bond strength. In the present study, the microtensile strength of G1 (without irrigation) and G2 (with irrigation) did not show significant differences. This finding is consistent with the results of two meta-analysis in 2018 [10,12], and the mechanisms of action of NaOCl on dental adhesion can explain this. NaOCl dissolves the exposed collagen network and creates a mineralized layer of dentin to which resin material can adhere, increasing the bond strength to dentin through a deproteinization process. This allows direct adhesion between adhesive resin and dentin without the hybrid layer that is usually created [1].

The present study compared three protocols prior to definitive restoration. The results showed that the microtensile strength of G3 and G4 was significantly higher than that of G1 (with no irrigation), and there were no differences between G2, G3, and G4 (with irrigation). However, the results of the three protocols appeared to show a tendency to increase bond strength (G4 > G3 > G2). Bonstein [13] compared a protocol of instrumentation of composite walls with another protocol of sandblasting the walls with aluminum oxide particles. The first protocol showed higher values because silane was applied after instrumentation with a drill and not in the second protocol. The CoJet^{TM} system (3M, St. Paul, MN, USA), which uses silica-coated alumina particles of the order of 30 μm, is indicated for composite repairs [14]. In addition to the sandblasting effect, the incorporation of silica into the substrate to be bonded contributes to a surface available for chemical bonding through the application of the silane [15,16]. Silane is a bifunctional molecule that acts as a binder and adhesion promoter and is chemically involved in the bonding of various restorative materials [14]. This makes the surface hydrophobic, leading to an improvement in the wettability of the composite [17]. Most studies show that the application of silanes helps to increase the bond strength compared to a simple application of the adhesive [18]. Similar results to our study were found in previous articles which concluded that sandblasting the surface with silica-coated alumina particles was a promising technique with higher bond strength values [19,20]. This can be explained by the action that the abrasive jet exerts on the substrate through its particles. By increasing the surface temperature and creating surface roughness through the impact of the alumina particles, micromechanical retention is increased. In addition to the increase in micromechanical retention, it was concluded that the application of silane to the access walls increased the bond strength by increasing the chemical adhesion, consistent with other studies [13,21].

5. Conclusions

Within the limitations of this in vitro study, irrigation with sodium hypochlorite does not affect the bond strength to dentin. The bond strength of the pre-endodontic restoration to dentin after endodontic irrigation reveals higher values when an adhesive protocol based on silicatization of the endodontic access walls is used.

Author Contributions: Conceptualization, I.C. and A.P.; methodology, M.R.; validation, I.C. and A.P.; formal analysis, I.C.F., I.C. and A.P.; investigation, M.R.; statistical analysis, L.P.; writing—original draft preparation, J.C.; writing—review and editing, A.P., I.C.F. and A.M.A.; presentation, S.S.; supervision, I.C., A.P. and A.M.A. All authors have read and agreed to the published version of the manuscript.

Funding: This research received no external funding.

Institutional Review Board Statement: Not applicable.

Informed Consent Statement: Not applicable.

Data Availability Statement: Data may be available upon request from the corresponding author.

Conflicts of Interest: The authors declare no conflict of interest.

References

1. Abuhaimed, T.S.; Neel, E.A.A. Sodium Hypochlorite Irrigation and Its Effect on Bond Strength to Dentin. *BioMed Res. Int.* **2017**, *2017*, 1930360. [CrossRef] [PubMed]
2. Kameyama, A.; Saito, A.; Haruyama, A.; Komada, T.; Sugiyama, S.; Takahashi, T.; Muramatsu, T. Marginal Leakage of Endodontic Temporary Restorative Materials around Access Cavities Prepared with Pre-Endodontic Composite Build-up: An in Vitro Study. *Materials* **2020**, *13*, 1700. [CrossRef] [PubMed]
3. Heydrich, R.W. Pre-Endodontic Treatment Restorations. A Modification of the "donut" Technique. *J. Am. Dent. Assoc.* **2005**, *136*, 641–642. [CrossRef] [PubMed]
4. Hargreaves, K.M.; Berman, L.H. *Cohen's Pathways of the Pulp*, 12th ed.; Elsevier: Amsterdam, The Netherlands, 2020.
5. Kshirsagar, S.; Aggarwal, S.; Mukhtar, A.; Gupta, P.; Rai, V.; Chawla, M. Pre Endodontic Build-up of a Grossly Destructed Tooth: A Case Report. *Int. J. Sci. Study* **2015**, *2*, 11.
6. *ISO/TS 11405:2015*; Dentistry—Testing of Adhesion to Tooth Structure. International Organization for Standardization: Geneva, Switzerland, 2015.
7. Armstrong, S.; Breschi, L.; Özcan, M.; Pfefferkorn, F.; Ferrari, M.; Van Meerbeek, B. Academy of Dental Materials Guidance on in Vitro Testing of Dental Composite Bonding Effectiveness to Dentin/Enamel Using Micro-Tensile Bond Strength (MTBS) Approach. *Dent. Mater.* **2017**, *33*, 133–143. [CrossRef] [PubMed]
8. Arashiro, F.N.; Figueiredo, J.L.G.; Guerisoli, D.M.Z.; Pereira, K.F.S.; Silva, P.G.; Yoshinari, G.H. Evaluation the Influence of Endodontic Irrigants on Bond Strength to Superficial Dentin of Bovine Teeth and Scanning Electron Microscope Analysis. *Rev. Odontol. Bras. Cent.* **2010**, *19*, 314–318.
9. Kharouf, N.; Pedullà, E.; La Rosa, G.R.M.; Bukiet, F.; Sauro, S.; Haikel, Y.; Mancino, D. In Vitro Evaluation of Different Irrigation Protocols on Intracanal Smear Layer Removal in Teeth with or without Pre-endodontic Proximal Wall Restoration. *J. Clin. Med.* **2020**, *9*, 3325. [CrossRef] [PubMed]
10. Hardan, L.; Bourgi, R.; Cuevas-Suárez, C.E.; Zarow, M.; Kharouf, N.; Mancino, D.; Villares, C.F.; Skaba, D.; Lukomska-Szymanska, M. The Bond Strength and Antibacterial Activity of the Universal Dentin Bonding System: A Systematic Review and Meta-Analysis. *Microorganisms* **2021**, *9*, 1230. [CrossRef] [PubMed]
11. Fortin, D.; Swift, E.J.; Denehy, G.E.; Reinhardt, J.W. Bond Strength and Microleakage of Current Dentin Adhesives. *Dent. Mater.* **1994**, *10*, 253–258. [CrossRef] [PubMed]
12. Bohrer, T.C.; Fontana, P.E.; Lenzi, T.L.; Soares, F.Z.M.; Rocha, R.D.O. Can Endodontic Irrigating Solutions Influence the Bond Strength of Adhesives to Coronal Dental Substrates? A Systematic Review and Meta-Analysis of In Vitro Studies. *J. Adhes. Dent.* **2018**, *20*, 481–494. [PubMed]
13. Bonstein, D.G. Evaluation of Varied Repair Protocols Applied to Aged Composite Resin. *J. Adhes. Dent.* **2005**, *7*, 41–49. [PubMed]
14. Matinlinna, J.P.; Lung, C.Y.K.; Tsoi, J.K.H. Silane Adhesion Mechanism in Dental Applications and Surface Treatments: A Review. *Dent. Mater.* **2018**, *34*, 13–28. [CrossRef] [PubMed]
15. Valente, L.L.; Sarkis-Onofre, R.; Gonçalves, A.P.; Fernández, E.; Loomans, B.; Moraes, R.R. Repair Bond Strength of Dental Composites: Systematic Review and Meta-Analysis. *Int. J. Adhes. Adhes.* **2016**, *69*, 15–26. [CrossRef]
16. Gutierrez, N.C.; Moecke, S.E.; Caneppele, T.M.F.; Perote, L.C.C.C.; Batista, G.R.; Huhtalla, M.F.R.L.; Torres, C.R.G. Bond Strength of Composite Resin Restoration Repair: Influence of Silane and Adhesive Systems. *J. Contemp. Dent. Pract.* **2019**, *20*, 880–886. [PubMed]
17. Fuentes, M.V.; Escribano, N.; Baracco, B.; Romero, M.; Ceballos, L. Effect of Indirect Composite Treatment Microtensile Bond Strength of Self-Adhesive Resin Cements. *J. Clin. Exp. Dent.* **2016**, *8*, e14–e21. [CrossRef] [PubMed]
18. Fornazari, I.A.; Brum, R.T.; Rached, R.N.; de Souza, E.M. Reliability and Correlation between Microshear and Microtensile Bond Strength Tests of Composite Repairs. *J. Mech. Behav. Biomed. Mater.* **2020**, *103*, 103607. [CrossRef] [PubMed]
19. Costa, T.R.F.; Ferreira, S.Q.; Klein-Júnior, C.A.; Loguercio, A.D.; Reis, A. Durability of Surface Treatments and Intermediate Agents Used for Repair of a Polished Composite. *Oper. Dent.* **2010**, *35*, 231–237. [CrossRef] [PubMed]
20. Celik, C.; Cehreli, B.S.; Bagis, B.; Arhun, N. Microtensile Bond Strength of Composite-to-Composite Repair with Different Surface Treatments and Adhesive Systems. *J. Adhes. Sci. Technol.* **2014**, *28*, 1264–1276. [CrossRef]
21. Bona, A.D.; Borba, M.; Benetti, P.; Cecchetti, D. Effect of Surface Treatments on the Bond Strength of a Zirconia-Reinforced Ceramic to Composite Resin. *Braz. Oral Res.* **2007**, *21*, 10–15. [CrossRef] [PubMed]

medical sciences forum

MDPI

Proceeding Paper

Kinematic Analysis Comparison between Normal and Cranial Cruciate Ligament Rupture Canine Gait Analysis: An Exploratory Study †

Paulo Miranda-Oliveira [1,2,3,*] and António Martinho Lopes [3,4]

1 Portuguese Athletics Federation (FPA), 2799-538 Oeiras, Portugal
2 School of Technology and Management (ESTG), Polytechnic of Leiria, 2411-901 Leiria, Portugal
3 Egas Moniz Center for Interdisciplinary Research (CiiEM), Egas Moniz School of Health & Science, 2829-511 Almada, Portugal; amlopes@egasmoniz.edu.pt
4 Centro de Investigação Interdisciplinar em Sanidade Animal (CIISA), Faculdade de Medicina Veterinária, Universidade de Lisboa, 1300-477 Lisboa, Portugal
* Correspondence: paoliveira@egasmoniz.edu.pt; Tel.: +351-919761503
† Presented at the 6th International Congress of CiiEM—Immediate and Future Challenges to Foster One Health, Almada, Portugal, 5–7 July 2023.

Abstract: This study aims to apply a kinematic analysis to characterize and compare a normal canine gait with a canine gait with cranial cruciate ligament (CrRL) rupture with free and open-source software. Two dogs walked ten times. A bidimensional kinematic analysis was performed. Spatiotemporal analysis showed significant differences between dogs. The dog with CrCL rupture obtained higher results for all parameters except stance and step time. Also, the stifle angle did not verify differences in absolute angle, but the signal showed differences in patterns between normal and abnormal gait. This study supports that software assisting clinicians' diagnosis with CrCl ruptures.

Keywords: canine gait; kinematic analysis; cranial cruciate ligament; exploratory study

Citation: Miranda-Oliveira, P.; Lopes, A.M. Kinematic Analysis Comparison between Normal and Cranial Cruciate Ligament Rupture Canine Gait Analysis: An Exploratory Study. *Med. Sci. Forum* **2023**, 22, 15. https://doi.org/10.3390/msf2023022015

Academic Editors: José Brito, Nuno Taveira and Ana I. Fernandes

Published: 9 August 2023

1. Introduction

CrCL rupture is one of the most common causes of dog lameness [1,2]. The etiopathogenesis of CrCL is not fully known, but regardless of the cause, it results in stifle joint instability and progressive degenerative joint disease [3]. Furthermore, this type of condition causes lameness, an increase in cranial tibial translation, and internal rotation and abduction of the tibia [4,5]. Gait analysis has evolved [6], it is well-established [7] and it is a way to perform clinical diagnosis for CrCl rupture [8]. Subjective analysis has allowed the identification of gait patterns, but the results, and thus the applications, are limited and different agreements between observers can be obtained [6,8,9]. Alternatively, objective gait analysis has emerged, allowing clinicians accurately study the canine gait cycle [6]. Several methods and instruments can be applied in gait analysis, such as kinetic and kinematic methods, and they are associated with instruments, such as force platforms and motion analysis systems, respectively [7,9]. The kinematic gait analysis is one the most applied, and it helps effectively in quantitative evaluation, measuring segments and joints and identifying gait asymmetry [8]. However, the current methods are usually expensive and require extensive data processing, so this study aims at an exploratory study which consists of applying a kinematic analysis to characterize and compare a normal canine gait with a canine gait with CrCL with free and open-source software. As a result, we could share with veterinaries a simple and useful tool for their daily work to follow and analyze their recovery of the canine gait.

2. Materials and Methods

2.1. Experimental Design Overview

The experiment conducted in the present study consisted of a comparison between a normal and an abnormal canine gait. Each dog performed ten trials with a natural speed in a developed outdoor space of two meters near the Egas Moniz Veterinary University Clinic. We used one high-speed camera and analyzed one step to determine the spatiotemporal parameters and the stifle angle using the motion analysis system Kinovea®.

2.2. Subjects

The study subjects were two dogs, one healthy Labrador Retriever and one Bouvier Bernouis with CrCL rupture, selected by convenience. The dogs' owners were informed of the objective and authorized the data collection. The CrCL rupture diagnosis was assessed by a cranial "drawer" and tibial compression test and conducted by an experienced veterinary orthopedic surgeon. The diagnosis was confirmed by Lateral radiography of the femur-tibio-patellar joint taken doing the tibial compression test and without tibial compression.

2.3. Measurement Protocol and Data Analysis

A veterinarian analyzed the subjects and placed three landmarks defining the stifle angle (Figure 1). The dogs performed a walk of three minutes before the analysis. Dogs started walking toward the outdoor calibrated zone of two meters with a natural but uncontrol speed, and each dog walked ten times. One high-speed camera (Casio Exilim® ZR200, Casio, Tokyo, Japan) collecting at 120 Hz was used during the bidimensional kinematic analysis. It was placed perpendicular to the right of the dog's line. Data processing for the high-speed was conducted using the Kinovea® software (version 0.9.5) to define the stance phase time, swing phase time, step time, step length and step speed, and tracking the ankle, stifle and grand trochanter landmarks defining the stifle angle. Step frequency ($SF = 1/ST$ [steps/s]) and step speed ($v_{step} = SL \times SF$ [m/s]) were calculated. Here, SF represents the step frequency; ST is the step time; v_{step} is the step speed and SL is the step length. Their results were organized using Microsoft Excel Office, and kinematic data processing was performed using the Spyder package in Python 3.7. Data from the high-speed camera were smoothed using the spectral power analysis [10], and a cut-off frequency of 9 Hz was applied.

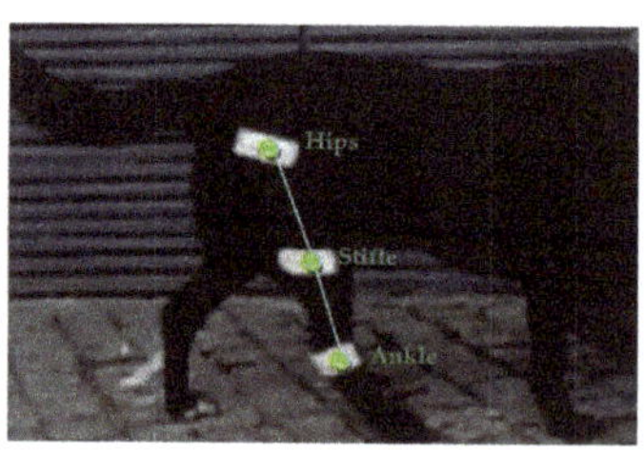

Figure 1. Segment model used to define the stifle angle.

2.4. Statistical Analysis

All walks (n = 20) performed by dogs were analyzed. Descriptive statistics were determined to describe the spatiotemporal data and stifle angle, with means and standard deviation values. Normality was calculated using the Shapiro–Wilk test ($p \leq 0.05$). The independent samples *t*-test was used to compare the proposed parameters, and when the normality was not verified, the Mann–Whitney test was applied [8].

3. Results

Table 1 provides the results of the descriptive and comparison statistical analysis for the spatiotemporal data and stifle angle parameters.

Table 1. Spatiotemporal data and stifle angle parameters descriptive and comparative analysis.

	CrCL		Normal		
	Mean ± SD		**Mean ± SD**		¥
Spatiotemporal parameters					
Stance phase time (s)	0.56 ± 0.09		0.46 ± 0.09		$p = 0.019$
Swing phase time(s)	0.26 ± 0.03		0.28 ± 0.15		$p = 0.150$
Step time (s)	0.82 ± 0.10		0.80 ± 0.09		$p = 0.051$
Stance phase (%)	68.00 ± 0.04		62.20 ± 0.05		$p = 0.005$
Swing phase (%)	32.00 ± 0.04		37.90 ± 0.05		$p = 0.005$
Step length (m)	0.73 ± 0.12		0.92 ± 0.03		$p < 0.001$
Step frequency (step/s)	1.23 ± 0.14		1.38 ± 0.17		$p = 0.048$
Step speed (m/s)	0.91 ± 0.22		1.27 ± 0.19		$p < 0.001$
Angular kinematic parameters					
Minimum stifle angle (degree)	−0.95 ± 1.88	*	−0.77 ± 1.48	*	$p = 0.912$
Maximum stifle angle (degree)	39.4 ± 7.45	*	40.3 ± 5.03		$p = 0.315$
Stifle angle amplitude (degree)	40.4 ± 7.38	*	41.4 ± 5.91		$p = 0.579$

*, *p*-value, non-normal distribution; ¥, *p*-value of the comparisons analysis between Abnormal and Normal dogs.

Figure 2 represents the stifle angular displacement in sagittal plane to a canine gait with CrCL (Figure 2a) and a normal canine gait (Figure 2b).

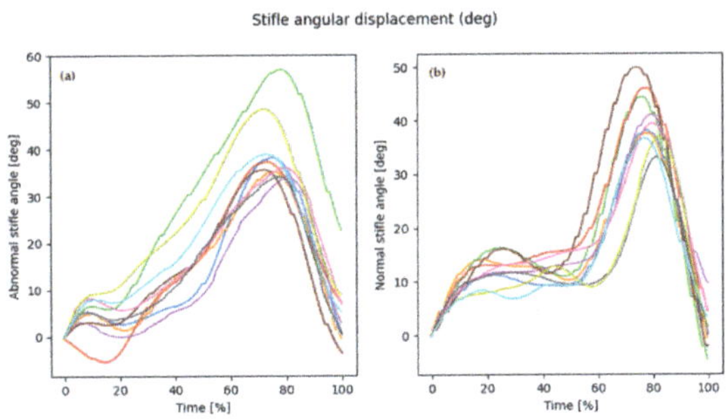

Figure 2. (**a**) Signal pattern to a canine gait with CrCL; (**b**) Signal pattern to normal canine gait; Each color line means each walk.

4. Discussion

In this study, we performed a bidimensional kinematic analysis to compare spatiotemporal parameters and the stifle angle between the normal and CrCL dog. When we analyzed Table 1, as expected, CrCL influenced canine gait by changing some parameters. All results were higher in normal canines except for the stance and step phase time following the literature [11] (p. 837). Specifically regarding the temporal parameters, our results were higher than Jenkins et al. (stride, stance, and swing phases were 0.575, 0.355 and 0.221 s, respectively) [12] (p. 81). However, they did not report the step speed, which is a factor that can change the temporal step parameters. Additionally, we verified significant differences in the stance phase time, stance and swing phase in percentage and step length, frequency, and speed (see Table 1). These results were divergent with Møller et al., which did not verify differences in the spatiotemporal parameters analyzed. However, they used a lower sample frequency (50 Hz) and a non-motion analysis system (Media Player Classic Home Cinema for Windows v.10.0) [8] (p. 66). Regarding the stifle angle (see Table 1), the amplitude did not verify differences between gaits and obtain angular displacement lower than Agostinho et al. (stifle angle: 52.48 ± 4.04 and 62.37 ± 6.53 deg), but they used a higher speed of 2.1 and 2.2 m/s, while we obtained a step speed around 1 m/s [9]. Lastly, regarding the stifle angle signal (Figure 2), when compared with the literature, we verified a similar but inverse signal [9,11], and signal differences between dogs, i.e., the gait to CrCL dog started flexion early (around 20% time gait phase). Our results followed the literature and showed that our proposal can be a useful diagnostic tool for veterinarians in their daily routine.

5. Conclusions

This study allowed us to characterize the normal and CrCL rupture canine gait with free and open-source software obtaining detailed information relevant to clinicians. Furthermore, this information should be confirmed with a higher sample and be suggested to veterinarians for inclusion in the follow-up after surgery to ensure that the dog restored all stifle joint range of motion.

Author Contributions: Conceptualization, P.M.-O. and A.M.L.; methodology, P.M.-O.; software, P.M.-O.; validation, P.M.-O. and A.M.L.; formal analysis, P.M.-O. and A.M.L.; investigation, P.M.-O.; resources, P.M.-O. and A.M.L.; data curation, P.M.-O.; writing—original draft preparation, P.M.-O. and A.M.L.; writing—review and editing, P.M.-O. and A.M.L.; visualization, P.M.-O. and A.M.L.; supervision, P.M.-O. and A.M.L.; project administration, P.M.-O. and A.M.L. All authors have read and agreed to the published version of the manuscript.

Funding: This research received no external funding.

Institutional Review Board Statement: Not applicable.

Informed Consent Statement: Not applicable.

Data Availability Statement: Not applicable.

Conflicts of Interest: The authors declare no conflict of interest.

References

1. Kowaleski, M.; Boudrieau, R.; Pozzi, A. Stifle Joint. In *Veterinary Surgery: Small Animal Expert Consult*; Elsevier: St. Louis, MO, USA, 2018; pp. 1071–1167, ISBN 978-0-323-32065-8.
2. Barnhart, M.; Bufkin, B.; Litsky, A. Biomechanical Comparison of Four Methods of Fixation of a Polymeric Cranial Cruciate Ligament in the Canine Femur and Tibia. *Vet. Comp. Orthop. Traumatol.* **2019**, *32*, 112–116. [CrossRef] [PubMed]
3. Niebauer, G.W.; Restucci, B. Etiopathogenesis of Canine Cruciate Ligament Disease: A Scoping Review. *Animals* **2023**, *13*, 187. [CrossRef] [PubMed]
4. Tinga, S.; Kim, S.E.; Banks, S.A.; Jones, S.C.; Park, B.H.; Pozzi, A.; Lewis, D.D. Femorotibial Kinematics in Dogs with Cranial Cruciate Ligament Insufficiency: A Three-Dimensional in-Vivo Fluoroscopic Analysis during Walking. *BMC Vet. Res.* **2018**, *14*, 85. [CrossRef] [PubMed]
5. Tashman, S.; Anderst, W.; Kolowich, P.; Havstad, S.; Arnoczky, S. Kinematics of the ACL-Deficient Canine Knee during Gait: Serial Changes over Two Years. *J. Orthop. Res.* **2004**, *22*, 931–941. [CrossRef] [PubMed]
6. Gillette, R.L.; Angle, T.C. Recent Developments in Canine Locomotor Analysis: A Review. *Vet. J.* **2008**, *178*, 165–176. [CrossRef] [PubMed]
7. Pálya, Z.; Rácz, K.; Nagymáté, G.; Kiss, R.M. Development of a Detailed Canine Gait Analysis Method for Evaluating Har-nesses: A Pilot Study. *PLoS ONE* **2022**, *17*, e0264299. [CrossRef] [PubMed]
8. Møller, J.H.; Vitger, A.D.; Poulsen, H.H.; Miles, J.E. Simple Video-Based Spatiotemporal Gait Analysis Is Not Better than Subjective Visual Assessment of Lameness in Dogs. *VCOT Open* **2021**, *4*, e65–e71. [CrossRef]
9. Agostinho, F.S.; Rahal, S.C.; Miqueleto, N.S.M.L.; Verdugo, M.R.; Inamassu, L.R.; El-Warrak, A.O. Kinematic Analysis of Labrador Retrievers and Rottweilers Trotting on a Treadmill. *Vet. Comp. Orthop. Traumatol.* **2011**, *24*, 185–191. [CrossRef] [PubMed]
10. Winter, D.A. *Biomechanics and Motor Control of Human Movement*, 3rd ed.; John Wiley & Sons: Hoboken, NJ, USA, 2005; ISBN 978-0-471-44989-8.
11. DeCamp, C.E. Kinetic and Kinematic Gait Analysis and the Assessment of Lameness in the Dog. *Vet. Clin. N. Am. Small Anim. Pract.* **1997**, *27*, 825–840. [CrossRef] [PubMed]
12. Jenkins, G.J.; Hakim, C.H.; Yang, N.N.; Yao, G.; Duan, D. Automatic Characterization of Stride Parameters in Canines with a Single Wearable Inertial Sensor. *PLoS ONE* **2018**, *13*, e0198893. [CrossRef] [PubMed]

medical sciences forum

MDPI

Proceeding Paper

ID NOW: A NAAT System Solution for the Rapid and Accurate Detection of SARS-CoV-2 with VTM Sampling †

Susana Bandarra [1], Célia Neto [1], Fátima Gil [1], Lurdes Monteiro [1,*] and Laura Brum [2]

1 Molecular Pathology Laboratory, SYNLAB Lisboa, Av. Columbano Bordalo Pinheiro, 75A, 1070-061 Lisbon, Portugal; susana.bandarra@synlab.pt (S.B.); celia.neto@synlab.pt (C.N.); fatima.gil@synlab.pt (F.G.)

2 Clinical Laboratories SYNLAB Portugal, Av. Columbano Bordalo Pinheiro, 75A, 1070-061 Lisbon, Portugal; laura.brum@synlab.pt

* Correspondence: lurdes.monteiro@synlab.pt; Tel.: +351-217-216-060

† Presented at the 6th International Congress of CiiEM—Immediate and Future Challenges to Foster One Health, Almada, Portugal, 5–7 July 2023.

Abstract: ID NOW™ COVID-19 is a rapid molecular test for the detection of SARS-CoV-2. According to its instructions for use, this point-of-care test should be performed on dry nasopharyngeal swab (NPS) specimens. However, this method completely consumes the swab, with the limitation that additional analyses cannot be performed if required. The aim of this work was to evaluate the analytical performance of the ID NOW™ COVID-19 using NPS sampled on a viral transport medium. When compared to a reference RT-PCR, the positive and negative percent agreement was 86% and 100%, respectively. False negatives were associated with high RT-PCR Ct values.

Keywords: point-of-care; SARS-CoV-2; rapid diagnostic

Citation: Bandarra, S.; Neto, C.; Gil, F.; Monteiro, L.; Brum, L. ID NOW: A NAAT System Solution for the Rapid and Accurate Detection of SARS-CoV-2 with VTM Sampling. *Med. Sci. Forum* **2023**, *22*, 13. https://doi.org/10.3390/msf2023022013

Academic Editors: José Brito, Nuno Taveira and Ana I. Fernandes

Published: 9 August 2023

1. Introduction

A real-time polymerase chain reaction (RT-PCR), the gold standard method for viral RNA identification, is highly effective. However, it can be time-consuming and requires specialized equipment and operator training. In some situations, a long time to result is not suitable for patient management, such as in emergency departments or for controlling airplane traveler infections during a pandemic.

Point-of-care (POC) tests based on RT-PCR or other nucleic acid amplification technologies are growing exponentially for the detection of human viral infections. Lower costs and developed technologies have had a real impact on viral diagnostics and patient management, not only because these easy-to-use assay kits allow for the decentralization of testing but, most importantly, because of the reduction in the sample-to-answer turnaround time. The COVID-19 pandemic over the last few years has provided living proof of this. Such systems have great potential for the diagnosis of SARS-CoV-2 or other viruses, especially in settings with limited resources or where PCR is not available.

The ID NOW™ COVID-19 assay was developed at the beginning of the pandemic and has been implemented in several countries. According to the manufacturer, the ID NOW™ COVID-19 should be performed on dry nasopharyngeal swabs (NPS) with the swab entirely consumed. A second swab is required in case of invalid results or if further analysis is required.

In the present study, we aimed to assess the analytical performance of the ID NOW™ COVID-19 using NPS sampled on viral transport media (VTM) compared to the reference (RT-PCR).

2. Materials and Methods

This study was conducted at the Molecular Pathology Unit of the SYNLAB Central Laboratory, Lisbon, Portugal. NPS were collected in the VTM between 4 August and

9 September 2021 from hospitals (emergency room, urgent care, and hospitalized patients) and community assessment centers (AC) and were sent to the laboratory for SARS-CoV-2 molecular detection. All NPS were first processed for diagnosis using the gold standard RT-PCR assay (Alinity m SARS-CoV-2, Abbott Molecular, Des Plaines, IL, USA) and were then tested using the Abbott ID NOW device. All samples were tested within 24 h of collection.

The ID NOW™ COVID-19 assay is an isothermal nucleic acid amplification system that specifically detects a unique region of the *RdRp* gene segment with fluorescently labeled molecular beacons and includes an internal control. The assay was performed on 107 NPS according to the manufacturer's instructions with the following modification: the use of 100 μL of NPS collected in VTM. All data were analyzed using Microsoft Excel 2016 (Microsoft Corporation, Redmond, WA, USA) and SPSS software v.28, (IBM, Chicago, IL, USA). For all statistical analyses, a p-value < 0.05 was considered statistically significant.

3. Results

3.1. Sample Population Analyses

A total of 107 NPSs were analyzed, including 53 samples from COVID-19 AC and 54 from hospital centers, 45.8% (49/107) females and 54.2% (58/107) males, aged <1–94 years (median 33 years), representing 21.5% of children and 78.5% of adults. In the RT-PCR (Alinity m SARS-CoV-2 assay), 98/107 (91.6%) were positive, with a range of cycle threshold (Ct) values from 11.9 to 33.1 (median 19.1, interquartile range [IQR] 16.3–26.9).

3.2. Analytical Performance of ID NOW™ COVID-19 Assay

The ID NOW™ COVID-19 assay detected 84 (85.7%; 95 CI 77.9–91.7) of the 98 positive NPS, with the remaining 14 testing as false negatives (14.3%; 95% CI 8.8–23.2) (Table 1). None of the nine negative SARS-CoV-2 samples gave a false positive result, with the ID NOW™ COVID-19 assay, and no invalid results were found in all samples. The overall, positive, and negative percent agreements were 86.9%, 85.7%, and 100%, respectively (Table 1). The agreement analysis comparing the performance of ID NOW showed that the agreement was considered moderate with a Kappa value of 0.502 (50.2%).

Table 1. Performances of ID NOW™ COVID-19 using clinical specimens versus reference RT-PCR.

	Alinity m RT-PCR		**Total**	**Percent Agreement**
	Detected	**Not Detected**		
ID NOW				
Positive	84	0	84	85.7%
Negative	14	9	23	100%
Total	98	9	107	86.9%

When stratifying by Ct, all NPS with a Ct $\leq$ 25.0 by RT-PCR were positive for ID NOW™ COVID-19, whereas only 18 (56.3%) of the 32 NPS displaying a Ct > 25.0 were positive. Of the 14 false negatives, half had a Ct between 25.0 and 30.0, and the other 7 were between 30.0 and 33.0. False negative cases (ID NOW negative/RT-PCR positive) had a higher Ct, suggesting a lower viral load. The mean Ct for concordant positive samples was 19.6 (95% CI, 18.4–20.8), ranging from 11.9 to 33.1, with a standard deviation of 5.6. The mean Ct for discordant samples was 29.6 (95% CI, 28.3–30.9), ranging from 25.4 to 32.9, with a standard deviation of 2.4.

The estimated diagnostic performance of the ID NOW™ COVID-19 assay is shown in Table 2.

Table 2. Estimated diagnostic performance ID NOW™ COVID-19.

Measurement	Diagnostic Performance
Sensitivity	85.7% (84/98); CI: 77.9–91.7
Specificity	100% (9/9)
Positive predictive value (PPV)	100% (84/84)
Negative predictive value (NPV)	39.1% (9/23); CI: 21.1–59.4

CI—95% Confidence interval.

The sensitivity of the ID NOW assay for RT-PCR-positive samples with Ct values less than 30.0 was 92.9% (78/84 cases) (Table 3).

Table 3. Sensitivity of ID NOW™ COVID-19 by cycle threshold.

Ct Value	Positive ID NOW/Positive RT-PCR	Sensitivity (%)
≤25	66/66	100%
25–30	12/19	63.2%; CI: 40.8–82.2
30–33	6/13	46.2%; CI: 21.6–72.1
Ct > cut off *	0/0	NA

* Ct values [33.7–38.7]; CI—95% Confidence interval.

4. Discussion

The development of POC assays improved access to diagnostic testing during the COVID-19 pandemic, which was advertised as rapid, accurate, and relatively easy to perform. However, caution is required because POC diagnostics have both advantages and potential pitfalls, including low sensitivity, as their use is recommended only for acute infections; the increased risk of and inappropriate use of the diagnostic tests; the misinterpretation of test results; and lack of quality control procedures when the diagnostics are removed from the specialized, controlled diagnostic laboratory environment.

In this study, we evaluated NPS sampled in VTM for the presence of SARS-CoV-2 RNA and found good performance (86% sensitivity and 100% specificity), with an overall agreement of 86.9, which was higher than that obtained in other studies [1]. In addition, some invalids have been previously reported with the ID NOW™ COVID-19 [2,3]. However, no invalids were obtained in the present study, suggesting that the rate of the invalids of ID NOW™ COVID-19 was probably low. These performances are slightly lower than conventional RT-PCR but much higher than those of antigenic tests [4]. Therefore, the ID NOW™ COVID-19, when available, could probably replace rapid antigenic tests. The ID NOW device provides a rapid qualitative result (positive, negative, uninterpretable) that does not require specialized interpretation.

Overall, our results showed an NPV of 39.1%, with false negatives occurring for samples with Ct values > 25.0. However, the NPV was 100% for Ct values > 33.0. This is in line with the majority of studies [1,5,6]. Additionally, the lowest performance was obtained for samples displaying Ct values between 30– and 33 with a positive percent agreement of 46.2%, which is slightly higher than that observed in other studies [1,5]. False negatives on ID NOW appear to be strictly related to the viral load, based on the distribution of false negative Ct values obtained and confirmed by the higher limit of detection of Alinity m (100 copies/mL vs. 3225 copies/mL for ID NOW).

In this study, the ID NOW™ COVID-19 assay demonstrated good performance for the detection of SARS-CoV-2 strains compared to the Alinity m SARS-CoV-2 RT-PCR assay. Approximately one year after the declaration of a state of emergency, this test was implemented in some hospital centers (acute care) to guarantee a turnaround time of 2 h and in Portuguese airports to respond quickly to the control of infected passengers, especially in a situation of stopovers between flights. All discrepant and invalid results were confirmed by RT-PCR in the central laboratory. This method allowed us to obtain the quick and appropriate response needed to minimize SARS-CoV-2 transmission.

5. Conclusions

In the case of the ID NOW™ COVID-19 assay, the use of NPS sampled on VTM has the great advantage of allowing repeat testing on the same sample without a significant loss of sensitivity. However, as for other POC assays, the results of this high-speed assay should be interpreted in a clinical and epidemiological context. In our opinion, POC assays are a promising tool for screening acute medical admissions with urgency to ensure the prompt treatment of patients or minimize nosocomial transmission.

Author Contributions: Methodology, C.N., F.G. and S.B.; investigation, data curation and writing—original draft preparation, S.B.; conceptualization, writing—review and editing, supervision, L.M.; Validation, L.M. and L.B.; Project administration, L.M. All authors have read and agreed to the published version of the manuscript.

Funding: This research received no external funding.

Institutional Review Board Statement: Ethical review and approval were waived for this study since this study was carried out on surplus samples that were sent to the laboratory as part of the molecular diagnosis of SARS-CoV-2 and under no circumstances would the use of samples in this study jeopardize the diagnosis.

Informed Consent Statement: Patient consent was not required as surplus samples that arrived at the laboratory with a clinician's request to detect SARS-CoV-2 were used.

Data Availability Statement: Not applicable.

Acknowledgments: We thank Lisbon Molecular Pathology Laboratory team for assistance with testing. The authors gratefully thank Abbott Diagnostics for providing reagents for the comparison assay.

Conflicts of Interest: The authors declare no conflict of interest. The manufacturer had no role in the design of the study; in the collection, analyses, or interpretation of data; in the writing of the manuscript; or in the decision to publish the results.

References

1. Srivastava, S.; Singh, P.; Malhotra, R.; Mathur, P. Comparison of Abbott ID NOW, a novel isothermal amplification based COVID-19 diagnostic method with RTPCR. *J. Virol. Methods* **2022**, *304*, 114521. [CrossRef] [PubMed]
2. Jin, R.; Pettengill, M.A.; Hartnett, N.L.; Auerbach, H.E.; Peiper, S.C.; Wang, Z. Commercial Severe Acute Respiratory Syndrome Coronavirus 2 (SARS-CoV-2) Molecular Assays: Superior Analytical Sensitivity of cobas SARS-CoV-2 Relative to NxTAG CoV Extended Panel and ID NOW COVID-19 Test. *Arch. Pathol. Lab. Med.* **2020**, *144*, 1303–1310. [CrossRef] [PubMed]
3. Zhen, W.; Smith, E.; Manji, R.; Schron, D.; Berry, G.J. Clinical Evaluation of Three Sample-to-Answer Platforms for Detection of SARS-CoV-2. *J. Clin. Microbiol.* **2020**, *58*, e00783-20. [CrossRef] [PubMed]
4. Khalid, M.F.; Selvam, K.; Jeffry, A.J.N.; Salmi, M.F.; Najib, M.A.; Norhayati, M.N.; Aziah, I. Performance of Rapid Antigen Tests for COVID-19 Diagnosis: A Systematic Review and Meta-Analysis. *Diagnostics* **2022**, *12*, 110. [CrossRef] [PubMed]
5. Aupaix, A.; Lazarova, E.; Chemais, M. A brief performance evaluation and literature review of Abbott ID Now COVID-19 rapid molecular-based test. *J. Virol. Methods* **2021**, *298*, 114293. [CrossRef] [PubMed]
6. Farfour, E.; Asso-Bonnet, M.; Vasse, M. The ID NOW COVID-19, a high-speed high-performance assay. *Eur. J. Clin. Microbiol. Infect Dis.* **2021**, *40*, 2045. [CrossRef]

medical sciences forum

MDPI

Proceeding Paper

Developing Communication Skills in Higher Education—The Use of the Pecha Kucha †

Ana Mano Azul [1], Luís Proença [1], Catarina Ramos [1], Sílvia Roda Couvaneiro [2] and João Couvaneiro [1,3,*]

1 Egas Moniz Center for Interdisciplinary Research (CiiEM), Egas Moniz School of Health & Science, 2829-511 Almada, Portugal; aazul@egasmoniz.edu.pt (A.M.A.); lproenca@egasmoniz.edu.pt (L.P.); cramos@egasmoniz.edu.pt (C.R.)
2 Departement of Education, Évora University, 7004-516 Évora, Portugal; silvia.couvaneiro@uevora.pt
3 Campus of Caparica, Interuniversity Centre for the History of Science and Technology (CIUHCT), 2829-516 Almada, Portugal
* Correspondence: jcouvaneiro@egasmoniz.edu.pt
† Presented at the 6th International Congress of CiiEM—Immediate and Future Challenges to Foster One Health, Almada, Portugal, 5–7 July 2023.

Abstract: In an Integrated Master of Science in Dentistry, it is relevant to develop communication skills. Through the Information and Communication Methodologies curricular unit, students learn to synthesize information, create engaging presentations, and use formats such as Pecha Kucha. The presentation model, with 20 slides of 20 s each, allows them to convey information clearly and concisely while efficiently managing time. Students were challenged to summarize scientific articles in this format, promoting communication skills. The activity was evaluated by the teachers, who gave feedback to the students.

Keywords: communication skills; short presentations; pecha kucha; dentistry

Citation: Azul, A.M.; Proença, L.; Ramos, C.; Couvaneiro, S.R.; Couvaneiro, J. Developing Communication Skills in Higher Education—The Use of the Pecha Kucha. *Med. Sci. Forum* **2023**, *22*, 16. https://doi.org/10.3390/msf2023022016

Academic Editors: José Brito, Nuno Taveira and Ana I. Fernandes

Published: 9 August 2023

1. Introduction

Public speaking and the public presentation of scientific content are important skills for academics and practitioners. However, failures, inconsistencies, and problems related to time management are frequent during public oral presentations. Not seldom do students find it difficult to synthesize, select, and organize information [1].

Given the need to contribute to the development of students' communication skills, a curricular unit (CU) of Information and Communication Methodologies was introduced in the first year of the study cycle. Among its contents, short communication formats for public speaking are explored, such as Pecha Kucha (PK). The presentation model was developed in 2003 by architects Mark Dytham and Astrid Klein. Its structure consists of a sequence of 20 slides, with automatic transitions every 20 s, for a total duration of 6 min and 40 s [2]. With very little text, it relies mainly on images, which should be understood as authentic visual metaphors. It was created with the purpose of making presentations more dynamic and engaging, and these benefits have already been demonstrated [3].

The goals of this activity were to lead students to develop public speaking and communication skills, such as the ability to synthesize and organize information, the structuring and creation of captivating presentations with an eminently visual narrative line, and the use of presentation production software. The assessment was carried out through direct observation, giving further feedback on the students' performance.

2. Materials and Methods

This was a descriptive, observational, and cross-sectional study. In the academic year 2022–2023, all the first-year students were organized in groups and challenged to prepare a presentation in the PK model summarizing a scientific article randomly assigned to each group. They then answered a survey in English (they are all speakers of Portuguese as their first language), made available online, with a set of questions/statements from the study "Do Students Learn Better with Pecha Kucha, an Alternative Presentation Format?" [4]. The survey includes 12 statements about the reasons for not including certain elements/information in the presentation (group 1), 17 statements about the reasons for doing so (group 2), and 7 statements aimed at assessing the level of confidence about the presentation in general (group 3). The statements in groups 1 and 2 are distributed over the following five areas: audience engagement, relevance of content, evidence-based evaluation, logistics, and credibility. For each statement, students were asked to indicate their level of agreement on a Likert-type scale, ranging from 1 to 5, where 1 corresponds to "played no role in my decision" and 5 corresponds to "definitely played a role".

3. Results

The data were analyzed using descriptive statistical methodologies. Of the 213 students enrolled, 83 (38.9%) participated by answering the survey. The average age is 19.9; 26.3% are male and 73.7% are female.

In group 1 of the survey, questions 4 (I decided not to present certain information because "it was too advanced and it might go over some students' heads"—area: audience engagement—Table 1) and 6 (I decided not to present certain information because "it was so advanced that even I was confused"—area: evidence-based evaluation—Table 1) revealed that students did not exclude elements or information in the presentation because of its complexity for the class or for the presenters. The main reasons for excluding information were related to the limited time available for the presentation. This situation was demonstrated in question 9 (I decided not to present certain information because "it wouldn't fit in the allotted time"—area: evidence-based evaluation—Table 1).

Table 1. Descriptive analysis (N, %) of questions from group 1—reasons for not including certain elements/information in the presentation (N = 83).

Score	Question 4		Question 6		Question 9	
	n	%	*n*	%	*n*	%
1—played no role in my decision	24	28.9	24	28.9	3	3.6
2	20	24.1	18	21.7	2	2.4
3	15	18.1	23	27.7	23	27.7
4	17	20.5	11	13.3	30	36.1
5—definitely played a role	7	8.4	7	8.4	25	30.1

Notes. Question 4—I decided not to present certain information because "it was too advanced and it might go over some students' heads"; Question 6—I decided not to present certain information because "it was too advanced and it might go over some students' heads"; Question 9—I decided not to present certain information because "it wouldn't fit in the allotted time".

Regarding group 2, on the reasons for including information in the presentation, question 13 (I decided to present certain information because "it was essential background information on the topic"—area: content relevancy—Table 2) reveals that the option for including information considered the need for contextualization. The same expression was found in the answers to question 17 (I decided to present certain information because "it was something that, when put together, would flow as a comprehensive story"—area: credibility—Table 2), which reveal that the option to include information also obeyed the establishment of a narrative line, an important dimension in the communicational process, emphasized throughout the CU lessons, particularly for the development of communicational competencies.

Table 2. Descriptive Analysis (N, %) of Questions from Group 2—reasons for including certain elements/information in the presentation (N = 83).

Score	Question 13		Question 17	
	n	%	*n*	%
1—played no role in my decision	2	2.4	2	2.4
2	4	4.8	6	7.2
3	16	19.3	14	16.9
4	17	20.5	26	31.3
5—definitely played a role	44	53.0	35	42.2

Notes. Question 13—I decided to present certain information because "it was essential background information on the topic"; Question 17—I decided to present certain information because "it was something that, when put together, would flow as a comprehensive story".

As for question 21 (I decided to present certain information because "it was something that all students in the class would be interested in knowing"), only 9.6% of the students (score 1 + 2) did not consider the inclusion of information relevant because it was interesting for the other students. As for question 25, I decided to present certain information because "it is the future direction where the science is heading" and only 10.8% of the students (score 1 + 2) did not take into account the evolutionary trend of science.

A cross-analysis of the questions in group 3 shows that after the presentation, the students feel confident to answer questions from a varied audience (high school students, dentists, family members, non-teaching staff, teachers, and even the president of the higher education institution).

In the last open-ended question, students were asked if they had any additional comments on this learning experience. Some of their answers include: "It was fun"; "it was impactful in my interaction skills"; "I think it improved my communication and synthesis skills"; "It's pretty hard to fit everything you need to say"; "it's perfect".

4. Discussion

The use of short presentation formats, such as PK, in meaningful and properly planned learning situations reveals benefits in terms of student motivation, involvement, information synthesis, public speaking and communication skills, innovation, and creativity [4].

In our study, the data suggest that students have self-awareness of the parameters that influenced the inclusion and exclusion of elements in their presentations. Nevertheless, in the future, to address possible limitations of the study, increasing the number of respondents may prove useful.

As suggested by authors such as Warmuth, PK presentations allow for better understanding of content and longer-lasting retention than simple traditional digital presentations such as PowerPoint or Keynote [5].

It has also been shown that the fast pace of the presentation facilitates the concentration of the speakers and the audience [2].

In the development of communicational skills, this type of methodology can be favorably applied in different scientific fields and other training contexts in higher education, and its usefulness has already been demonstrated in scientific areas such as Dentistry [6], Psychology [2], and Nursing [7], among others.

Author Contributions: Conceptualization, J.C. and A.M.A.; methodology, J.C. and A.M.A.; data curation, L.P.; writing—original draft preparation, J.C. and S.R.C.; writing—review and editing, J.C., A.M.A., L.P. and C.R. All authors have read and agreed to the published version of the manuscript.

Funding: This research received no external funding.

Institutional Review Board Statement: Not applicable.

Informed Consent Statement: Informed consent was obtained from all subjects involved in the study.

Data Availability Statement: Data is available upon reasonable request from the corresponding authors.

Conflicts of Interest: The authors declare no conflict of interest.

References

1. Pedaste, M.; Mäeots, M.; Siiman, L.A.; de Jong, T.; van Riesen, S.A.N.; Kamp, E.T.; Manoli, C.C.; Zacharia, Z.C.; Tsourlidaki, E. Phases of inquiry-based learning: Definitions and the inquiry cycle. *Educ. Res. Rev.* **2015**, *14*, 47–61. [CrossRef]
2. Beyer, A.M. Improving Student Presentations. *Teach. Psychol.* **2011**, *38*, 122–126. [CrossRef]
3. Ave, J.S.; Beasley, D.; Brogan, A. A Comparative Investigation of Student Learning through PechaKucha Presentations in Online Higher Education. *Innov. High. Educ.* **2020**, *45*, 373–386. [CrossRef]
4. Liao, M.-K.; Lewis, G.; Winiski, M. Do Students Learn Better with Pecha Kucha, an Alternative Presentation Format? *J. Microbiol. Biol. Educ.* **2020**, *21*, 21.3.66. [CrossRef] [PubMed]
5. Warmuth, K.A.; Caple, A.H. Differences in Instructor, Presenter, and Audience Ratings of PechaKucha and Traditional Student Presentations. *Teach. Psychol.* **2020**, *49*, 224–235. [CrossRef]
6. Nichani, A. Life after death by power point: PechaKucha to the rescue? *J. Indian Soc. Periodontol.* **2014**, *18*, 127. [CrossRef]
7. Byrne, M.M. Presentation innovations: Using Pecha Kucha in nursing education. *Teach. Learn. Nurs.* **2016**, *11*, 20–22. [CrossRef]

medical sciences forum

MDPI

Proceeding Paper

Raman Spectroscopy Applied to Blue and Green Ink Discrimination—A Pilot Study †

Daniela Eusébio [1,2], Vlad Tatarescu [1,3], Márcia Vieira [4], Carlos Família [1,2] and Alexandra Bernardo [1,2,*]

1 Laboratório de Ciências Forenses e Psicológicas Egas Moniz, Monte de Caparica, 2829-511 Almada, Portugal; danielaeusebio99@icloud.com (D.E.); vlad.tatarescu@gmail.com (V.T.); carlosfamilia@gmail.com (C.F.)
2 Egas Moniz Center for Interdisciplinary Research (CiiEM); Egas Moniz School of Health & Science, 2829-511 Almada, Portugal
3 Mestrado em Tecnologias Laboratoriais em Ciências Forenses, Egas Moniz School of Health and Science, Monte de Caparica, 2829-511 Almada, Portugal
4 Department of Conservation and Restoration and LAQV-REQUIMTE, Faculty of Sciences and Technology, NOVA University of Lisbon, 2829-511 Almada, Portugal; mc.vieira@campus.fct.unl.pt
* Correspondence: abernardo@egasmoniz.edu.pt
† Presented at the 6th International Congress of CiiEM—Immediate and Future Challenges to Foster One Health, Almada, Portugal, 5–7 July 2023.

Abstract: The requirement for the preservation of evidence in court continues to pose a challenge to forensic document analysis, where the answer to the discrimination of writing utensils is recurrent in requested questions. This study aims to contribute to the identification of the blue and green inks of handwriting instruments using Raman spectroscopy. Based on the Raman spectra obtained and after processing with Spectragryph software, a predictive model was built using KNIME software, where a discriminative power of 72.7% was obtained.

Keywords: questioned document; inks; Raman spectroscopy; handwriting instruments

Citation: Eusébio, D.; Tatarescu, V.; Vieira, M.; Família, C.; Bernardo, A. Raman Spectroscopy Applied to Blue and Green Ink Discrimination—A Pilot Study. *Med. Sci. Forum* **2023**, *22*, 29. https://doi.org/10.3390/msf2023022029

Academic Editors: José Brito, Nuno Taveira and Ana I. Fernandes

Published: 14 August 2023

1. Introduction

The analysis of the ink of common writing instruments is one of the most relevant areas of modern criminalistics in document fraud since the specific characteristics of inks can answer identification, comparison, and/or dating questions in requested forensic examinations [1–3]. This work aims to contribute to the discrimination of blue and green inks from common writing instruments through a non-destructive technique in which the full analysis of Raman spectra is exploited with Knime software in order to build a predictive model.

2. Materials and Methods

Eleven pens (five blue and six green) were selected from a collection of pens purchased on commercial surfaces. After signal Raman screening: pen 8: Unilakubo—SG100 Rubber Grip 0.7 mm; pen 10: Bic Crystal; pen 14: Bic—unidentifiable model; pen 32: Uni-ball Eye micro 0.5 mm; pen 39: Stabilo Sensor F 0.3 mm; pen 3: Papermate—Inkjoy 1.0 mm; pen 9: Bic Crystal; pen 51: Bic Crystal Soft Medium; pen 53: Papermate Inkjoy 1.0 mm; pen 54: Staples Stick Pens; pen 55: FlairZing—Color Flair 0.7 mm. The ink of each pen has been deposited in notebooks, created with current paper (Navigator) in circular areas with different stroke overlaps in each quadrant, in order to optimize the acquisition conditions. Raman spectra of the ink circular area deposits were acquired using a Raman Mira DS spectrometer equipped with a laser with an excitation wavelength of 785 nm (100 mW) and a spectral range of 200–2300 nm. The equipment provides a spectral resolution of 8–10 cm^{-1} and a measuring spot of 0.042–2.5 mm. The detection technique used is Orbital Raster Scan (ORS), where a tightly focused beam (0.04 mm) scans a relatively larger area compared to the conventional Raman scanning, averaging the information. It has a universal fixture with three positions that allow the surface of objects to be analysed at three different distances: in contact and at 3 mm and 5 mm.

After acquiring 3 replicates of the Raman spectra for the pen inks, the respective baseline was adjusted using Spectragryph software version 1.2.15 [4]. The baseline of the spectra was adjusted 2 times. The Raman spectra files of each ink were split into 2 databases to create a predictive model using KNIME Analytics Platform software version 4.6.0 [5]: (a) the training database (which contained the data from 2 replicates for each of the dyes, the averages of the triplicates and the average between each two replicates) and the (b) validation database (which contained 1 of the replicates that were not present in the training database).

3. Results

The visual analysis of the Raman spectra of the pen inks studied revealed differentiating bands that could contribute to the construction of a predictive model, Figure 1.

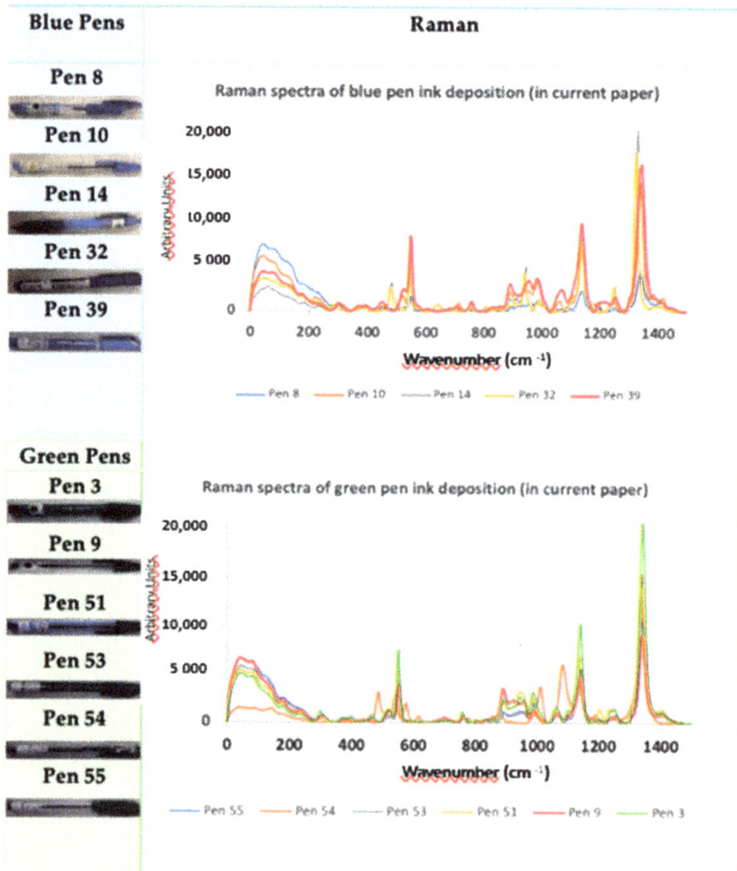

Figure 1. Raman spectra of blue and green ink pens.

After obtaining the Raman spectra on the handwriting instrument inks, the predictive model was built using Knime software to evaluate the discriminative ability of the model, Figure 2.

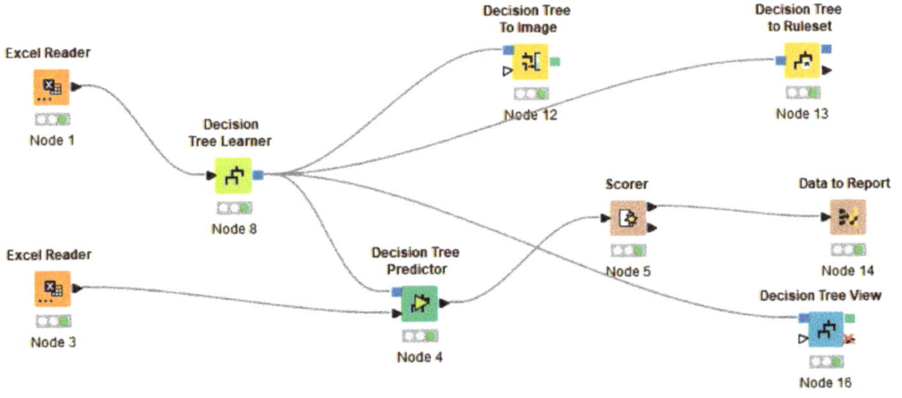

Figure 2. Predictive model performed in KNIME Analytics Platform software version 4.6.0 for the analysis of ink samples from handwriting instruments.

The following nodes were used in the model: Decision Tree Learner, Decision Tree Predictor, Scorer, Decision Tree View, Decision Tree To Image, Decision Tree To Ruleset, and Data to Report.

According to Berthold's instructions [5], in the nodes called Excel Reader (Nodes 1 and 3), the files intended to be processed by the predictive model were inserted: in node 1, the document of the training database, and in node 3, the test database. Decision Tree Learner (node 8) parameters were set, and a decision tree was generated from the training file (node 1). The Decision Tree Predictor (node 4) draws on the tree of node 8 to predict the class value for new patterns, in this case, for the "test" file. The Scorer, node 5, analyses the test and training files inserted in the previous nodes and provides the accuracy and error of the analysis. The following nodes (12, 13, 14, and 16) serve only to visualize the decision trees on an image.

After creating the Decision Tree Learner, a second tree was generated with the same nodes, but with the objective of validation in real context. In node 1, the same training file was inserted (for each technique), and in node 3, the validation file containing the data of the one replicate that was not present in the training file was inserted.

The predictive model created achieved a discriminatory power of 72.7% and an error of 27.3%.

4. Discussion

Despite the reduced number of replicates in each database, the present technique achieved values not much lower than those obtained in other studies with other techniques such as Micro-ATR-FTIR combined with chemometrics in which the discrimination power did not exceed 86.67% [6] and 89.99% [7], or in studies where multispectral imaging coupled with FCM was used and achieved discrimination percentages of 76% [8] and 85.7% [9].

5. Conclusions

The present pilot study allowed the development of a predictive model based on pen ink Raman spectra with a high discriminatory power of 72.7%.

Author Contributions: Conceptualization, A.B.; methodology, A.B. and M.V.; software, D.E. and C.F.; investigation, D.E. and V.T.; writing—original draft preparation, D.E.; writing—review and editing, A.B., M.V. and C.F.; supervision, A.B., M.V. and C.F. All authors have read and agreed to the published version of the manuscript.

Funding: This research received no external funding.

Institutional Review Board Statement: Not applicable.

Informed Consent Statement: Not applicable.

Data Availability Statement: Data are contained within the article.

Acknowledgments: Laboratório de Ciências Forenses e Criminais Egas Moniz and Departamento de Conservação e Restauro Faculdade de Ciências e tecnologias da Universidade Nova de Lisboa.

Conflicts of Interest: The authors declare no conflict of interest.

References

1. Cantu, A.; Ink Analysis, A. *Wiley Encyclopedia of Forensic Sciences*, 2nd ed.; John Wiley & Sons, Ltd.: Hoboken, NJ, USA, 2009; pp. 375–379. [CrossRef]
2. Ellen, D. *The Scientific Examination of Documents*, 2nd ed.; Taylor & Francis: Boca Raton, FL, USA, 1997.
3. Ortiz-Herrero, L.; Bartolomé, L.; Durán, I.; Velasco, I.; Alonso, M.L.; Maguregui, M.I.; Ezcurra, M. DATUVINK pilot study: A potential non-invasive methodology for dating ballpoint pen inks using multivariate chemometrics based on their UV–vis-NIR reflectance spectra. *Microchem. J.* **2018**, *140*, 158–166. [CrossRef]
4. Menges, F. Spectragryph. 2019. Available online: http://spectroscopy.ninja (accessed on 27 April 2021).
5. Berthold, M. KNIME Analytics Platform v4.0.2. 2019. Available online: https://www.knime.com/ (accessed on 27 April 2021).
6. Lee, L.C.; Othman, M.R.; Pua, H. Systematic assessment of attenuated total reflectance-fourier transforms infrared spectroscopy coupled with multivariate analysis for forensic analysis of black ballpoint pen inks. *Malays. J. Anal. Sci.* **2012**, *16*, 262–272.
7. Nam, Y.S.; Park, J.S.; Lee, Y.; Lee, K.B. Application of micro-attenuated total reflectance fourier transform infrared spectroscopy to ink examination in signatures written with ballpoint pen on questioned documents. *J. Forensic Sci.* **2014**, *59*, 800–805. [CrossRef] [PubMed]
8. Khan, M.J.; Yousaf, A.; Khurshid, K.; Abbas, A.; Shafait, F. Automated forgery detection in multispectral document images using fuzzy clustering. In Proceedings of the 13th IAPR International Workshop on Document Analysis Systems, Viena, Austria, 24–27 April 2018; Volume 2018, pp. 393–398. [CrossRef]
9. Asri, M.N.M.; Yahya, N.N.; Nor, N.A.M.; Desa, W.N.S.M.; Ismail, D. Towards establishing a non-destructive technique for forensic ink analysis involving Raman spectroscopy with chemometric procedures. *AIP Conf. Proc.* **2019**, *2155*, 1–7. [CrossRef]

medical sciences forum

MDPI

Proceeding Paper

Facilitating NGS-Based Screening of Genetic Disorders Using -AI-Driven Bioinformatics †

Ricardo Pais [1,2,*], Amanda Carneiro [2], Yolanda Zendzela [2], Yosra Sdiri [2], Tidiana Rodrigues [2], Maria Guilhermina Moutinho [2], Tabisam Khan [1] and Markella Mikkelsen [1,*]

1 Molmart Ltd., Manchester M1 7ED, UK; info@molmart.co.uk
2 Egas Moniz Center of Interdisciplinary Research (CiiEM), Egas Moniz School of Health and Science, 2829-511 Almada, Portugal; amanda.sc@outlook.pt (A.C.); zenzelayolanda@yahoo.com (Y.Z.); sdiriyosra4@gmail.com (Y.S.); tidianelothrodrigues@gmail.com (T.R.); gmoutinho@egasmoniz.edu.pt (M.G.M.)
* Correspondence: rjpais@molmart.co.uk (R.P.); mmikkelsen@molmart.co.uk (M.M.)
† Presented at the 6th International Congress of CiiEM—Immediate and Future Challenges to Foster One Health, Almada, Portugal, 5–7 July 2023.

Abstract: Next-Generation Sequencing (NGS) is used as a diagnostic strategy for identifying pathogenic genetic variants in children and adults. However, the analysis is complex, requiring specialized bioinformaticians, and it can take weeks to finalize one study. This has been a limiting factor for the application of NGS in the screening of populations for rare genetic diseases. In this work, we show two case studies, where we applied an AI-driven bioinformatics framework in a diagnostic and a preventive scenario, respectively. The AI analysis was accurate and substantially faster than using conventional bioinformatics tools. Our results support the concept that AI-driven bioinformatics is a scalable solution for rendering accurate results and enabling a more widely available genetic screening for rare diseases.

Keywords: genomics; bioinformatics; rare diseases; artificial intelligence; pre-conception

Citation: Pais, R.; Carneiro, A.; Zendzela, Y.; Sdiri, Y.; Rodrigues, T.; Moutinho, M.G.; Khan, T.; Mikkelsen, M. Facilitating NGS-Based Screening of Genetic Disorders Using -AI-Driven Bioinformatics. *Med. Sci. Forum* **2023**, 22, 39. https://doi.org/

Academic Editors: José Brito, Nuno Taveira and Ana I. Fernandes

Published: 17 August 2023

1. Introduction

Whole Exome Sequencing (WES) using Next-Generation Sequencing (NGS) is a clinically accepted diagnostic technology for the identification of pathogenic genetic variants in children and adults [1]. Finding gene-function-disruptive variants (SNPs and INDELs) in sequences is fundamental in determining the cause of the genetic disease and for genetic counselling consultations. Additionally, the application of this method at the pre-conception stage can also enable parents to make informed decisions regarding the possible birth of children with a particular genetic disease. Databases such as ClinVar and OMIM have been accumulating information on an ever-increasing number of new pathogenic variants [2]. In these databases, gene–disease associations have also been growing over time, leading to more than eight thousand having been already reported [3]. Public and private healthcare facilities are beginning to use these data as a front-line tool over conventional techniques to diagnose pediatric rare genetic diseases [1,4]. However, the analysis of WES using bioinformatics is complex and requires specialist skills and training, hence it can take several weeks from sample to diagnosis [5]. The relative complexity associated with the high labor intensity is a substantial bottleneck in the field, leading to a heavy cost in human resources. This has been a limiting factor for the screening and prevention of rare diseases in the general population. Artificial Intelligence (AI) is considered to be a solution for automating complex analysis and decision-making [6]. In this work, we present two case studies where we applied an AI-driven bioinformatics framework in a diagnostic and a preventive scenario, respectively.

2. Methodology

2.1. Clinical Samples and Sequencing

Saliva samples were collected in DNA/RNA saliva collection tubes (GeneFix™, Isohelix) using the commercial ExoMart and SureMart kits from MolMart Ltd., Manchester, United Kingdom. Relevant clinical data were submitted by the referring clinician into the MolMart online form for kit activations (https://molmartgenomics.com, accessed on 27 October 2022). WES was performed by NGS using the Illumina platform. The exome library was prepared with Agilent's SureSelect V6+UTR-post kit.

2.2. Bioinformatics

Variant Calling Files (VCF) were generated from FASTQ files using a standard bioinformatics pipeline [7,8]. BWA (Burrows–Wheeler Alignment Tool) software version 0.7.12 and reference human genome version hg38 were used for read mapping and alignment. Variant calling and variant annotation of genetic modifications was made using GATK (Genome Analysis Toolkit) software version 3.4.0 and SnpEff version 4.1, respectively. The MolMart Artificial Intelligence Analyst (MAIA) was used for pathogenic gene variant candidate identification and ranking on the clinical observations of the Variant Calling Files (VCF). Clinical observation matching and pathogenic scoring were performed by MAIA, considering both experimental evidence on databases and sequence predictions.

3. Results

We applied an AI-driven bioinformatics framework to analyze two case studies, one a diagnostic (Case Study 1) and the other a preventive scenario (Case Study 2).

3.1. Case Study 1

An 8-month-old infant was referred for genetic testing with hypotonia, delayed development, hepatosplenomegaly and strabismus. We applied AI-driven bioinformatics on the sequenced exome containing about 114,000 gene variants, taking into account the clinical phenotype (Figure 1). From all gene variants, the AI took ~5 s to identify a total of 757 putative pathogenic variants, where only 15 had high-scoring matches on disease database annotations that related to the clinical observations. Furthermore, the top-ranked variant (Figure 1) was the one chosen by independent molecular geneticists as causative of the phenotype by manually checking in the OMIM and ClinVar databases.

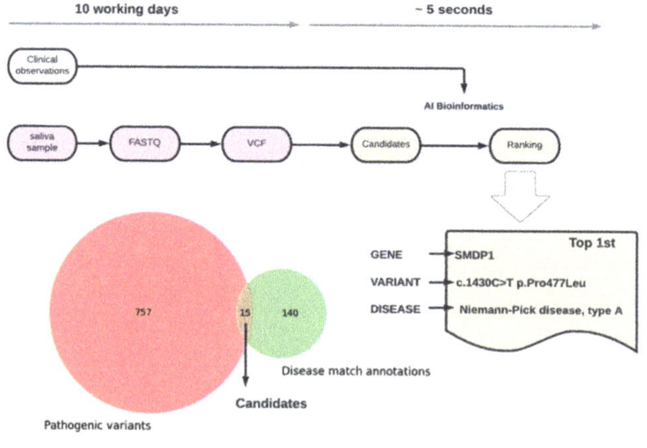

Figure 1. Overview of the bioinformatics analysis pipeline and final outcome on case study 1.

3.2. Case Study 2

In this case, we screened a healthy couple at the pre-conception stage for their potential risk of having a child affected by a genetic disease. We applied AI-driven bioinformatics on the male and female exomes containing about 112,000 and 113,000 gene variants, respectively (Figure 2). The AI took ~12 s to identify six putative pathogenic gene variants that can be transmitted from both males and females. From these, only one raised some concern based on strong gene–disease association evidence, with an estimated probability of 23% of having a child with mannose binding deficiency.

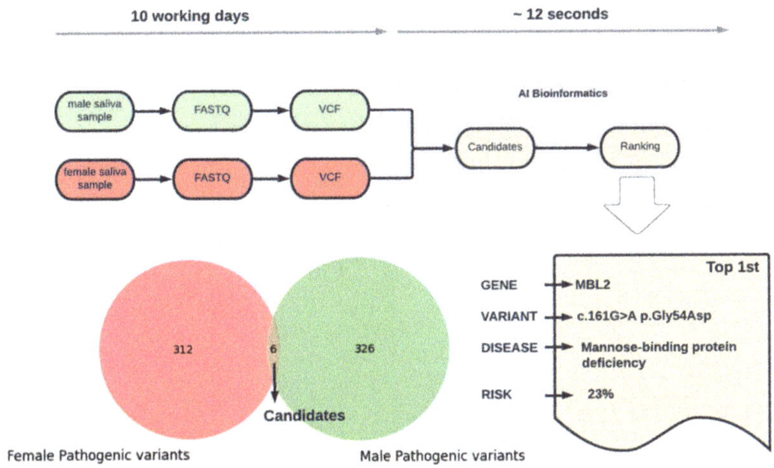

Figure 2. Overview of the bioinformatics analysis pipeline and final outcome on case study 2.

4. Conclusions

The case studies shown here demonstrate that AI-driven bioinformatics analysis is substantially faster than conventional bioinformatics tools and platforms. Furthermore, our results support the concept that AI-driven bioinformatics is an accurate and scalable solution which can make population-wide genetic screening for rare diseases possible.

Author Contributions: Conceptualization, R.P., M.M. and T.K.; methodology, R.P., M.M. and T.K.; formal analysis, R.P., A.C., Y.Z., Y.S. and T.R.; investigation, R.P., A.C., Y.Z., Y.S., T.R. and M.G.M.; data curation, R.P., A.C., Y.Z., Y.S., T.R. and M.G.M.; writing—original draft preparation, R.P.; writing—review and editing, M.M.; supervision, R.P., M.G.M. and M.M. All authors have read and agreed to the published version of the manuscript.

Funding: This research received no external funding.

Institutional Review Board Statement: The study was conducted in accordance with the Declaration of Helsinki, and approved by the Institutional Review Board (or Ethics Committee).

Informed Consent Statement: All patients have given their consent to use their clinical and genomic data for research and clinical purposes.

Data Availability Statement: Medical and genomic data are not publicly available due to privacy restrictions.

Conflicts of Interest: The authors declare no conflict of interest.

References

1. Frésard, L.; Montgomery, S.B. Diagnosing Rare Diseases after the Exome. *Mol. Case Stud.* **2018**, *4*, a003392. [CrossRef] [PubMed]
2. Pereira, R.; Oliveira, J.; Sousa, M. Bioinformatics and Computational Tools for Next-Generation Sequencing Analysis in Clinical Genetics. *J. Clin. Med.* **2020**, *9*, 132. [CrossRef] [PubMed]

3. Fridman, H.; Yntema, H.G.; Mägi, R.; Andreson, R.; Metspalu, A.; Mezzavila, M.; Tyler-Smith, C.; Xue, Y.; Carmi, S.; Levy-Lahad, E.; et al. The Landscape of Autosomal-Recessive Pathogenic Variants in European Populations Reveals Phenotype-Specific Effects. *Am. J. Hum. Genet.* **2021**, *108*, 608–619. [CrossRef] [PubMed]
4. Thareja, G.; Al-Sarraj, Y.; Belkadi, A.; Almotawa, M.; Ismail, S.; Al-Muftah, W.; Badji, R.; Mbarek, H.; Darwish, D.; Fadl, T.; et al. Whole Genome Sequencing in the Middle Eastern Qatari Population Identifies Genetic Associations with 45 Clinically Relevant Traits. *Nat. Commun.* **2021**, *12*, 1250. [CrossRef] [PubMed]
5. Richards, S.; Aziz, N.; Bale, S.; Bick, D.; Das, S.; Gastier-Foster, J.; Grody, W.W.; Hegde, M.; Lyon, E.; Spector, E.; et al. Standards and Guidelines for the Interpretation of Sequence Variants: A Joint Consensus Recommendation of the American College of Medical Genetics and Genomics and the Association for Molecular Pathology. *Genet. Med.* **2015**, *17*, 405–424. [CrossRef] [PubMed]
6. Mann, M.; Kumar, C.; Zeng, W.F.; Strauss, M.T. Artificial Intelligence for Proteomics and Biomarker Discovery. *Cell Syst.* **2021**, *12*, 759–770. [CrossRef] [PubMed]
7. Pirooznia, M.; Kramer, M.; Parla, J.; Goes, F.S.; Potash, J.B.; McCombie, W.R.; Zandi, P.P. Validation and Assessment of Variant Calling Pipelines for Next-Generation Sequencing. *Hum. Genom.* **2014**, *8*, 14. [CrossRef] [PubMed]
8. Roy, S.; Coldren, C.; Karunamurthy, A.; Kip, N.S.; Klee, E.W.; Lincoln, S.E.; Leon, A.; Pullambhatla, M.; Temple-Smolkin, R.L.; Voelkerding, K.V.; et al. Standards and Guidelines for Validating Next-Generation Sequencing Bioinformatics Pipelines: A Joint Recommendation of the Association for Molecular Pathology and the College of American Pathologists. *J. Mol. Diagn.* **2018**, *20*, 4–27. [CrossRef] [PubMed]

medical sciences forum

MDPI

Proceeding Paper

Predicting Cancer Prognostics from Tumour Transcriptomics Using an Auto Machine Learning Approach †

Ricardo Jorge Pais [1,2,*], Filipa Lopes [3], Inês Parreira [3], Márcia Silva [3], Mariana Silva [3] and Maria Guilhermina Moutinho [2]

1 Bioenhancer Systems, Office 63 182-184 High Street North, East Ham, London E6 2JA, UK
2 Egas Moniz Center for Interdisciplinary Research, Egas Moniz School of Health & Science, 2829-511 Almada, Portugal; gmoutinho@egasmoniz.edu.pt
3 Egas Moniz School of Health & Science, 2929-511 Almada, Portugal; filipa.lopes2000@gmail.com (F.L.); ines.parreira03@gmail.com (I.P.); marciasofyfs@gmail.com (M.S.); marianarodriguessasilva123@gmail.com (M.S.)
* Correspondence: rjpais@bioenhancersystems.com
† Presented at the 6th International Congress of CiiEM—Immediate and Future Challenges to Foster One Health, Almada, Portugal, 5–7 July 2023.

Abstract: Cancer prognostics using tumour transcriptomics is a promising precision medicine approach for helping decisions during cancer treatment. However, currently used cancer prognostic biomarkers still have low predictive power. This work tested the potential of applying machine learning (ML) algorithms for generating patients' survival prognostics on lung, breast, and kidney tumour transcriptomics datasets. We evaluated the performance of models generated by ML and reported their optimal sensitivity, specificity, accuracy, and computed ROC-AUC. The results support the potential for applying auto ML approaches for the future development of cancer prognostics tools based on transcriptomics data.

Keywords: bioinformatics; cancer prognostics; machine learning; transcriptomics

Citation: Pais, R.J.; Lopes, F.; Parreira, I.; Silva, M.; Silva, M.; Moutinho, M.G. Predicting Cancer Prognostics from Tumour Transcriptomics Using an Auto Machine Learning Approach. *Med. Sci. Forum* **2023**, *22*, 6. https://doi.org/10.3390/msf2023022006

Academic Editors: José Brito, Nuno Taveira and Ana I. Fernandes

Published: 8 August 2023

1. Introduction

Cancer is a worldwide flagellum, leading to millions of deaths per year. The success of cancer treatments often depends on the choice of the correct treatment [1]. Treatment success is associated with tumour heterogenicity and genetic factors. Cancer prognostic biomarkers are considered a promising personalized medicine approach for helping decision-making during cancer treatment [2]. Cancer prognostic biomarkers still have low predictive power, explaining only 25% to 75% of the cases [2]. Transcriptomics is an affordable and accurate high-throughput methodology often described as a promising precision medicine approach that enables the quantification of gene expression levels of multiple genes [3]. The application of machine learning (ML) frameworks on transcriptomics data is thought to have the potential to identify biomarker signatures for a binary classification (yes/no) of patient's survival with predictive power [4,5]. However, this approach is still not applied as a solution for treatment prognostics.

Multiple ML algorithms can be applied in bioinformatics to combine biomarkers to improve models' predictive capacity [5,6]. Further, there is an infinite possibility of models that can arise from these algorithms due to all possible parameter combinations making it hard and labour-intensive to find optimal models. Auto ML approaches have proven useful for the optimal model generation with reasonable computational effort and provide a much faster route to achieving better-performing models [7,8]. This work used an auto ML approach to test the potential of ML for generating transcriptomics-based cancer prognostic predictors for lung, breast, and kidney cancers transcriptomics datasets. Here, the model's performance was evaluated and reported their optimal sensitivity, specificity, accuracy, and computed ROC-AUC.

2. Methods

2.1. Data Collection

Transcriptomics data were collected from the 2021 updated Human Protein Atlas database records, which contained mRNA expression (FPKM) of 200 genes from 1075 anonymized cancer patients [9,10]. The TCGA transcriptomics data of breast, lung and renal cancers biopsies were downloaded from these sources. Metadata including patients' age, sex, survival time after biopsy and time of death were also collected from this source.

2.2. Dataset Construction

From collected transcriptomics data, 58 genes were selected. These 58 genes are key components of signalling pathways involved in the regulation of epithelial-to-mesenchymal transition, which plays a critical role in metastasis acquisition [11]. From the collected metadata, we select only the transcriptomes associated with patients who have been reported to survive over 5 years after the diagnostics (good prognostics) or with a reported death within the first 2 years (poor prognostics). The sample numbers of the final curated datasets used in this work are summarized in Table 1.

Table 1. Cancer transcriptomics datasets and their sample numbers.

Tumour Biopsy	Number of Patients	Good Prognostics	Poor Prognostics
Breast	239	199	40
Lung	325	94	231
Kidney	318	210	108

2.3. Auto Machine Learning Framework

Models were generated using the Tree-Based Pipeline Optimization Tool (TPOT). TPOT is an open-source software package developed in Python for an automated generation of ML-derived predictive models [7,8]. TPOT relies on genetic programming to generate predictive models with optimal performance, testing multiple ML algorithms and modelling parameters [8]. The open-source TPOT version 0.11.7 was installed under Python 3.9 anaconda distribution. TPOT auto ML pipeline was implemented and run under the Jupyter notebook environment. All scripts were run on a MacBook pro with a 2.4 GHz 8-Core Intel Core i9 processor.

2.4. Model Generation and Evaluation

The TPOT Classifier method was used for model training, testing and optimization. It was set up to perform 100 generations with a population size of 50 randomly selected ML algorithms. Optimization criterium was set to find the optimal Receiver Operating Curve (ROC) given by the Area Under the Curve (AUC) value. A random selection of 50% of the data were used in training and the remaining for testing [12]. Models' final performance was computed by generating predictions using the selected models on all data selected for training, and then calculating the accuracy, sensitivity, specificity and ROC-AUC [13].

3. Results

We applied the auto ML framework (TPOT) on the curated datasets of breast, lung, and kidney tumour transcriptomics (Table 1) for generating patient survival prognostics. TPOT ran for about 1 h on each dataset, generating and testing an average of 2.6 models/second, evaluating a total of 10,000 different variants that use and combine distinct ML algorithms (e.g., Random Forest, Naïve Bayes, Neural Networks, and many others). The best-performing algorithms selected were substantially distinct among cancer types with different performances (Table 2).

Table 2. Generated optimal predictive models and their associated performances.

Tumour Biopsy	Algorithm Pipeline	SEN *	SPE	AUC	ACC
Breast	Multinomial Naïve Bayes with Random Forest	94%	58%	53%	84%
Lung	KNeigbours with Random Forest	59%	83%	48%	52%
Kidney	Normalized Random Forest	94%	66%	70%	71%

* SEN (sensitivity); SPE (specificity); AUC (Area Under the Curve); and ACC (accuracy).

The results obtained (Table 2) showed that predicted models generated for breast and kidney cancer prognostics performed with very good sensitivity (SEN = 94%). However, these models had poor specificity (SPE < 66%), which indicates a huge tendency to generate false positives if applied to predict the survival of a patient with a tumour [13]. In contrast, the predicted model obtained for lung cancer prognostics had a reasonable specificity (SPE > 83%) but with a poor sensitivity (SEN = 59%), indicating a high tendency for generating false negatives [13]. Further, the obtained ROC-AUCs showed that only the kidney cancer prognostic model has good predictive power (AUC > 70%) with reasonable accuracy (ACC > 70%), indicating that only this model, among all, generates robust predictions not given by chance [13].

The performances obtained from the models generated by TPOT (Table 2) also showed that the currently available ML algorithms are not enough to generate high-performance models on our cancer transcriptomics datasets. This low performance may be explained by cancer heterogenicity, missing key regulatory genes on the dataset, or confounding variables associated with the clinical data (age, gender, ethnicity, death reasons, and treatment choices).

4. Conclusions

This study demonstrated that the auto ML approach is a powerful methodology for the fast and systematic generation of predictive models that can be applied in cancer prognostics from tumour transcriptomics. Here, we illustrated the ML approach application with three types of cancers that showed promising performances, particularly for kidney cancer. Moreover, the results in this work support the idea of technical challenges in this modelling framework that justify future work for improving either the data or the tools to generate predictive models.

Author Contributions: Conceptualization, R.J.P. and M.G.M.; methodology, R.J.P.; formal analysis, R.J.P., F.L., I.P., M.S. (Mariana Silva) and M.S. (Márcia Silva); investigation, F.L., I.P., M.S. (Mariana Silva) and M.S. (Márcia Silva); data curation, F.L., I.P., M.S. (Mariana Silva) and M.S. (Márcia Silva); writing—original draft preparation, R.J.P.; writing—review and editing, M.G.M.; supervision, R.J.P. and M.G.M. All authors have read and agreed to the published version of the manuscript.

Funding: This research received no external funding.

Institutional Review Board Statement: The study was conducted in accordance with the Declaration of Helsinki, and approved by the Institutional Review Board (or Ethics Committee) of Bioenhancer Systems Ltd. on the 21 February 2023.

Informed Consent Statement: All data involving humans used in this study were anonymous and retrieved from public databases, where informed consent was given.

Data Availability Statement: Publicly available datasets on breast, lung and renal cancer transcriptomics were analysed in this study. The original collected data (rna_cancer_sample.tsv.zip) can be found here: https://www.proteinatlas.org/about/download (accessed on 12 September 2021). The curated datasets used in the study can be found here: https://www.bioenhancersystems.com/resources (accessed on 12 September 2021). Medical data are not publicly available due to privacy restrictions.

Conflicts of Interest: The authors declare no conflict of interest.

References

1. Hanahan, D.; Weinberg, R.A. Hallmarks of Cancer: The Next Generation. *Cell* **2011**, *144*, 646–674. [CrossRef] [PubMed]
2. Šutić, M.; Vukić, A.; Baranašić, J.; Försti, A.; Džubur, F.; Samaržija, M.; Jakopović, M.; Brčić, L.; Knežević, J. Diagnostic, Predictive, and Prognostic Biomarkers in Non-Small Cell Lung Cancer (NSCLC) Management. *J. Pers. Med.* **2021**, *11*, 1102. [CrossRef] [PubMed]
3. Van Allen, E.M.; Robinson, D.; Morrissey, C.; Pritchard, C.; Imamovic, A.; Carter, S.; Rosenberg, M.; McKenna, A.; Wu, Y.M.; Cao, X.; et al. A Comparative Assessment of Clinical Whole Exome and Transcriptome Profiling across Sequencing Centers: Implications for Precision Cancer Medicine. *Oncotarget* **2016**, *7*, 52888–52899. [CrossRef] [PubMed]
4. De Ridder, D.; De Ridder, J.; Reinders, M.J.T. Pattern Recognition in Bioinformatics. *Brief. Bioinform.* **2013**, *14*, 633–647. [CrossRef] [PubMed]
5. Mann, M.; Kumar, C.; Zeng, W.F.; Strauss, M.T. Artificial Intelligence for Proteomics and Biomarker Discovery. *Cell Syst.* **2021**, *12*, 759–770. [CrossRef] [PubMed]
6. Pais, R.J. Predictive Modelling in Clinical Bioinformatics: Key Concepts for Startups. *BioTech* **2022**, *11*, 35. [CrossRef] [PubMed]
7. Le, T.T.; Fu, W.; Moore, J.H. Scaling Tree-Based Automated Machine Learning to Biomedical Big Data with a Feature Set Selector. *Bioinformatics* **2020**, *36*, 250–256. [CrossRef] [PubMed]
8. Olson, R.S.; Urbanowicz, R.J.; Andrews, P.C.; Lavender, N.A.; Kidd, L.C.; Moore, J.H. Automating Biomedical Data Science Through Tree-Based Pipeline Optimization. In *Lecture Notes in Computer Science (Including Subseries Lecture Notes in Artificial Intelligence and Lecture Notes in Bioinformatics)*; Springer: Cham, Switzerland, 2016; Volume 9597, pp. 123–137.
9. Pontén, F.; Jirström, K.; Uhlen, M. The Human Protein Atlas-a Tool for Pathology. *J. Pathol.* **2008**, *216*, 387–393. [CrossRef] [PubMed]
10. Uhlen, M.; Oksvold, P.; Fagerberg, L.; Lundberg, E.; Jonasson, K.; Forsberg, M.; Zwahlen, M.; Kampf, C.; Wester, K.; Hober, S.; et al. Towards a Knowledge-Based Human Protein Atlas. *Nat. Biotechnol.* **2010**, *28*, 1248–1250. [CrossRef] [PubMed]
11. Pais, R.J. Simulation of Multiple Microenvironments Shows a Pivot Role of RPTPs on the Control of Epithelial-to-Mesenchymal Transition. *Biosystems* **2020**, *198*, 104268. [CrossRef] [PubMed]
12. Swan, A.L.; Mobasheri, A.; Allaway, D.; Liddell, S.; Bacardit, J. Application of Machine Learning to Proteomics Data: Classification and Biomarker Identification in Postgenomics Biology. *Omics J. Integr. Biol.* **2013**, *17*, 595–610. [CrossRef] [PubMed]
13. Dankers, F.J.W.M.; Traverso, A.; Wee, L.; van Kuijk, S.M.J. Prediction Modeling Methodology. In *Fundamentals of Clinical Data Science*; Springer International Publishing: Cham, Switzerland, 2019; pp. 101–120.

medical sciences forum

MDPI

Proceeding Paper

Impact of Acid Rain with Different pH Values in *Monstera deliciosa* Plants †

Afonso Ramos [1,*], Sandra Almeida [1], Luísa Ganço [1,2], Patricia Costa [1] and António Pereira [1]

1 Colégio Atlântico, International Baccalaureate Diploma Level, 2840-167 Seixal, Portugal; sandraalmeida@colegioatlantico.pt (S.A.); lu.ganco@egasmoniz.edu.pt (L.G.); patriciacosta@colegioatlantico.pt (P.C.); antoniopereira@colegioatlantico.pt (A.P.)

2 Egas Moniz Center for Interdisciplinary Research (CiiEM), Egas Moniz School of Health & Science, Campus Universitário, 2829-511 Almada, Portugal

* Correspondence: a.afonsoramos@colegioatlantico.pt; Tel.: +351-924338326

† Presented at the 6th International Congress of CiiEM—Immediate and Future Challenges to Foster One Health, Almada, Portugal, 5–7 July 2023.

Abstract: The present research experimentally outlines the reaction of a *Monstera deliciosa* plants to the effects of acid rain at various pH levels. Sulfuric acid was used to simulate acid rain, with prepared solutions having four different pH values. Paper chromatography showed that the plant's pigment begins splitting more and the chlorophyll further deteriorated. The plants also started to lose their ability to be soluble. Through spectrophotometry, plants with the lowest pH levels of acid were losing their capacity to absorb light and radiation more quickly since their absorbance values were lower.

Keywords: acid rain; *Monstera deliciosa*; sulphuric acid; pH; paper chromatography; chlorophyll; spectrophotometry

Citation: Ramos, A.; Almeida, S.; Ganço, L.; Costa, P.; Pereira, A. Impact of Acid Rain with Different pH Values in *Monstera deliciosa* Plants. *Med. Sci. Forum* **2023**, *22*, 21. https://doi.org/10.3390/msf2023022021

Academic Editors: José Brito, Nuno Taveira and Ana I. Fernandes

Published: 11 August 2023

1. Introduction

Acid rain refers to a variety of ways in which acids are released into the atmosphere. This research evaluates the consequences of acid rain on living and deceased *Monstera deliciosa* plant, simulating acid rain with sulphuric acid with different pH levels to intensify the solutions' acidity, examining how *Monsteras* react to different acidic environments. *Monstera deliciosa* plants can grow in a soil that has a pH level between 5.0 and 7.5. Although *Monsteras* can thrive in acidic soil, they do not meet the criteria for a true acid-loving plant [1]. *Monstera* plants were chosen to be tested as they can grow fast indoors; therefore, when changing their desire environment, speedy reactions will occur [2]. The analysis was carried out by performing paper chromatography and spectrophotometry. The impacts of acid rain on plants are well-studied, particularly those that flourish in aquatic habitats, are the reason why an indoor plant was chosen to be studied.

Plants with the most acidic solutions were predicted to show less resistance to sulphuric acid solutions. The plants receiving the solutions with the lowest pH levels, through paper chromatography, will show a lesser range of pigments with weaker colours, as chlorophyll is being deteriorated. Through spectrophotometry, *Monstera* plants will lose their ability to absorb radiation at different wavelengths, showing lower absorbance values. From 330 nm to 420 nm, absorbance values tend to decrease. At 430 nm, there should be an increase in the absorbance registered, since chlorophyll A absorbs the light in the blue wavelengths, with absorption peaks at 430 nm [3].

2. Procedure and Results

Five *Monstera deliciosa* plants were used for the experimental protocol. Acid rain was simulated with sulphuric acid at 98% concentration with a density of 1.92 gmL^{-1}, a strong acid, and distilled water solution. Then, 4 mL of sulphuric acid was used and a total of 1000 mL of distilled water. Sulphuric acid with distinctive pH values was made in different solutions, using four volumetric flasks (each for its respective pH), to reproduce the differences between the acidity of acid rain. pH was measured using the *Texas TI-Nspire CX II* calculator pH meter and a universal pH meter. Paper chromatography and spectrophotometry were carried out to analyse the effectiveness of acid rain, using the *Zuzi* spectrophotometer model 4111 RS and cuvettes. Each plant was assigned a specific sulphuric acid solution, the same solution with the same pH being poured during the entire experiment. One of the plants was not given water for a week before initiating the experience to represent a deceased and not well-treated plant (identified as "No treatment"). The procedure was applied three days per week—Monday, Wednesday, and Friday—for three weeks. The temperature was measured using the temperature meter of the *Texas TI-Nspire CX II* calculator, being proved not to affect the results. Then, 10 mL was poured on the plants each day, the solution being doubled at day ten since all plants kept showing some resistance.

2.1. Paper Chromatohraphy

2.1.1. First Day of Experiment

Paper chromatography was performed before applying the acid solutions as they were all under the same conditions, besides the "No treatment" plants. A leaf of each plant was taken and placed in its respective and identified mortar (one for each plant being treated with its sulphuric acid solution), where 5 mL of di-ethylether was added, crushing the leaves with the solvent using a pestle. A 6 × 8 cm chromatography paper was cut, 1 cm margin was left at the bottom for the paper to be dipped in hexane. Five drops of each preparation, which was suspended vertically in the solvent chamber only touching the levels of the solvent. Hexane was used since it is a potent non-polar solvent, due to the lack of electronegativity in the H-C bond, repelling polar components and attracting non-polar components. Figures 1 and 2 correspond to the performing of paper chromatography twice regarding the *Monstera deliciosa* plants.

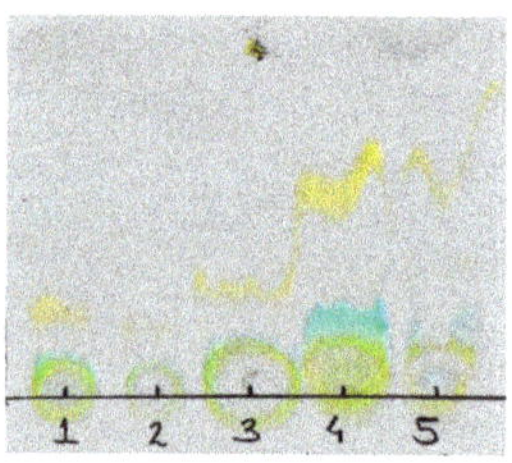

Figure 1. Chromatography 1st trial.

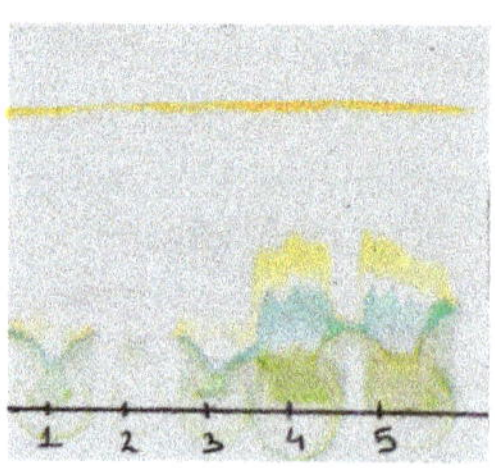

Figure 2. Chromatography 2nd trial.

2.1.2. Ninth Day of Experiment

While analysing the three chromatograms, Figures 3–5, the yellow pigment has a stronger presence, meaning the chlorophyll keeps deteriorating as the sulphuric acid mixture is applied. A larger chromatography paper was cut, 14 × 7 cm, to have a larger distance between the samples extracted from the plants.

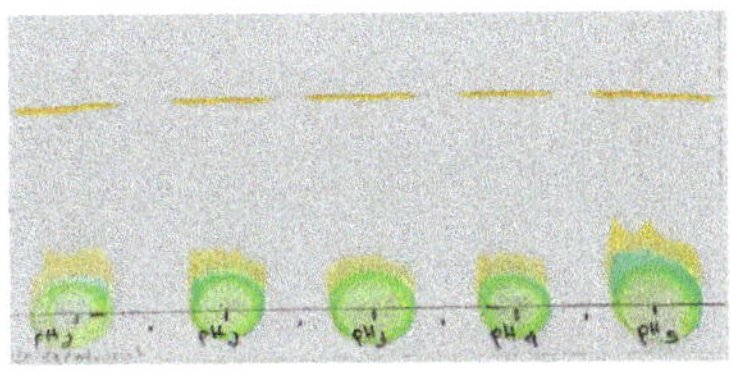

Figure 3. First trial at 9th day.

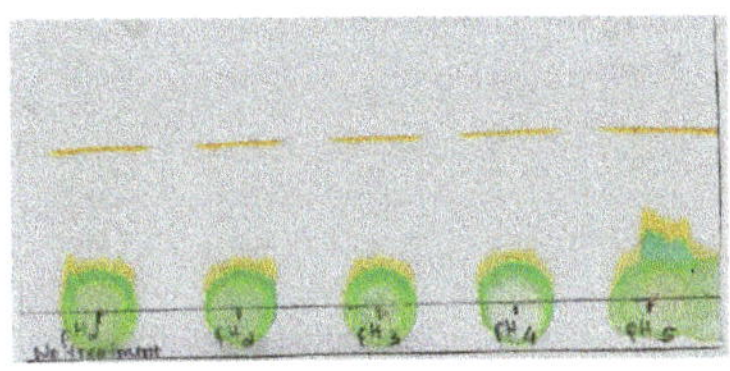

Figure 4. Second trial at 9th day.

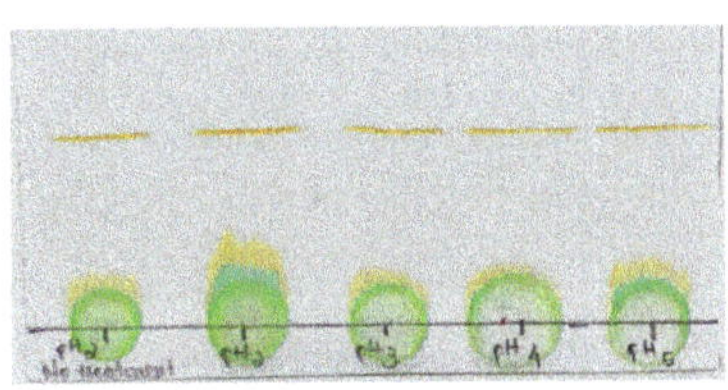

Figure 5. Third trial at 9th day.

2.1.3. Eighteenth Day of Experiment

All plants showed yellow leaves, as only yellow leaves were used this day. It may be assumed that sulphuric acid solutions caused, in all plants, almost the total deterioration of the chlorophyll pigment. This degradation caused the leaves to turn yellow.

2.2. Spectrophotometry

2.2.1. Tenth Day of Experiment

The first spectrophotometry was performed after four applications of the 10 mL acid solutions. The leaves were crushed with distilled water using a mortar and pestle, passing it, with a Pasteur pipette, into a funnel where there was filter paper to retain all the impurities, draining the clean extraction of the leaves. This extraction was placed in a cuvette and then inserted into the spectrophotometer. The values read by spectrophotometer were obtained by the wavelengths present in the blue region of the spectrum, from 330 nm to 520 nm. Analysing Figure 6 with all plants, generally the absorbance decreases from 330 nm to 420 nm. At 430 nm, there is an increase in the absorbance registered, which goes according to the literature. Contrary to chromatography, this experiment shows data that are capable for distinguishing how the different pH values of the sulphuric acid affected the plants. As shown in Figure 6, there is an increase in the absorbance values as the pH increases. Verifying the hypothesis, the lower the pH values, the more damage the sulphuric acid solution will cause to the plant. The "no treatment" plant also showed a much lower absorbance value.

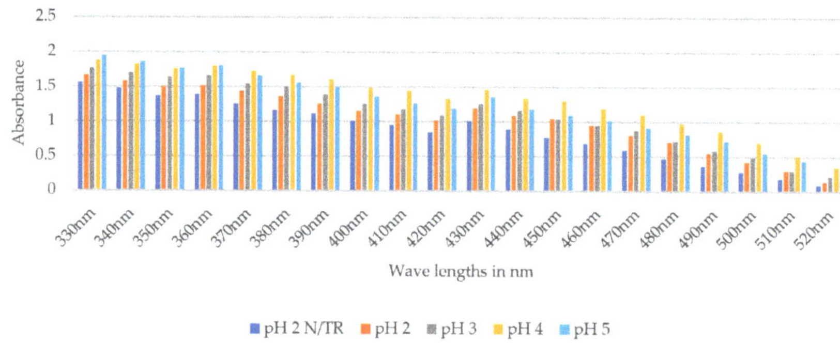

Figure 6. Spectrophotometry of *Monstera deliciosa* plants after 4 applications of the 10 mL sulphuric acid solutions.

2.2.2. Nineteenth Day of Experiment

Second spectrophotometry was performed where four times the 20 mL solution had already been applied. More accurately in Figure 7, all values have decreased in comparison to Figure 6. The doubling of the solution applied to 20 mL guided the plants to lower values of absorbance, meaning a greater loss of their capabilities of absorbing light and radiation. The plant being applied with the highest pH value showed results that were kept almost identical, with few decreases to the ones before, which proves the affirmation that these plants are more likely to grow in a soil where the pH is between 5.0 and 7.5. In the same way, there is an increase in absorbance from 420 nm to 430 nm, leading to a peak representing each plant. Also, there are negative values in all trials for the plants with pH values of 2. Physically, negative values should not be presented, as they are not possible to reach or analyse, but chemically, they signify that the amount of light that passes through the sample is more intense than the amount of light that passes through the reference.

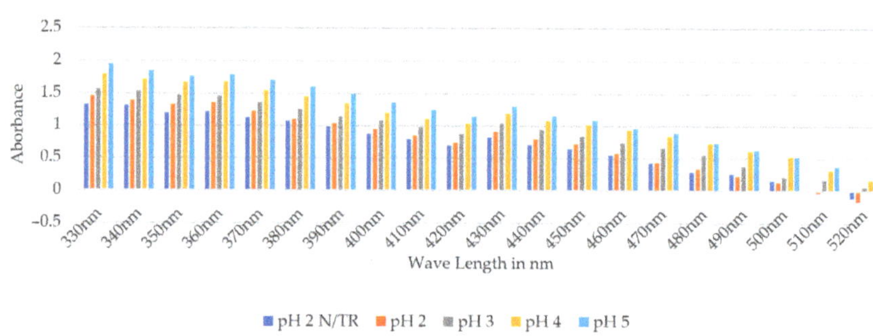

Figure 7. Spectrophotometry of *Monstera deliciosa* plants after 4 applications of the 20 mL sulphuric acid solutions.

3. Discussion and Conclusions

Via paper chromatography, because chlorophyll is degrading, plants receiving solutions with the lowest pH level solutions will exhibit a smaller variety of pigments and weaker colours.

Through spectrophotometry, by having quantitative data to investigate, plants suffering from the lowest pH sulphuric acid solution were losing their capabilities of absorbing light and radiation more quickly by showing lower absorbance values.

Therefore, by paper chromatography and spectrophotometry, this investigation has demonstrated how acid rains of various pH levels affect Monstera plants.

Author Contributions: A.R. and S.A. conceived, designed and performed the experiments; A.R. and L.G. wrote the paper; P.C. project administration; A.P. funding acquisition. All authors have read and agreed to the published version of the manuscript.

Funding: The project was financed by Colégio Atlântico (private institution). CiiEM has also provided support through Project UIDB/04585/2020, funded by FCT.

Institutional Review Board Statement: Not applicable.

Informed Consent Statement: Not applicable.

Data Availability Statement: The data sets used and/or analysed during the current study are available from the corresponding author.

Conflicts of Interest: The authors declare no conflict of interest.

References

1. Akin, C. What pH Level Is Best for Monstera Plants? Monstera Plant Resource, 29 July 2021. Available online: https://monsteraplantresource.com/what-ph-level-is-best-for-monstera-plants/ (accessed on 7 July 2023).
2. Crane, J.H.; Carlos, F.B. Monstera Growing in the Florida Home Landscape. *EDIS* **2006**, *2006*, 1. [CrossRef]
3. Photosynthesis—Photosynthetic Spectrum—Excite Europe. Available online: www.exciteled.eu/spectrum-photosynthesis (accessed on 7 July 2023).

MDPI
St. Alban-Anlage 66
4052 Basel
Switzerland
www.mdpi.com

Medical Sciences Forum Editorial Office
E-mail: msf@mdpi.com
www.mdpi.com/journal/msf

www.ingramcontent.com/pod-product-compliance
Lightning Source LLC
Chambersburg PA
CBHW061434210926
51850CB00008B/1514
* 9 7 8 3 0 3 6 5 9 1 7 4 2 *